Liquid, Gaseous and Solid Biofuels - Conversion Techniques

Liquid, Gaseous and Solid Biofuels - Conversion Techniques

Editor

Azat Solak

scitus academics

Liquid, Gaseous and Solid Biofuels - Conversion Techniques
Edited by **Azat Solak**

ISBN: 978-1-68117-509-6
Library of Congress Control Number: 2016942997

www.scitusacademics.com
Box No. 4766, 616 Corporate Way,
Suite 2, Valley Cottage,
NY 10989

Printed in United States of America on Acid Free Paper

Preface

A biofuel is a gas produced from prehistoric biological matter through contemporary biological processes such as anaerobic digestion and agriculture, rather than a gas produced by geological processes such as those engaged in the development of fossil fuels such as petroleum and coal. Bioenergy is biofuels-based energy. Biofuels can also be derived from forestry, agricultural or fishery products or municipal waste, as well as by-products and waste from agro-industry, food industry and food service. In general, bioenergy accounts for about 10% of the world's total energy demand. Traditional unprocessed biomass, such as fuel wood, charcoal and animal dung, accounts for almost all of this and is the primary source of power for many people in developing countries who use it primarily for heating and cooking liquid, solid and gaseous biofuels.

This book emphasizes the state-of-the-art development for the production of liquid and gaseous biofuels that chemists and technologists should benefit from. More innovative and efficient converting technologies now enable biofuels to be extracted from materials such as wood, plants and waste. Bioenergy is made up of solid, liquid or gaseous fuels, perhaps. Liquid fuels can be used in the current stock of road, rail and air transport networks, as well as in electricity generators for engines and turbines. Gaseous and solid fuels can be used to create electricity from direct or possibly indirect turbine power plants designed for use. About 85% of the liquid biofuels are currently produced in the form of bioethanol, the main producers being Brazil and the USA as well. In the European Union, biodiesel production is primarily concentrated. Commercial technology for conversion of biomass improves in terms of resource efficiency and mitigation of environmental impacts and financial results. This book will appeal to researchers, industry professionals and academics.

Table of Contents

CHAPTER 1

Biofuels and Co-Products Out of Hemicelluloses

Ariadna Fuente-Hernández, Pierre-Olivier Corcos, Romain Beauchet and Jean-Michel Lavoie

Industrial Research Chair on Cellulosic Ethanol (CRIEC), Département de Génie Chimique et de Génie Biotechnologique, Université de Sherbrooke, Sherbrooke, Québec, Canada

1. INTRODUCTION

Second generation biofuels are based on the utilisation of non-edible feedstock for the production either of ethanol to be inserted in the gasoline pool or of biodiesel to be inserted in the diesel pool. Ethanol is usually produced out of fermentation of C6 sugars (although other approaches does exist, see [1]) and the latter came, in first generation ethanol, from starch. In second-generation ethanol, the source of carbohydrate considered is usually cellulose, which, in turns, is obtained from lignocellulosic biomass. Recent work by Lavoie*et al.* [2] have depicted an overview of many types of lignocellulosic biomass and in most cases, cellulose, although a major component, is not the only one and is accompanied by lignin, hemicelluloses, extractives and, in case of agricultural biomass, proteins. High grade biomass (as wood chips, sugar cane or even corn) are usually very expensive (more than 100 USD/tonne) because, in most part, of the important demand related to those feedstock in industries and this is why cellulosic ethanol is more than often related to residual biomass. The latter includes but is not limited to residual forest and agricultural biomass as well as energy crops. In all cases, although the feedstock is rather inexpensive (60-80 USD/tonne), it is composed of many different tissues (leaves, bark, wood, stems, etc.) making its transformation rather complex [3].

Industrialisation of second-generation biofuel requires specific pre-treatment that should be as versatile as efficient in order to cope with the economy of scale that has to be implemented in order to make such conversion economical.

The whole economics of cellulosic ethanol relies first on ethanol, which has a commodity beneficiates from a quasi-infinite market as long as prices are competitive. Assuming average cellulose content of 45-55 % (wt) in the lignocellulosic biomass, the ethanol potential of lignocellulosic biomass would range between 313-390 L per tonne of biomass converted. With an actual market price of 0.48 USD per liter the value of this ethanol would range between 150-187 USD per tonne of biomass processed. Since the latter is more expensive to process (first isolation of cellulose then hydrolysis of cellulose) and considering the fact that the feedstock is itself expensive, there is a necessity to get an added value out of the remaining 55-45 % (wt) content. This residual carbon source is composed mostly of hemicelluloses and of lignin. The latter is a very energetic aromatic-based macromolecule, that has a high calorific value explaining why many processes converting such biomass (as some pulp and paper processes) relies on the combustion of lignin to provide part of the energy for the industry. It could also serve as a feedstock for the production of added-value compounds and although the subject is very pertinent to the field, it is out of the scope of this review, which focuses mostly on C5 sugars derived from hemicelluloses.

Conversion of the carbohydrates is of course an important part of the process although; isolation of hemicellulose for the lignocellulosic matrix is also crucial for such an approach and in consequence should also be briefly assessed. For years now, the pulp and paper industry have worked with lignocellulosic substrates and they have over the year developed many techniques allowing isolation of hemicelluloses. Chemical processes as soda pulping and kraft pulping allows isolation of both lignin and hemicellulose whilst protecting the cellulosic fibres in order to produce the largest amount of pulp possible per ton of biomass. Nevertheless, in both chemical processes previously mentioned, the hemicellulose are rather difficult to reach since they are mixed with a variety of organic and inorganic compounds including lignin as well as the chemicals that were used for the pulping process. During the last decades, the pulp and paper industry have started to look toward other processes that could allow a preliminary removal of hemicelluloses in order to avoid a complicated and expensive isolation after a chemical pulping process.

Amongst the techniques used for prehydrolysis, treatments with hot water catalyzed or not have been investigated in details in literature. As an example, Schild*et al.* [4] performed a preliminary extraction with water (via auto-hydrolysis) or with alkaline water prior to soda pulping in order to recuperate the hemicellulose prior to pulping. Similar testing was also performed on northern spruce with pressurised hot water in the presence of sodium bicarbonate [5]. Hot water extractions were also performed at temperature around 170 °C at different pH (the latter were adjusted with a phthalate buffer) and these experiments showed that control of pH was crucial in order to extract more of the hemicelluloses (up to 8 % wt on original biomass) [6]. Hot water extractions at similar temperature range have also been performed on maple [7] as well as on sugarcane bagasse [8]. Overall the hot water pretreatment may be a very promising approach for isolation of hemicelluloses although reported rates did not go far over 10 % because of the necessity to preserve the cellulosic fibres in order to avoid losses for papermaking. Acid catalyst has also been used as pretreatment to remove hemicellulose prior to pulping as reported by Liu*et al.* [9]. Utilisation of sulphuric acid, although very efficient to remove hemicellulose may also have an impact on cellulose thus reducing the pulp production rates.

Another process that could lead to isolation of hemicellulose is the organosolv process, which is to a certain extent comparable to classical chemical pulping in that sense that the technique allows simultaneous removal both for lignin and hemicelluloses. However, instead of using only an aqueous mixture of ions, the process relies on the utilisation of a combination of ions (usually alkaline) in a 50/50 mixture of aqueous organic solvent. In most cases, the solvent is methanol for obvious economic reasons although other solvents as butanol and certain organic acids have also been investigated to the same purposes. Recent work by Wang*et al.* [10]have shown that in an organosolv process using different solvent as well as different catalyst with poplar, sodium hydroxide was shown to be the best catalyst for hemicellulose removal from the pulp. Recent work by Brosse *et al.* [11] also showed that for *Miscanthus Gigantheus*, an ethanol organosolv process combined with an acid catalyst (sulphuric) lead to removal of most of the hemicelluloses and lignin from the original biomass.

Finally, another approach that could lead to isolation of hemicellulose from a lignocellulosic matrix is steam processes. This technique relies on impregnation of the feedstock with water (either catalyzed or not) then treatment under pressure at temperature ranging from 180-230 °C for a certain period of time after which pressure is relieved suddenly thus creating an “explosion” of the feedstock. Such process could lead, depending on the operating condition, to the isolation of either hemicellulose or lignin in two steps or in a single step. Our team has demonstrated the feasibility of both processes for different substrates [12-14].

Independently of the substrate or the technique used for the isolation of the hemicelluloses, conversion of lignocellulosic biomass, either for the production of paper or for the production of biofuels requires a complete utilization of the carbon compound found in biomass. Once the hemicelluloses are isolated from the original feedstock, they can undergo different types of transformation leading to different added value compounds that could lead to increase the margin of profit for the industries in the field.

Hemicelluloses account for 15-35 % of lignocellulosic biomass dry weight [2] and they are usually composed of different carbohydrates as well as small organic acids as acetic and formic acid. Glucose and xylose are often the most abundant sugars in hemicelluloses hydrolysis although mannose, arabinose and galactose might also be present in lower concentrations. The carbohydrate compositions of some lignocellullosic biomass are shown in Table 1. Whilst the C6 sugars could easily be fermented to ethanol following detoxification of the mixture, C5 sugars remains hard to convert to ethanol, mostly because classical yeasts don't metabolise them and the genetically modified organism that ferment C5 sugars are usually slower than classical organisms used in the production of etanol from C6 sugars. Nevertheless, even if ethanol production may remain a challenge, other alternatives could be considered, both on the chemical and on the microbiological point of view, to allow conversion of C5 sugar into added value products.

Carbohydrates tend to react in acidic, basic, oxidative or reductive mediums and therefore, numerous do arise for the conversion of C5 sugars. Although many options are available, this review will focus solely on 4 different pathways: acid, base, oxidative, and reductive. Each of these pathways could be inserted in an integrated biorefinery process where each of the fractions could be isolated and upgraded to high value compounds (see Figure 1).

Table 1. Carbohydrate composition of some lignocellulosic biomass.

Components (wt%)	Energy crops		Agricol residues		Forest residues	Coniferous
	Switchgrass [15]	Miscanthus [16]	Wheat Straw [15]	Corn Stover [17]	Aspen [18]	Loblolly Pine [19]
Glucan	38.5	55.5	39.2	36.2	52.4	36
Xylan	26.3	12.4	24.6	20.1	14.9	7.5
Galactan	1.16	-	-	1.45	2.2	2.5
Mannan	0.13	-	-	-	2.3	8.2
Arabinan	3.41	-	1.9	3.0	0.9	1.6

In this review, emphasis will be made on the recent work made for each of these conversion pathways both on the chemical and on the biochemical pathways. The review will focus on these 4 approaches also for their generally simple nature that would make them adaptable to an industrial context. These results will be compared to classical fermentation processes to produce ethanol with different types of organisms that can metabolise C5 sugars.

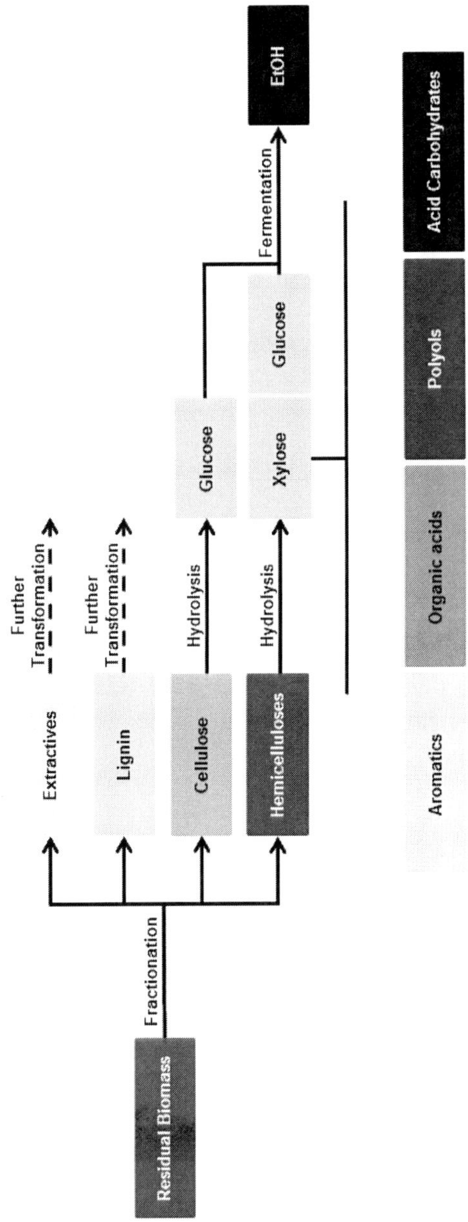

Figure 1. Potential utilization of hemicelluloses in an optimized conversion process for residual lignocellulosic biomass where C6 sugars are converted to ethanol, lignin and extractives to other added value products.

2. CONVERSION OF XYLOSE UNDER AN ACID CATALYST

2.1. The Chemical Pathway

Either in cyclic or aliphatic form, xylose then tends to dehydrate thus leading to the production of furfural whilst losing three molecules of water. Although this approach could explain the formation of furfural, it is not the sole options and many detailed reports have shown, by correlating the intermediaries with the actual structure, could be formed by many approaches depending on the reactant as reported by Marcotullio *et al.* [20] using halogen ions and proceeding only via the aliphatic form or as reported by Nimlos *et al.* [21] either via an aliphatic or a cyclic pathway (D-xylopyranose). Many different types of acid catalyst, either Brønsted or Lewis have been tested for the production of furfural. Although most of the acids reported in literature have been efficient so far for the production of the targeted molecule, one of the major side-reaction of furfural is polymerisation which influences the conversion rates and the selectivity of most of the processes reported in literature. An example of the abundance of research on this specific conversion is shown in Table 2 for different dehydration reactions under acid catalyst.

For these reactions, the temperature is generally between 140-240 °C under proportional pressure allowing the mixture to remain liquid. Many researches also use a co-solvent, often toluene in order to isolate furfural from the aqueous mixture. The reason why toluene is so popular to this purpose is mostly related to the fact that toluene has affinity for fufural thus inhibiting its polymerization.

Heterogeneous catalyst has been proven to be very efficient for the process [22,23] although polymerisation tend to reduce the surface activity thus leading to a short-term deactivation of the catalyst. On the other hand, homogeneous catalyst was also shown to be efficient but at this point the whole technique relies on how the organic solvent is dispersed in the aqueous mixture. Reducing the size of the organic solvent particles in water (or vice-versa) to the maximum should allow the best transfer between the aqueous phase to the organic phase, assuming of course that furfural has suitable affinity for the solvent and that the partition coefficient favours the solvent.

TABLE 2. Molar conversion to furfural in relationship with the catalyst used for the dehydration of xylose to furfural under acid catalyst.

Catalyst	Conversion	Reference
H-Mordenite	98%	[22]
Sulphonic acid/Silica surface	99%	[23]
1-methylimidazole	91%	[24]
KI, KCl (dilute acid)	88%	[20]
NaCl, H_2SO_4	83%	[25]
1-alkyl-3-methylimidazolium	84%	[26]
NaCl, HCl	78%	[27]
Aluminium chloride Hexahydrate	76%	[28]
Amberlyst 70	75%	[29]
Zeolite H-Beta	74%	[30]
MCM-22, ITQ-2	70%	[31]
$FeCl_3$	71%	[32]
Nafion	60%	[33]
Keggin type acids	62%	[34]
Vanadyl pyrophosphate	53%	[35]

Production of furfural itself is of course of significant interest because, amongst many factors, this chemical is commonly used in the industry as a solvent (mostly in oil chemistry). The average world production for furfural is 250 000 t/y and the actual market price evolves around 1000 USD/t [36] with recent market value reported to be closer to 1600 USD/tonne [37]. Furfural can also be a gateway to other products that could be used either as biofuels or as biomolecules. Example of such would be furfuryl alcohol via partial reduction of furfural (see Figure 2 below).

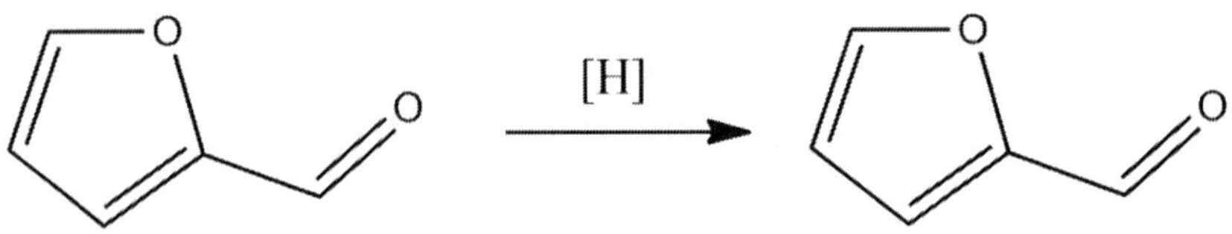

Figure 2. Reduction of furfural to furfuryl alcohol.

Furfuryl alcohol is also of interest since it is used as resins, adhesives and wetting agent, it has been mentioned that most of the 250 Kt/y of the furfural production is oriented toward production of furfuryl alcohol. The market value of this compound has been reported to be around 1800-2000 USD/tonne [38] and many reports in open literature mentions high selectivity for the conversion of furfural with iridium and ruthenium catalyst [39], rhodium [40], iron [41] and with zirconium oxide [42].

Another possible target for the transformation of furfural is for the production of 2-methyltetrahydrofuran (Me-THF) (see Figure 3). The latter is actually accredited as an additive for fuel and therefore, the possible market is virtually very important. It is also used in the petroleum industry to replace tetrahydrofuran (THF) that usually comes from non-renewables.

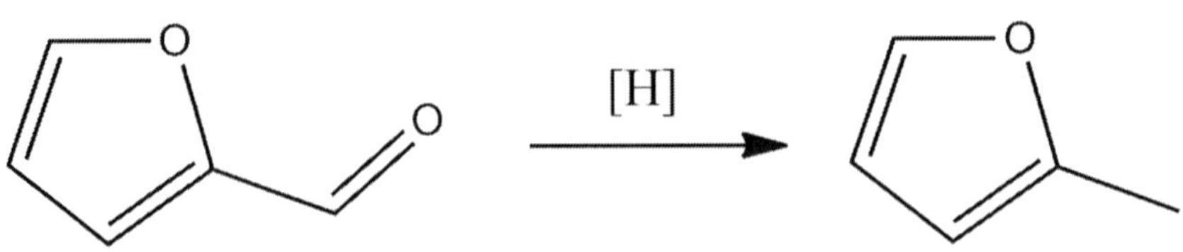

Figure 3. Reduction of furfural to 2-methyltetrahydrofuran.

Reduction of furfural to Me-THF seems to represent an important challenge since there is fewer reports mentioned in literature on the subject, as compared, as an example, to the reduction of furfural to furfuryl alcohol. Wabnitz *et al.* [43, 44] patented a one and two step process allowing conversion of furfural to Me-THF under a palladium-based catalyst and a mixture of palladium and copper oxide and chromium oxide as for the two step process. Lange [45] patented a process using palladium and titanium oxide whilst Zheng et al. [46] worked with a copper alloy. Value for Me-THF could be estimated from the price of THF which is around 3000 USD/tonne [47] and the gap between the value of furfural and Me-THF could justify the process although hydrogen value can be estimated to be around 4.5 USD/Kg (estimated with the actual price of natural assuming reforming of the latter).

Another potentially interesting approach for a transformation of furfural would be decarboxylation to furan. The general process is depicted in Figure 4 below.

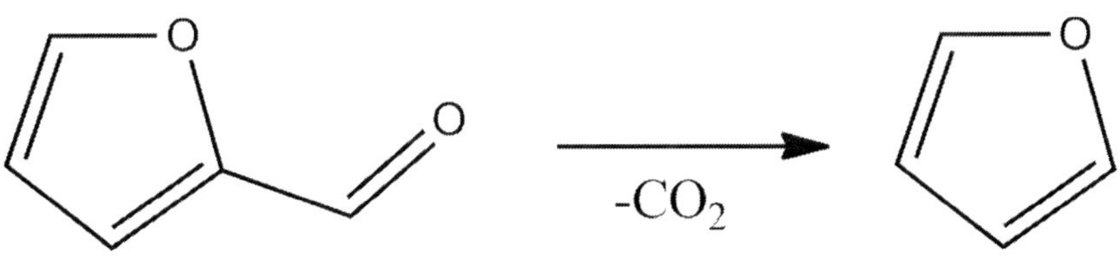

Figure 4. Decarboxylation of furfural to furan.

Many researches have focused on decarboxylation including work by Zhang *et al.* [48] who mentioned decarboxylation with potassium-doped palladium, and Stevens *et al.* [49] who reported conversion with copper chromite in supercritical CO_2.

Results reported in literature show that xylose, under an acid catalyst, tend invariably to dehydrate to furfural thus limiting the possibilities for side-products in such specific conditions. The acids could be Brønsted or Lewis type, all lead to the production of furfural furthermore when temperature are raised above 150 °C.

2.2. The Biological Pathway

Although furfural is a very common route for the conversion of xylose under an acid catalyst, furfural itself is rarely related to microorganisms in that sense that it is often considered as an inhibitor instead of a metabolite.

Nevertheless, to the best of our knowledge, no report mentioned a biological conversion of xylose to furfural.

3. CONVERSION OF XYLOSE UNDER A BASE CATALYST

3.1. The Chemical Pathway

The interaction between xylose and bases, either Brønsted or Lewis, is rather less reported in the literature when compared to the acid conversion of xylose to furfural indicated in the previous section. Many very different reactions have been reported as in the case of Popoff and Theander [50] that have quantified the cyclic compounds produced after a base-catalyzed reaction of pure D-xylose at 96 °C for 4 hours. The produced compounds are rather peculiars in comparison to other work made on the subject (see Figure 5) since most of the reported compounds are aromatics. The presence of aromatics may be a result that the reaction time was long and the isomerisation that was required in order to induce such reaction was efficient. Johansson and Samuelson [51] tested the effect of alkali treatments (NaOH) on birch xylan and contrarily to the previous research; they found that the treatment led to the production of a variety of organic acids. Testing on untreated xylene showed that most of the organic acids were already obtained from xylans and the most distinctive impact was observed after a 2 day test at 40 °C where the concentrations of L-galactonic and altronic acids increased significantly which could be related to a less severe treatment of xylans that also include C6 sugars.

El Khadem *et al.* [52] studied the effect of xylose conversion in an alkali medium at low temperatures (room) and for long periods (1-4 weeks) and one of the interesting features of his work was that the process did lead to the epimerization of sugars, but furthermore, it leads to the production of C6 sugars most probably from a reverse aldol reaction. Among the sugars that were formed during the reaction, conversion of xylose was shown to be more efficient to lyxose (18 %) and arabinose (15 %) with a decrease observed for most of the compounds between 1 and 4 weeks (see Figure 6). A vast majority (more than 50 %) of xylose remains on its original form and the reaction leads to the production of 1 % glucose and 2.5 % of sorbose, both are C6 sugars.

Xylose, as the other carbohydrates, is converted to smaller organic acids when reacted with a strong alkali medium. As an example, Jackson *et al.* [53] have demonstrated that the conversion of xylose to lactic acid could reach 64 % (molar) accompanied by glyceric acid. Although they did not used xylose but rather ribose and arabinose, they were able to reach conversions between 35-43 % into lactic acid using potassium hydroxide as catalyst under microwave irradiation [54]. Rahubadda *et al.* [55] have provided a mechanism for the conversion of xylose to lactic acid under a base catalyst. The simplified pathway is depicted in Figure 7 below.

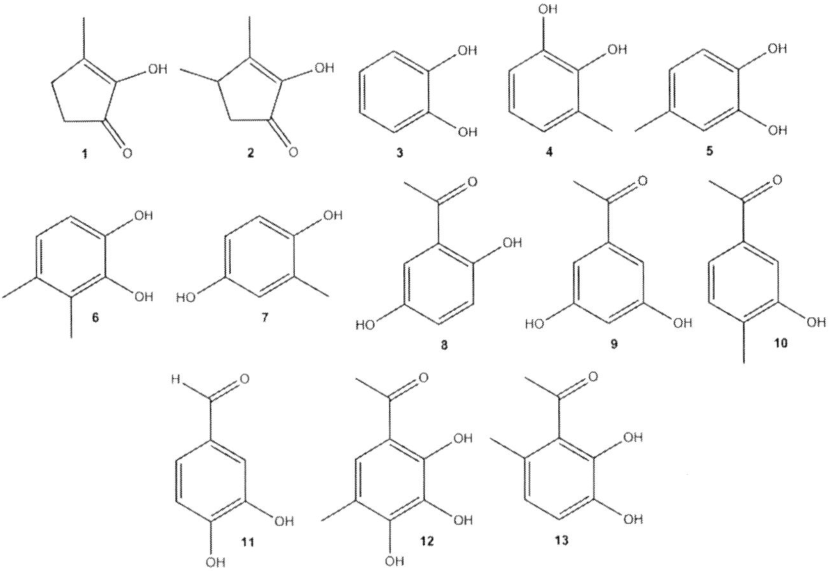

Figure 5. Cyclic and aromatics obtained from the based-catalysed treatment of D-xylose under a sodium hydroxide catalyst where (1) 2-hydroxy-3-methylcyclopent-2-enone; (2) 2-hydroxy-3,4-dimethylcyclopent-2-enone; (3) pyrocatechol; (4) 3-methylbenzene-1,2-diol; (5) 4-methylbenzene-1,2-diol; (6) 3,4-dimethylbenzene-1,2-diol; (7) 2-methylbenzene-1,4-diol; (8) 1-(2,5-dihydroxyphenyl)ethanone; (9) 1-(3,5-dihydroxyphenyl)ethanone; (10) 1-(3,4-dihydroxyphenyl)ethanone; (11) 3,4-dihydroxybenzaldehyde; (12) 1-(2,3,4-trihydroxy -5-methylphenyl)ethanone; (13) 1-(2,3-dihydroxy-6-methylphenyl)ethanone.

They mentioned in this report that methylglyoxal is most probably derived from glyceraldehyde as depicted in Figure 8 below. The possible reaction leading to methylglyoxal may involve an E2 reaction on C2 leading to removal of the hydroxyl group on C3 then a keto-enol rearrangement to methylglyoxal.

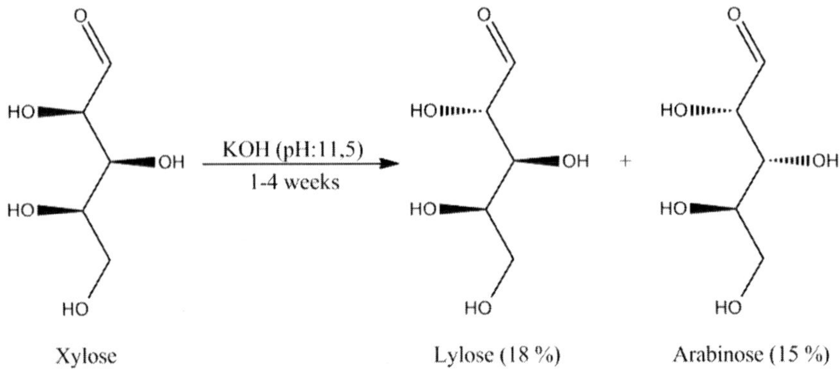

Figure 6. Major epimerisation products from 1-4 week reaction of D-xylose in a pH 11.5 KOH solution at room temperature.

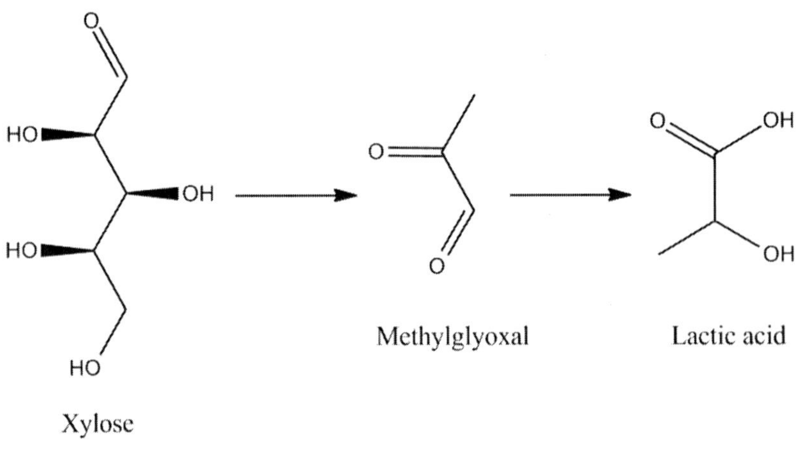

Figure 7. Conversion of D-xylose to lactic acid via the methylglyoxal pathway.

Onda *et al.* [56] achieved a conversion rate of more than 20 % when using xylose as a feedstock with a carbon-supported platinum catalyst in alkali solution. In a recent report by Ma *et al.* [57], it was shown that using model compounds, different carbohydrates tend to convert into lactic acid at different levels. Fructose was shown to be more effectively converted to lactic acid than glucose and finally than xylose. The work also showed a correlation between the amount of catalyst (varying from 1-3 % wt.) of NaOH, KOH and $Ca(OH)_2$ respectively. Part of the work by Aspinall *et al.* [58] was aimed at the non-oxidative treatment of xylans from different substrates using sodium hydroxide as solvent. The reaction was performed at room temperature for 25 days and amongst the products that emerged from this reaction, a majority was acidic and lactic acid as well as formic acid were the two major products. Other work by Yang *et al.* [59] showed that higher temperature treatments of xylose (200 °C) in a $Ca(OH)_2$ solution produced about 57 % (mol.) of lactic acid with 2,4-dihydroxybutanoic acid in second with 10 % (mol.). The same conversion patterns were observed by Raharja *et al.* [60] with production rates for lactic acid above 50 %.

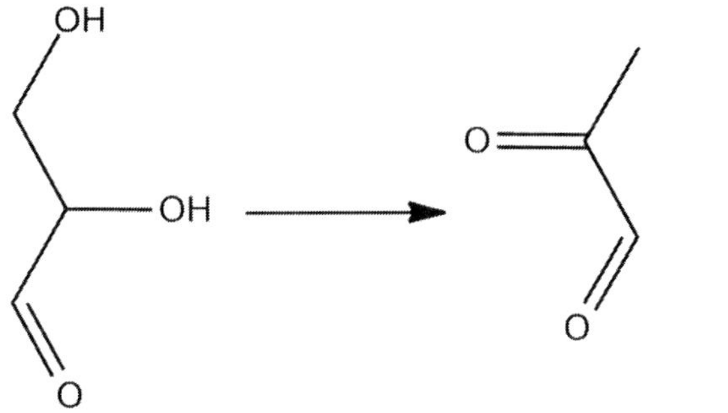

Figure 8. Conversion of glyceraldehyde to methylglyoxal.

3.2. The Biological Pathway

Amongst the different options for the conversion of xylose reported in the previous chapter, production of lactic acid via the microbial route is a vastly studied field [61-63] since currently, all of the production of lactic acid at an industrial scale in the world is biologically based. Traditionally, the concept evolves around fermenting carbohydrate-based syrup by homolactic organisms, mostly lactic acid bacteria (LAB). The most common carbohydrate-based substrates used to this purpose may be molasses, corn syrup, whey, sugarcane or even beet bagasse. Highly efficient LAB includes*Lactobacillus delbrueckii*, *L. amylophilus*, *L. bulgaricus* and *L. leichmanii*. Mutant *Aspergillus niger*has also been reported to be effective at an industrial scale [64]. LAB have the particularity to possess an homofermentative metabolism producing only lactic acid as extracellular waste product, instead of the heterofermentative pathway yielding by-products such as aldehydes, organic acids and ketones. The catabolic pathway yielding lactic acid is essentially the same across all organisms; the pyruvate intermediate is converted to lactic acid by a lactate dehydrogenase (LDH). Thus for hexose sugars, the theoretical yield is 2 moles of lactate per mole of sugar (or 1g sugar for 1g lactate). This enzymatic catalysis has the advantage over its chemical counterpart to be stereospecific: both L-lactate-dehydrogenase (L-LDH) and D-lactate-dehydrogenase (D-LDH) exist, generating either L-lactate or D-lactate respectively [65]. Both are NAD-dependant (nicotinamide adenine dinucleotide) and may be found alone or together in wild lactate-producing microbial strains. Since optical purity of lactate is a major requirement for the lactate industry, research focuses on stereospecificity as much as yields and productivity [61,66-70].

An efficient lactate producer has to display specific attributes, mainly the adaptability to low-cost substrates, high selectivity of desired enantiomer (L, D or both), high optimal temperature for decreased contamination risks, low pH tolerance and high performances (yield and productivity). LAB display appreciable performances, but lack a low pH tolerance, which implies uses of a pH control apparatus during the fermentation process. LAB optimal pH is near neutral, but the pKa of lactic acid being 3.8, an alkali agent, usually $Ca(OH)_2$, must be used thus generating calcium lactate. After typical batch fermentation, the medium is acidified with H_2SO_4 therefore regenerating and purifying the lactic acid [64]. Another drawback of LAB is their requirement for a complex growth medium, since they are auxotroph for certain amino acids and vitamins [71]. In order to overcome this problem, many fungi were also investigated for lactate production. Strains of *Rhizopus*, *Mucor* and *Monilla sp.* have shown potential whilst other fungi even displayed amylolytic activity, which could lead to a direct starch-to-lactate conversion [72-74].

Most researches still focuses on hexose conversion, and research group have optimized strains and process strategies in order to obtain high lactate titers, yields and productivities. Ding and Tan [75] developed a glucose fed-batch strategy using *L. casei* and generating up to 210 g/L of lactic acid with a 97 % yield. Chang *et al.* [76] proposed a continuous high cell density reactor strategy yielding a titer of 212.9 g/L and productivity of 10.6 g/L/h with *Lb. rhamnosus*. Dumbrepatil *et al.* [77] created a *Lb. delbrueckii* mutant by ultraviolet (UV) mutagenesis producing 166 g/L with productivity of 4.15 g/L/h in batch fermentation. Genetically engineered non-LAB biocatalysts yet have to match the performances of highly efficient wild LAB. In fact, *C. glutamicum*, *S. cerevisiae* and *E. coli*recombinant have been developed, but with limited success [61].

The search for lignocellulose-to-lactate biocatalysts have led to the discovery of many strains of pentose-utilizing LAB. *Lb. pentosus ATCC8041* [78, 79], *Lb. bifermentans DSM20003* [80], *Lb. brevis*[81], *Lb. Plantarum* [82], *Leuconostoc lactis* [83, 84], and *E. mundtii QU 25* [85, 86]. Lactic acid produced from xylose per say has been investigated by few [84,85, 87, 88], but with mitigated results, mainly due to the fact that the pentose-utilizing LAB do not perform as well in pentoses as in hexoses-rich metabolism. This phenomenon is most likely due to the fact that pentoses are metabolized in the PK pathway (phosphoketolase), thus for a given strain, even if hexoses are fermented through an homofermentative route, pentose will yield heterofermentative products (i.e. acetic and lactic acid) [78, 89].

Nevertheless, Tanaka *et al.*[84] have shown that in addition to the PK, *L. lactis* could metabolize xylulose-5-phosphate (X5P), an intermediate pentose catabolite, through the pentose phosphate pathway (PPP). The theoretical yield through the PPP is 5 moles of lactate for 2 moles of pentoses, but through the PK it decreases to 1:1 [61], thus, the conversion advantage of the PPP is obvious. Okano *et al.* [87,89] demonstrated this approach by creating a pentoses-utilizing *Lb. plantarium* recombinant in which the native L-lactate dehydrogenase (L-LDH) gene was disrupted, leaving only the homologous D-lactate dehydrogenase (D-LDH) active. However, this strain produced both acetic and D-lactic acid; hence the PK gene (*xpk1*) was substituted by a heterologous transketolase (*tkt*) from *L. lactis*, thereby shifting heterolactic fermentation to a homolactic one.

Modification of yeast strains in order to achieve xylose-to-lactate conversion has also been investigated, as an example Ilmen *et al.* [90] expressed the L-LDH gene from *L. helveticus* in *P. stipitis* and was able to reach a titer of 58 g/L of lactate with a yield of 58 %. These results were obtained despite the fact that no effort had been made to silence the native PDC/ADH (pyruvate decarboxylase/alcohol dehydrogenase) ethylic pathway, consequently 4.5 g/L of ethanol was simultaneously produced as the endogenous PDC rivalled against the recombinant L-LDH for pyruvate. Tamakawa *et al.* [88] went further by transforming *C. utilis*, disrupting the native *pdc1* gene, and expressing heterologous LDH, XR (xylose reductase), XDH (xylitol dehydrogenase) and XK (xylulokinase) enzymes. Furthermore, to prevent the redox imbalance, they increased the XR's NADH (reduced nicotinamide adenine dinucleotide) affinity by site-directed mutagenesis. In batch culture this recombinant was able to yield titers up to 93.9 g/L of lactate at a yield of 91 %. Table 3 shows the most recent and most efficient strains developed for lactic acid production, both from hexoses and pentoses.

Lactic acid seems to be, on the biological as well as on the chemical point of view the best possible compound that could be derived from a based-catalysed reaction of xylose. Racemic mixtures of lactic acid (most probably derived from chemical synthesis) can be evaluated to 1150 USD/tonne [100] whilst the pure isomer was reported to have a price market around 1750 USD/tonne [101]. As in many cases, the price will vary proportionally with purity of the compound. Utilisation of lactic acid on the market is mostly related to polymers, food, pharmaceutical and detergents. The annual world demand for the compound should reach a little more than 367 Ktonnes/year by 2017 [102].

Table 3. Lactic acid concentration (LA), time of fermentation (Tf), yield and production rate for the most common microorganisms used for the biological conversion of xylose to lactic acid

Strain	Gen Eng Str	Medium	Process	LA (g/L)	Tf (h)	Yield (g/g)	Prd (g/L/h)	Ref
E. mundtii QU 25	-	Cellobiose	Batch	119	106	0.83	1.12	[86]
	-	Xylose	Batch	86.7		0.84	0.9	[85]
	-	Glucose/ cellobiose	Batch	35.1	15	0.91	2.99	[86]
Lactobacillus sp. RKY2	-	Wood hydrolysates*	Continuous w/cell recycling	27	-	0.9	6.7	[91]
Strain	**Gen Eng Str**	**Medium**	**Process**	**LA (g/L)**	**Tf (h)**	**Yield (g/g)**	**Prd (g/L/h)**	**Ref**
Lb. bifermentas DSM 20003	-	Wheat bran hydrolysates	Batch	62.8	60	0.83	1.17	[80]
Lb. casei NCIMB 3254	-	Cassava bagasse	Batch SSF**	83.8	60	0.96	1.4	[92]
Lb. delbrueckii Uc-3	UV mutagenesis	Cellobiose	Batch	90	40	0.9	2.25	[93]
		Molasse	Batch	166	40	0.95	4.15	[77]
Lb. lactis RM 2-24	UV mutagenesis	Cellobiose	Batch	80	48	0.8	1.66	[94]
		Cellulose	Batch SSF	73	48	0.73	1.52	
Lb. plantarum ΔldhL1-xpk1::tkt	Disruption of endogenous LDH gene. Replacment of endogenous PK (*xpk1*) gene with heterologous *tkt* to redirect the PK pathway to the PPP.	Arabinose	Batch	38.6	28	0.82	1.37	[89]
Lb. plantarum ΔldhL1-xpk1::tkt-Δxpk2	Idem as above. Disruption of 2nd PK gene (*xpk2*) to terminate acetate production.	Xylose	Batch	41.2	60	0.89	0.67	[87]
Lb. rhamnosus ATCC 7469	-	Paper Sludge	Batch SSF	73	168	0.97	0.45	[95]
Lb. rhamnosus ATCC 9595	-	Apple pomace	Batch	32.5	6	0.88	5.41	[96]

Strain	Gen Eng Str	Medium	Process	LA (g/L)	Tf (h)	Yield (g/g)	Prd (g/L/h)	Ref
(CECT288)								
L. lactis IO-1	-	Xylose	Batch	33.3	-	0.68	-	[84]
S. cerevisiae recombinant	Replacement of native *pdc1* and *pdc5* by heterologous bovine L-LDH gene.	Glucose	Batch	82.3	192	0.83	0.43	[97]
S. cerevisiae recombinant	Disruption of *pcd1* and *adh1* genes. Expression of bovine L-LDH.	Glucose	Batch	71.8	65	0.74	1.1	[98]
K. lactis	Disruption of PDC and PDH genes. Expression of bovine L-LDH gene.	Glucose	Semi-Batch	60	500	0.85	0.12	[99]
C. utilis	Disruption of endogenous PDC gene.Expression of heterologous of heterologous LDH, XR, XDH and XK. XR gene site-specific mutation for preferential NADH cofactor utilization	Xylose	Batch	66.7	78	0.79	1.62	[88]
			Batch	93.9	78	0.91	2.18	
P. stipitis	Expression of LDH from *L. helveticus*.	Xylose	Batch	58	147	0.58	0.39	[90]

* *No xylose consumption occurred*

***SSF = simultaneous saccharification and fermentation*

4. CONVERSION OF XYLOSE UNDER REDUCING CONDITIONS

4.1. The Chemical Pathway

Xylose, as all the other carbohydrates that can be isolated from lignocellulosic biomass, has a carbonyl function that is susceptible to transformations, including reduction. One of the most common compounds that can be derived from xylose is xylitol, a pentahydroxy chiral compound as depicted in Figure 9.

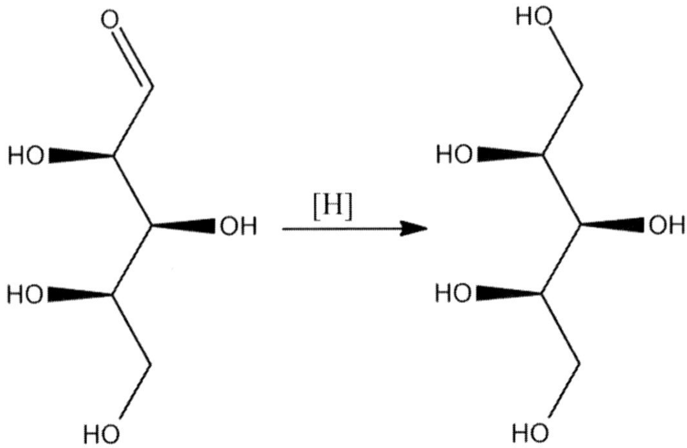

Figure 9. Simplified conversion of D-xylose to D-xylitol.

Amongst the most reported catalysts in the literature are nickel and Raney nickel. According to Wisniak *et al.* [103] they are good catalysts for the production of xylitol from xylose with total conversion at 125 °C and 515 psi. In the same year, the authors published the use of ruthenium, rhodium and palladium for the reduction of xylose [104] concluding that the efficiency of those metals was declining in the order Ru>Rh>Pd at temperatures around 100-125 °C under pressure. Mikkola *et al.* [105, 106] also used nickel as a catalyst by ultrasonic process that generated close to 50 % conversion of xylose to xylitol. From this process was reported that an important problem was the deactivation of the catalyst. Utilisation of nickel also led to the publication of two patents, one in 2003 [107] and another in 2007 [108]. In the case of the first, the concept relied on the isomerization of D-xylose to L-xylose prior to catalytic reduction under a nickel catalyst.

Ruthenium as well as ruthenium-based compounds has also been reported as catalysts for the reduction of xylose to xylitol. Ruthenium has been operated at temperatures between 90 °C and 110 °C under pressure using ruthenium supported either on silica [109] or on carbon [110]. Conversion rates for the latter have been reported to reach 35 % to xylitol for the latter with coproduction of glycerol and ethylene glycol. Ruthenium chloride ($RuCl_3$) has also been reported as a catalyst for the reduction of xylose to xylitol [111, 112].

Treatment of carbohydrates at a higher severity leads to the hydrogenolysis, implying not only the carbonyl compounds being reduce to alcohol but a breakage of the carbon-carbon bonds in the original carbohydrate. Recent work [113] shows that temperature above 250 °C and pressure between 600-1000 psi, can lead to conversion of xylose to ethylene glycol, propylene glycol and glycerol, as depicted inFigure 10 below.

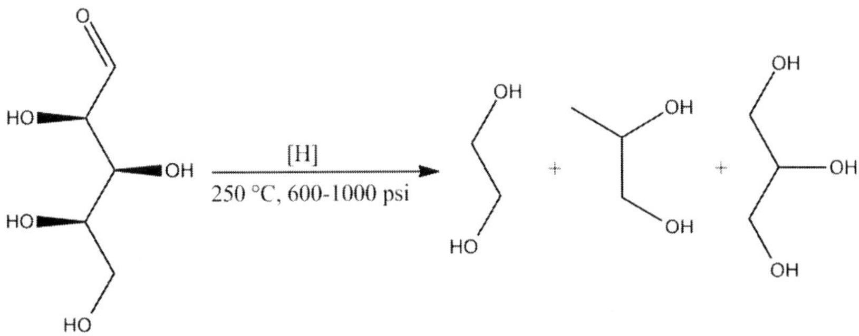

Figure 10. Simplified conversion of D-xylose to ethylene glycol, propylene glycol and glycerol as reported by Crabtree *et al.* [113].

Production of ethylene glycol and glycerol has also been reported by Guha *et al.* [110] as a side product of their xylitol production. Hydrogenolysis of xylitol is a logical suite for reduction of xylose and specific work has been reported using different catalytic systems and experimental setups. As an example, it was recently reported [114] that xylitol could be converted into a mixture of polyols and different other products as formic acid and lactic acid as well as xylitol, which, according to the previously mentioned work in this chapter, is given when xylose is submitted to a noble metal catalyst under hydrogen. In this specific case, the catalyst was platinum supported on carbon under a base-catalyzed matrix. Chopade *et al.* [115] also presented a patent reporting the conversion of carbohydrates (including xylose) into polyols using a ruthenium catalyst as did Dubeck and Knapp in 1984 [116].

In 2010 it was reported the use of nickel as a catalyst for hydrogenolysis of xylose [117] whilst Kasehagen [118] reported hydrogenolysis of carbohydrates under a nickel-iron-copper catalyst using a matrix of alkali salts with glycerol as the main product. The effects of nickel was studied by Wright [119] but this time using tungsten as a co-catalyst. Finally, there is a report about hydrogenolysis of carbohydrates under a rhenium catalyst [120].

4.2. The Biological Pathway

Only a few bacteria have been shown to naturally produce xylose as a metabolite. It has been showed [121] that a bacteria belonging to the genus *Gluconobacter* was able to produce xylitol from arabitol by way of a membrane-bound D-arabitol deshydrogenase (AraDH), followed by a soluble XDH. Rangaswamy *et al.* [122] isolated strains of *Serratia*, *Cellulomonas* and *Corynebacterium* species that were able both to grow and produce xylitol with xylose as sole carbon source, although the reported yields were very low. In early work [123, 124], it was found that both *Corynebacterium* and*Enterobacter liquefaciens* strains were able to grow and produce xylitol from xylose although gluconate had to be present as cosubstrate. Nevertheless, studies using wild bacterial strains for xylitol production are scares [122, 125-127]. In most metabolic pathways, bacteria go through direct xylose to xylulose conversion via isomerisation, bypassing the xylitol intermediate. Subsequently, xylulose is phosphorylated in X5P and can be metabolized by most prokaryotes and eukaryotes via the PPP, or the PK pathway in the case of heterolactic bacteria (Figure 11) [128].

Although the fact that yeast and fungi are generally more efficient xylitol producers than bacteria is widely recognized [129], certain highly productive species such as *Candida* are actually known for their pathogenic nature [130]. Moreover, construction of recombinant yeasts by introduction of xylose reduction pathway in GRAS species such as *S. cerevisiae* have been accomplish, although these recombinant still have to match the productivities found using non-GMO organisms (genetically modified *organisms*) [131-134]. Bacterial species on the other hand present high yields, fast metabolism and many GRAS (generally recognized as safe) species with recombinant strains often display higher efficiencies than their non-altered counter-part [135].

It was found that the catabolic rate of xylose is usually enhanced by the presence of a co-substrate such as glucose [136, 137]. However, most organisms preferentially use glucose to any other sugars due to allosteric competition in sugar transport and/or repression of other carbon catabolites [138, 139]. Thus, a suitable biocatalyst would have to simultaneously metabolize both substrates. This functionality was achieved in *E. coli* [140]by replacing the putative cAMP-dependent receptor protein (CRP) with a cAMP-independent mutant, which also expressed a plasmid-based xylose transporter. Similarly, some authors [125] used this approach as well as inserting the heterologous XR gene and silencing the endogenous xylose isomerase (XI). Alternatively, heterologous XR and XDH may be introduced and the putative XK (*xylB* gene) silenced.

Other well suited candidates for such a bioconversion would be LAB, offering the advantage of an energy metabolism completely independent of their limited biosynthetic activity, thus their glycolysis pathways may be engineered without disturbing other key structural pathways [129]. By introduction of yeast XR gene, as well as a heterologous xylose transporter in *L. lactis,* they showed that bacterial productivity and yield might reach those of the best yeasts. Even if all xylose is not consumed when in high initial concentration, the non-pathogenic and anaerobic nature of *L. lactis* is a notable advantage.

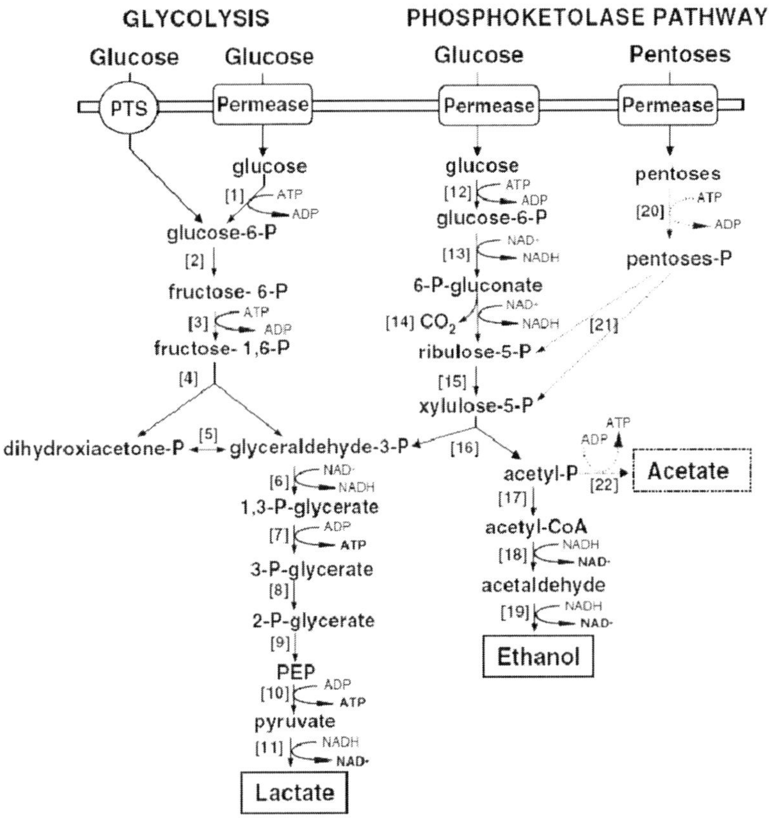

Figure 11. Glycolysis and phosphoketolase (pentose phosphate) pathways in lactic acid bacteria (1) glucokinase, (2) phosphoglucose isomerase, (3) phosphofructokinase, (4) fructose 1,6-bisP aldolase, (5) triose-phosphate isomerase, (6) glyceraldehyde-3P dehydrogenase, (7) phosphoglycerate kinase, (8) phosphoglycerate mutase, (9) enolase, (10) pyruvate kinase, (11) lactate dehydrogenase, (12) hexokinase, (13) glucose-6P dehydrogenase, (14) 6-phosphogluconate dehydrogenase, (15) ribulose-5P 3-epimerase, (16) xylulose-5P phosphoketolase, (17) phosphotransacetylase, (18) acetaldehyde dehydrogenase, (19) alcohol dehydrogenase; (20) pentose kinase, (21) pentose phosphate epimerase or isomerase, (22) acetate kinase. *CoA* coenzyme A.

Early work done on *Corynebacterium glutamicum* showed another alternative for the production of xylitol but the necessity of inserting gluconate as co-substrate for NADPH (nicotinamide adenine dinucleotide phosphate) regeneration rendered the application non economical [122,124]. Sasaki *et al.*[141] developed a *C. glutamicum* recombinant achieving simultaneous co-utilization of glucose/xylose. This was done by introducing the pentose transporter area in *C. glutamicum* chromosomal DNA (deoxyribonucleic acid). *C. glutamicum* is a noticeable candidate for its non-pathogenic and gram-positive nature, as well as its extensive use for amino and nucleic acid industrial synthesis [142, 143]. It was established [135] that xylitol productivity may be improved by disabling the xylitol import system (ptsF gene) and suggested that more work done on xylitol export system and redox balance may yield further improvements. Nevertheless, their CtXR7 *C. glutamicum* recombinant attained a productivity of 7.9 g/L/h and final xylitol concentration of 166 g/L after 21 h (see Table 4). This was achieved by (to date this is considered the best xylitol bacterial producer):

- introduction homologous pentose transporter (*araE*);
- disruption of the native lactate deshydrogenase (*ldhA*);
- expression of single-site mutant XR from *C. tenuis*;
- disruption of XK native gene (*xylB*);
- disruption of phosphoenolpyruvate-dependent fructose phosphor-transferase (*ptsF* gene; PTS^{fru}).

As previously discussed for ethanol, the redox imbalance that often occurs from XR/XDH preferential use of NADPH/NAD+ cofactors is a key factor for xylitol accumulation in the cell. In most yeast studied, it has been shown that XR has a marked preference for NADPH, while XDH has a quasi-unique specificity for NAD+ [126]. The main exception being *P. stipitis* who shows a nearly by-specificity for NAD(P)(H) for its XR and *P. tannophilus* whose XDH shows a higher activity with NADP+ than NAD+ [158] proposed a theoretical maximum xylitol yield in yeasts of 0.905 mol of xylitol per mol of xylose when NADH was efficiently used as cofactor by the XR or under aerobic condition where the NADH can be oxidized back to NAD+ in the respiratory chain. Otherwise, under anaerobic conditions, the theoretical yield drops to 0.875. These yields follow the equations (1) and (2) below respectively:

$$126\ \text{xylose} + 3\ O_2 + 6\ \text{ADP} + 6\ P_i + 48\ H_2O \rightarrow 114\ \text{xylitol} + 6\ \text{ATP} + 60\ CO_2 \qquad (1)$$

$$48\ \text{xylose} + 15\ H_2O \rightarrow 42\ \text{Xylitol} + 2\ \text{ethanol} + 24\ CO_2 \qquad (2)$$

Table 4. Overview of the different strains allowing conversion of xylose to xylitol including yields, fermentation time (Tf), production (Prd) and the process strategy.

Strain	Genetic Engineering Strategy	Yield g/g	Xylose g/L	Xylitol (g/L)	Tf (h)	Prd (g/l/h)	Process Strategy	Reference
Candida athensensis SB18	-	83%	250	207.8	175	1.15	Batch limited O_2	[144]
		87%	300	256.5	250	0.97	Fed Batch limited O_2	
		79%	200	151.71	156	0.97	Batch limited O_2	
C. tropicalis ASM III	-	93%	200	130	120	1.08	Batch limited O_2	[145]
Candida sp. 559-9	-	99%	200	173	121	1.44	Batch limited O_2	[146]
C. tropicalis KCTC 10457	-	87%	200	172	48	3.66	Batch limited O_2	[147]
C. tropicalis KFCC 10960	-	93%	270	251	55	4.56	Fed Batch	[148]
C. tropicalis KCTC 10457	-	90%	260	234	48	4.88	Fed Batch	[147]
C. guilliermondii	-	73%	250	-	-	-	Fed Batch limited O_2	[149]
C. tropicalis	-	82%	750	189	58	4.94	Fed Batch/ Cell recyicing/ Glucose cosubstrate/ limited O_2	[150]
C. tropicalis	-	69%	100	-	-	5.7	Cell recycling/ limited O_2	[151]
C. tropicalis	-	85%	214	182	15	12	cell recycling/ limited O_2	[147]
S. cerevisiae	Expression heterologous XR gene from *P. stipitis*.	95%	190	-	-	0.4	Fed batch/ Glucose cosubstrate	[152]
Corynebacterium glutamicum CtXR7	Expression of *araE* pentose transporter gene. Disruption of *ldhA*. Single site mutation of heterologous XR gene. Disruption of *xylB*& PTS^{fru} genes.	-	120	166	21	7.9	Fed batch/ Glucose cosubstrate/ 40g/L dry cell	[135]
D. hansenii NRRL Y-7426	-	38%	45	19.7	72	0.274	Batch/ Detoxified grape marc hydrolysates	[153]
S. cerevisiae	Overexpression ALD6 & ACS1 genes.Expression of *P. stipitis* XR gene.	~100%	20	91.3	60	1.76	Fed batch/ Glucose cosubstrate	[154]
Lactobacillus brevis NZ9800	Expression of *P. stipitis* XR gene Expression of *Lb. brevisxylT* symporteur.	~100%	160	75	41	2.72	Fed batch/ Glucose cosubstrate	[129]
C. tropicalis	-	83%	80	96.5	120	1.01	Fed batch/ Corn Cob hydrolysates/limited O_2	[155]
*C. tropicalis*SS2	Xylitol-assimilation deficient strain by chemical mutagenesis.	93%	100	220	70	3.3	Fed batch/ aerobic	[156]
C. trolpicalis JH030	-	71%	45	31.1	80	0.44	Batch/ Rice straw hydrolysates	[157]

Owing the better yield both in xylitol and ATP (adenosine triphosphate) under oxygen-limited xylitol production, aeration is a crucial parameter. As a general trend, xylitol production increases when oxygen is allowed in the medium under a certain threshold concentration [159]. This preference is yeast specific since for *P. stipitis* it is reported that the absence of dissolved oxygen is needed for optimal xylitol production; while *P. tannophilus* reaches maximum yields under anoxic conditions [160, 161].

Many strains of *S. cerevisiae* have been transformed for xylose utilization in the early 90's. As for xylose-to-xylitol, Hallborn *et al.* [152] reported a highly efficient conversion of xylose to xylitol (95 % of theoretical). It has been suggested that the incapacity of *S. cerevisiae* to rapidly replenish its NADPH pool from its PPP during xylose metabolism is what causes the metabolic bottleneck [162,163]. This is mainly due to the fact that xylose is a non-preferred carbon substrate for *S. cerevisiae* and do not provide sufficient energy for growth and metabolism [164].

C. tropicalis is a candidate of choice for xylitol production among the few native strains reported as the best xylitol producers to date (see Table 4) and this research for native strains and genetically engineered recombinant is still under way today [155-157]. As in *S. cerevisiae*, the PPP is the major NADPH biosynthesis pathway and efforts have been made to increase its flux. Ahmad *et al.* [165] recently successfully increased the metabolic flux toward PPP for NADPH regeneration, thereby enhancing xylitol production of the original strain by 21 %. This was done by disrupting XDH putative gene, and over-expressing homologous glucose-6-phosphate dehydrogenase (G6PDH) and 6-phosphogluconate dehydrogenase (6-PGDH). Table 4 summarize the best xylitol producing strains found in the literature up to date.

Reduction of xylose either at low or at high severity thus producing either xylitol or polyols (including glycerol) is a process driven by the price of hydrogen. On the other hand, the market for small polyols as ethylene or propylene glycol may generate more opportunity than the xylitol market. Xylitol market value is between 3650 and 4200 USD/tonne [166] whilst ethylene glycol is reported at a market price of 980-1500 USD/tonne [167] and propylene glycol at 1500-1700 USD/tonne [168]. The market for each of the previously mentionned compound is around 100 Ktonnes/y for xylitol [169], 19 Mtonnes/y for ethylene glycol [170] and 1.4 Mtonnes/y for propylene glycol. Although the market for smaller polyols may seem to be larger, as an example conversion of xylose to ethylene glycol and propylene glycol would require 3 times as much hydrogen if compared to xylitol. Since the price for hydrogen can be estimated roughly at 4.5-5 USD/Kg, the very concept of polyols production relies on the efficiency of the hydrogenolysis process therefore explaining why many of the reported litterature in this chapter are patents.

5. CONVERSION OF XYLOSE UNDER OXIDIZING CONDITIONS

5.1. The Chemical Pathway

Oxidation of xylose has been numerously reported in the literature although focus interest, both on the biological as well as chemical point of view has been focused toward a simple oxidation of xylose to xylonic acid (see Figure 12).

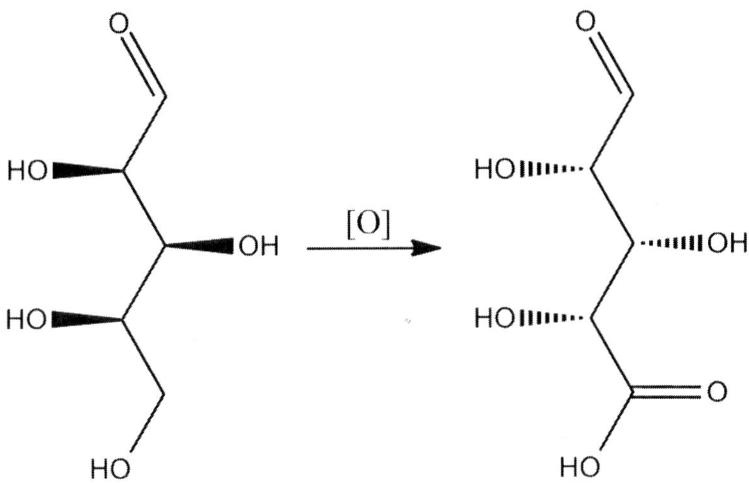

Figure 12. Simplified conversion of xylose to xylonic acid

Oxidation of xylose has been reported for a variety of different metallic catalyst including gold for high conversion rates [171]. Using a process performed a little higher than room temperature in a basic pH for 1 hour, they were able to reach a 78 % conversion of xylose to xylonic acid. Using comparable catalyst, Pruesse *et al.* [172] were able to reach 99 % selectivity with a conversion rate of 21 mmol/min/g (Au) in a continuous reactor. Nevertheless, contrarily to Bonrath, Pruesse and co-worker used a mixture of gold and palladium to perform this oxidation and temperature slightly higher (60 °C as compared to 40 °C).

Copper has also been indirectly investigated for the conversion of xylose to xylonic acid in that sense that Van der Weijden *et al.* [173] used C5 sugars (including xylose) for the reduction of copper sulfate in wastewater with very promising results. Although emphasis was not put on the carbohydrate itself, results showed that the reduction of copper from (II) to elemental was possible yet economical at larger scale. Xylonic acid was also observed as by-product of xylose oxidation using chlorine, as a side reaction of lignin oxidation. In this work [174], the concentration of xylonic acid increased by a factor of 40 after the chlorination process. Interesting enough, the xylitol concentration also increased, which might lead to the conclusion that oxidation, was probably not the sole factor here and that side reactions as the Cannizarro reaction between two xylose molecules could have been occurring. Jokic *et al.* [175] showed that it was possible up to an efficiency of 80 % to convert xylose simultaneously to xylonic acid and xylitol using electrotechnologies. Such process could be to a certain extent compared to the Cannizarro reaction where the original aldehyde is acting as redox reagent.

Further oxidation of xylose leads to a trihydroxydiacid, more specifically xylaric acid as depicted in Figure 13 below.

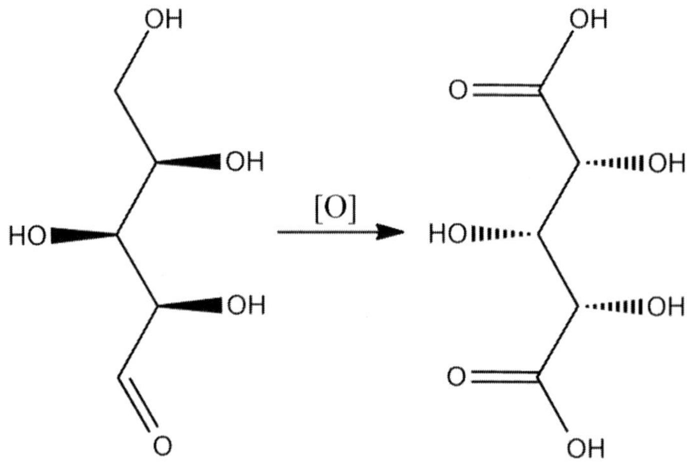

Figure 13. Simplified scheme for the conversion of xylose xylaric acid

Conversion of C5 sugars and to a smaller extent of xylose into aldaric acids has been described in literature in a few reports. Kiely *et al.* [176] reported that a conversion up to 83 % xylose into 2,3,4-trihydroxyglutaric acid was achievable in a reaction mixture composed of nitric acid and $NaNO_2$. The side product of this reaction was reported to be disodium tetrahydroxysuccinate. Conversion of xylose to xylaric adic was also reported [177] using oxygen under a platinum catalyst all of this in an alkali promoted medium. Comparable conversion process [178] was obtained without any alkali, though still performed the reaction in water at 90 °C under 75 psi of oxygen. The conversion for this process was 29 %. Fleche *et al.* [179] reported a maximum conversion of 58% once again using platinum supported on alumina.

Severer oxidizing conditions leads to a breakage of the carbon-carbon bonds in the carbohydrate molecule leading to the production, mostly, of small organic acids as formic and acetic acid on glucose [180]. A simplified scheme of such a reaction is presented in Figure 14 below:

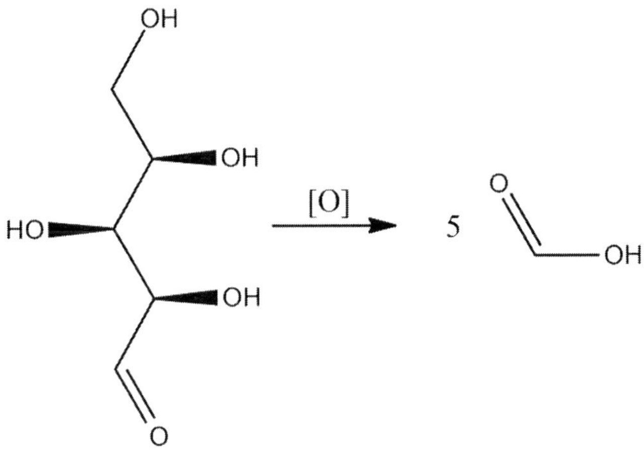

Figure 14. Simplified scheme for the conversion of xylose to formic acid under more severe oxidizing conditions.

An example of sever oxidation of xylose in a mixture of hydrogen peroxide and ammonium hydroxide have been recently reported [181] with a conversion of 96 % at room temperature for 1 h. Similar conversion of xylose was reported [182] for a process using oxygen and a molybdenum and vanadium catalyst. The reaction was done for 26 h at 353 K and 30 bar for a conversion of up to 54 % into formic acid with carbon dioxide as by-product.

5.2. The Biological Pathway

Xylonic acid synthesis from xylose has been reported for *Acetobacter* sp. [183], *Enterobacter cloacea*[184], *Erwinia* sp. [185, 186], *Fusarium lini* [187], *Micrococcus* sp. [188], *Penicillium corylophilum*,*Pichiaquer cuum* [185], *Pseudomonas* sp. [189, 190], *Pullularia pullulans* [191], *Gluconobacter* and *Caulobacter* [192, 193].

In metabolic pathways, xylose is converted to xylonate via 2 key enzymes. First, a xylose dehydrogenase (XD) oxidizes xylose to D-xylono-1,4-lactone (xylonolactone) using either NAD+ or NADP+ as cofactor. This reaction is followed by the hydrolysis of xylonolactone to xylonate either spontaneously or by an enzyme with lactonase activity [194, 195]. It is hypothesized that *Pseudomonas*and *Gluconobacter sp.* both carry a membrane-bound pyrroloquinoline quinine (PQQ)-dependent XD and a cytoplasmic one [195, 196]. Stephens *et al.* [193] recently proposed a full xylose catabolic pathway for *C. crescentus*. Note that a similar pathway was proposed for arabinose yielding L-arabonate [197]. As shown in Figure 15, the proposed metabolic pathway for *C. crescentus* shows that xylonate is an intermediate in catabolic reactions that is quite different from the XI or XR/XDH previously discussed which were more intensively studied.

Researches on highly efficient microbial xylonic acid production are scarce compared to biofuels or xylitol. Even if the identification of xylonate producing species began as early as 1938 [187], the first attempt to isolate a possible industrial biocatalyst was done by Buchert *et al.* [185], who identified *P. fragi ATCC4973* as a potentially high efficiency xylonate producer (92 % of initial sugar converted to xylonic acid with initial xylose concentration of 100 g/L). In further work, *P. fragi* and *G. oxydans*showed yields of over 95 % but the low tolerance of those native strains to inhibitors tends to be problematic for industrial uses [192]. As discussed above, the metabolic pathways implied by xylonate have been investigated in the recent years [193,196]. The first recombinant microorganism engineered for the industrial production of xylonate was done by Toivari *et al.* [198]. By introducing the heterologous *Trichoderma reesei xyd1* gene (coding for the NADP+ dependant XD) in *S. cerevisiae*, they were able to obtain up to 3.8 g/L xylonate with 0.036 g/L/h productivity and 40 % yield. Nygard*et al.* [195] engineered *K. lactis* by introducing *T. reesei xyd1* and deleting the putative *xyl1* gene coding for the XR. Up to 19 g/L xylonate where produced when grown on a xylose (40 g/L) and galactose (10.5 g/L) medium. The native ability of fast xylose uptake was an advantage, but high intracellular xylonate concentration was observed, which may indicate difficulties with product export. Liu *et al.* [199] used similar approach engineering *E. coli* by disrupting the native xylose metabolic pathways of XI and XK (as shown in Figure 16). The native pathway of xylonate was also blocked by disrupting xylonic acid dehydratase genes. The XD from *C. crescentus* was introduced and 39.2 g/L of xylonate from 40 g/L of xylose in minimal medium was obtained at high productivity 1.09 g/L/h. From these results it is clear that research is at its genesis and significant efforts will be required for the creation of a highly productive and effective xylonate production biocatalyst.

At this point it is rather hard to verify the potential or the economic value of oxidation products from xylose. Complete oxidation to formic acid could be the most suitable approach at this point since the market for xylonic and xylaric acid is not as well defined as for the simple methanoic acid with its actual market value between 750-950 USD/tonne [200] and an annual world demand suspected to reach 573 Ktonnes in 2012 [201]. Conversion of xylaric acid into glutaric acid (pentanedioic acid) would lead to a very interesting market as a plasticizer but dehydration or reduction of the three central hydroxyl groups may be a challenge that could be winning at lab scale although a multiple synthesis pathway would be very difficult to reach economic at an industrial level.

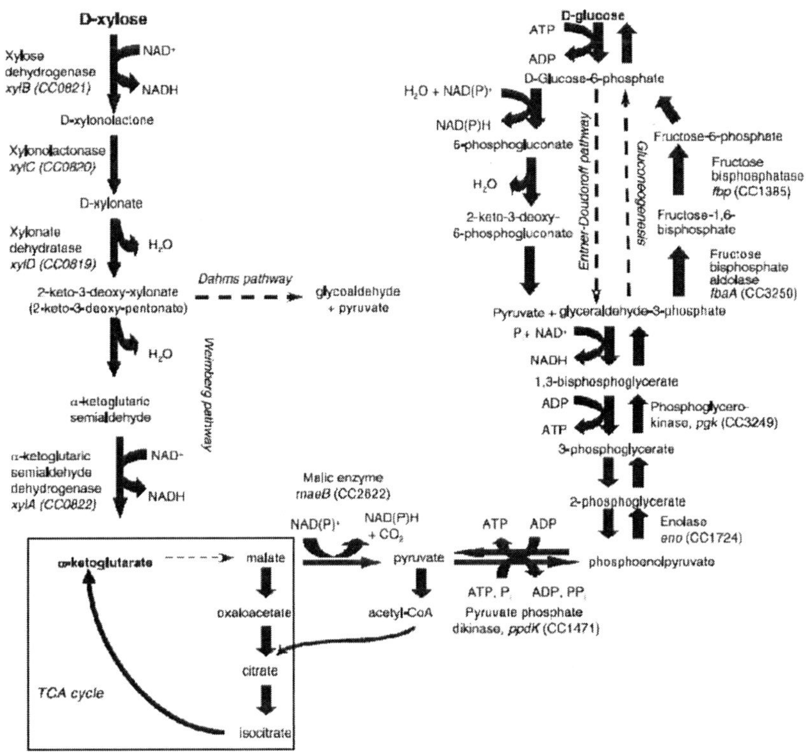

Figure 15. Proposed pathway ford-xylose metabolism in *C. crescentus* [193].

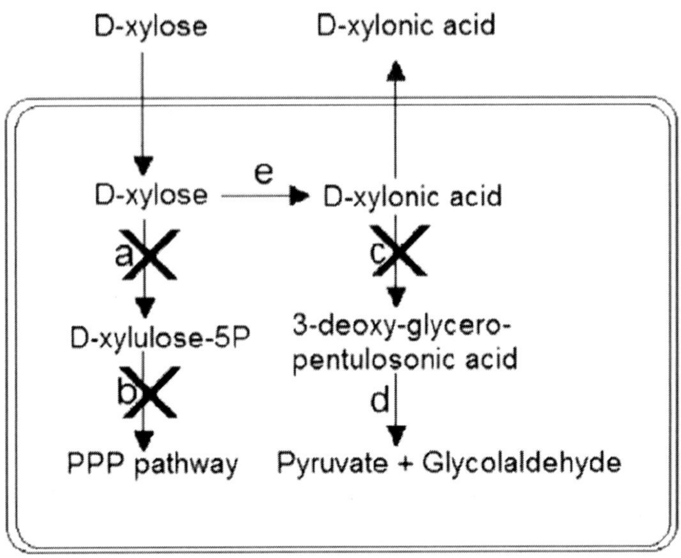

a: D-xylose isomerase (*E. coli*)
b: D-xylulose kinase (*E.coli*)
c: D-xylonate dehydratase (*E. coli*)
d: Aldolase (*E. coli*)
e: D-xylose dehydrogenase (*C. crescentus*)

Figure 16. D-xylose and D-xylonic acid metabolic pathways in *E. coli*. The symbol X denotes that the gene is disrupted.

6. CONCLUSION

Second-generation ethanol or "cellulosic ethanol" relies on the utilisation of lignocellulosic biomass as a source of carbohydrates via the "bio" conversion route (keeping in mind that other pathway, as thermocatalytic pathways, may also lead to cellulosic ethanol). Production of ethanol thus requires isolation of cellulose from lignocellulosic matrix, then hydrolysis of cellulose to glucose prior to fermentation. Both of the previously mentioned steps represent challenges for industry, but the whole economic of the process is perhaps the most challenging part of cellulosic ethanol production. Cellulose is usually available in lignocellulosic biomass in the 45-60 % range which, assuming a perfect conversion implies production of 300-400 L/tonne of lignocellulosic biomass processed. At an actual price of 0.48 USD/L, each ton of biomass has a potential value of about 150-200 USD/tonne of biomass processed.

The conversion of lignocellulosic biomass is rather more complex and to a certain extent more expensive than starch-based feedstock as corn and therefore, one can assume that the conversion price is going to be higher than classical or first generation ethanol production. Keeping that fact in mind, the conversion of cellulose to glucose itself is a major technological challenge since it either requires enzymes, ionic liquids or strong acids that are rather expensive to buy or expensive to recycle and since it is of outmost importance for the production of the ethanol, technology is to a certain extent limited by this reality.

The remaining carbon content of lignocellulosic biomass is also an important factor to be considered. Since the maximum production of ethanol from the total feedstock could vary around 300-400 L per tonne, there is at this point a necessity to generate co-products from the biomass in order to make this whole process economic at the end thus coping for technological problem as conversion of cellulose to glucose. Lignin is one of the most abundant macromolecule on earth bested only by cellulose. The aromatic nature of lignin is a challenge for ethanol production but not for added value compounds as aromatic monomers that could displace actual monomers used in the polymer industry that are usually obtained from non-renewable materials.

Hemicelluloses are also an important part of the lignocellulosic biomass. Hemicelluloses, contrarily to cellulose that is characterized by an amorphous and a crystalline part, are highly ramified and easy to hydrolyse. Usually, a simple diluted alkali solution, acidic solution or even hot water can allow conversion of hemicellulose to simple sugars. The major problem with hemicellulose is the heterogeneous composition including but not limited to small acids and a variety of C6 and C5 sugars. Whilst the C6 sugars could be easily fermented to ethanol, pending reduction of the organic acids and other inhibitors, the C5 sugars require speciality yeasts for fermentation.

Other than the classical fermentative pathway, C5 sugars can as well be converted, biologically as well as chemically into a wide variety of added value products and "green" compounds. In this paper, we have identified 4 pathways for the conversion of C5 sugars but more specifically xylose, a common carbohydrate in biomass hemicelluloses.

Reaction of xylose under an acid catalyst is probably one of the most investigated fields in this domain. The target for this conversion being furfural, a well-known chemical as well as precursor for other compound as furan, Me-THF, THF and furfuryl alcohol, a reactant used in the polymer industry. The best approach for the conversion of xylose furfural, to the best of our knowledge, is chemical as no microorganism allowing conversion of C5 sugars to furfural has been identified so far. The conversion of xylose to furfural was reported to reach more than 95 % for both heterogeneous and homogeneous catalyst. On the other hand, the selectivity toward furfural is not always as efficient since the latter undergoes polymerisation in acidic medium, which often also leads to deactivation of the catalyst.

A basic catalyst leads to a conversion of C5 sugars to lactic acid although this pathway as not been deeply investigated in the literature. Lactic acid is a compound well in demand on the market but the limitations for the chemical transformation is the lack of stereospecificity of the products. Conversion of xylose under a base catalyst leads to the production of a racemic mixture of D- and L-lactic acid and thus reducing the market value of the product, particularly if the polymer industry is targeted. On the other hand, the biological conversion of xylose to lactic acid is a well-known and extensively reported process for which the production was reported to reach 6.7 g/L/d for genetically modified organisms as, in this specific case, *Lactobacillus sp. RKY2*. According to the reports, the production of lactic acid would be more efficient by the biological approach since it can lead to a stereospecific and a higher market value.

Reduction of xylose can lead to many different products including xylitol for lower severity up to diols as ethylene glycol and propylene glycol at higher severity. It is ambiguous to determine at this point if either the chemical or the biological pathway is more efficient for the production of xylitol since reports on both pathways have shown promising results. The main problem with the xylitol market is that although it is increasing, it is fairly small and therefore it is harder to fit in a new production of xylitol. On the other hand, a more severe reduction of xylose, leading to diols, could be a very interesting opportunity for the production of ethylene glycol and propylene glycol, two very important products in the chemical industry. The downside of this approach would be the production of glycerol as a side-product.

Finally, oxidation of xylose is, at this point, the approach with the lower potential for a rapid commercialisation since the market for xylonic acid and xylaric acid is hard to size at present. The conversion process, both chemical and biological seems to have significant potential in terms of scalability but the end usage is not well defined at this point. The best option would be to produce glucaric acid from xylaric acid, which could be used as a plasticizer. On the other hand, such a process, overall rather complicated, would add a significant cost for a product that would land in the commodity range.

ACKNOWLEDGEMENT

We would like to acknowledge Enerkem, Greenfield Ethanol, CRB Innovations and the Ministry of Natural Resources of Quebec for financial support of the Industrial Chair in Cellulosic Ethanol.

REFERENCES

1. Lavoie JM, Marie-Rose S, Lynch D. Non-homogeneous residual feedstocks to biofuels and chemicals via the methanol route.Biomass Conversion and Biorefinery 2012; 1-6, ISSN: 2190-6823.

2. Lavoie JM, Beauchet R, Berberi V, Chornet M. Biorefining Lignocellulosic Biomass via the Feedstock Impregnation Rapid and Sequential Steam Treatment. Biofuel's engineering process technology.Intech publisher 2011; 685-714, ISBN: 978-953-307-480-1.

3. Lavoie JM, Chornet M, Chornet E. Biomass refineries: relationships between feedstock and conversion approach. Energy and sustainability II, WIT press 2009; 35-48,

4. Schild G, Sixta H. Sulfur-free dissolving pulps and their application for viscose and lyocell.Cellulose 2011;18(4) 1113-1128,

5. Song T, Pranovich A, Bjarne H. Characterisation of Norway spruce hemicelluloses extracted by pressurised hot-water extraction (ASE) in the presence of sodium bicarbonate. Holzforschung 2011;65(1) 35-42, ISSN: 0018-3830.

6. Song T, Pranovich A, Holmbom B. Effects of pH control with phthalate buffers on hot-water extraction of hemicelluloses from spruce wood. Bioresource technology 2011 ; 102 (22) 10518-10523;ISSN: 0960-8524.

7. Duarte GV, Ramarao BV, Amidon TE. Polymer induced flocculation and separation of particulates from extracts of lignocellulosic materials. Bioresource technology 2010;101(22) 8526-8534, PMID: 20605092.

8. Peng F, Ren JL, Xu F, Bian J, Peng P, Sun RC. Comparative study of hemicelluloses obtained by graded ethanol precipitation from sugarcane bagasse. Journal of agricultural and food chemistry 2009;57(14) 6305-6317,

9. Liu W, Hou Q, Mao C, Yuan Z, Li K. Hemicelluloses Prior to Aspen Chemithermomechanical Pulping: Pre-Extraction, Separation, and Characterization. Journal of Agricultural and Food Chemistry 2012;60(19) 4880-4885,

10. Wang K, Yang H, Guo S, Tang Y, Jiang J, Xu F, Sun RC. Organosolv fractionation process with various catalysts for improving bioconversion of triploid poplar. Process Biochemistry 2012;47(10) 1503-1509,

11. Brosse N, El Hage R, Sannigrahi P, Ragauskas A. Dilute sulphuric acid and ethanol organosolv pretreatment of Miscanthus x Giganteus. Cellulose Chemistry and Technology 2010;44(1-3) 71-78,ISSN:05769787.

12. Lavoie JM, Capek-Menard E, Gauvin H, Chornet E. Production of pulp from Salix viminalis energy crops using the FIRSST process. Bioresource Technology 2010;101(13) 4940-4946,

13. Lavoie JM, Beauchet R, Berberi V, Chornet M. Biorefining quasi homogeneous biomass via the Feedstock Impregnation Rapid and Sequential Steam Treatment. Biofuel's engineering process technology 2011; Intech publishing 685-715, ISBN:978-953-307-480-1.

14. Lavoie JM, Beauchet R. Biorefinery of Cannabis sativa using one-and two-step steam treatments for the production of high quality fibres. Industrial Crops and Products 2012;37(1) 275-283,.

15. Pasangulapati V, Ramachandriya KD, Kumar A, Wilkins MR, Jones CL, Huhnke RL. Effects of cellulose, hemicellulose and lignin on thermochemical conversion characteristics of the selected biomass. Bioresource Technology. 2012;114 663–669, ISSN: 0960-8524.

16. Sathitsuksanoh N, Zhu Z, Zhang YHP. Cellulose solvent- and organic solvent-based lignocellulose fractionation enabled efficient sugar release from a variety of lignocellulosic feedstocks. Bioresource Technology 2012;117 228–233, ISSN: 0960-8524

17. Xu J, Zhang X, Cheng JJ. Pretreatment of corn stover for sugar production with switchgrass-derived black liquor. Bioresource Technology 2012;111 255–260, ISSN: 0960-8524

18. Jun A, Tschirner UW, Tauer Z. Hemicellulose extraction from aspen chips prior to kraft pulping utilizing kraft white liquor. Biomass and bioenergy 2012;37 229-236, ISSN:09619534.

19. Rana D, Rana V, Ahring BK. Producing high sugar concentrations from loblolly pine using wet explosion pretreatment. Bioresource Technology 2012;121 61–67, ISSN: 0960-8524

20. Marcotullio G, de Jong W. Furfural formation from D-xylose: the use of different halides in dilute aqueous acidic solutions allows for exceptionally high yields. Carbohydrate Research 2011;346(11) 1291-3, ISSN:1873-426X.

21. Nimlos MR, Qian X, Davis M, Himmel ME, Johnson DK. Energetics of xylose decomposition as determined used quantum mechanics. J. Phys. Chem. A.2006;110(42) 11824-11838, ISSN:1089-5639.

22. Lessard J, Morin JF, Wehrung JF, Magnin D, Chornet E. High Yield Conversion of Residual Pentoses into Furfural via Zeolite Catalysis and Catalytic Hydrogenation of Furfural to 2-Methylfuran. Top Catal. 2010;53(15-18) 1231-1234, ISSN:10225528.

23. Bugrayev A, Al-Haq N, Okopie RA, Qazi A, Suggate M, Sullivan AC, Wilson JRH. Covalently linked ethylmercaptophenyl sulfonic acid and ethylmercaptobenzyl sulfonic acid silica materials - Synthesis and catalytic activity. Journal of Molecular Catalysis A: Chemical 2008;280(1-2) 96-101, ISSN: 1381-1169.

24. Tao F, Song H, Chou L. Efficient process for the conversion of xylose to furfural with acidic ionic liquid. Canadian Journal of Chemistry 2011;89(1) 83-87, DOI: 10.1139/V10-153.

25. Rong C, Ding X, Zhu Y, Li Y, Wang L, Qu Y, Ma X, Wang Z. Production of furfural from xylose at atmospheric pressure by dilute sulfuric acid and inorganic salts.Carbohydrate Research 2012;350(1) 77-80, ISSN:1873-426X.

26. Lima S, Neves P, Antunes MM, Pillinger M, Ignatyev N, Valente AA. Conversion of mono/di/polysaccharides into furan compounds using 1-alkyl-3-methylimidazolium ionic liquids. Appl. Catal. A: Gen 2009;363(1-2)93-99, DOI: 10.1016/j.apcata.2009.04.049.

27. Gürbüz EI, Wettstein SG, Dumesic JA. Conversion of Hemicellulose to Furfural and Levulinic Acid using Biphasic Reactors with Alkylphenol Solvents. ChemSusChem 2012;5(2) 383-387,.

28. Yang Y, Hu CW, Abu-Omar MM. Synthesis of Furfural from Xylose, Xylan, and Biomass Using AlCl3 6H2O in Biphasic Media via Xylose Isomerization to Xylulose. ChemSusChem 2012;5(2) 405-410,

29. Agirrezabal-Telleria I, Larreategui A, Requies J, Güemez MB, Arias PL. Furfural production from xylose using sulfonic ion-exchange resins (Amberlyst) and simultaneous stripping with nitrogen. Bioresource Technology 2011;102(16) 7478-7485,

30. Lima S, Antunes MM, Fernandes A, Pillinger M, Ribeiro MF, Valente AA. Catalytic cyclodehydration of xylose to furfural in the presence of zeolite H-Beta and a micro/mesoporous Beta/TUD-1 composite material. Applied Catalysis, A: General 2010;388(1-2) 141-148, ISSN: 0926-860X.

31. Antunes MM, Lima S, Fernandes A, Pillinger M, Ribeiro MF,Valente AA. Aqueous-phase dehydration of xylose to furfural in the presence of MCM-22 and ITQ-2 solid acid catalysts. Applied Catalysis, A: General 2012; 417-418 (29) 243-252,

32. vom Stein T, Grande PM, Leitner W, Dominguez de Maria P. Iron-Catalyzed Furfural Production in Biobased Biphasic Systems: From Pure Sugars to Direct Use of Crude Xylose Effluents as Feedstock. ChemSusChem 2011;4(11) 1592-1594, DOI:10.1002/cssc.201100259.

33. Lam E, Majid E, Leung AC, Chong JH, Mahmoud KA, Luong JH. Synthesis of Furfural from Xylose by Heterogeneous and Reusable Nafion Catalysts. ChemSusChem2011;4(4) 535-541,.

34. Dias AS, Pillinger M, Valente AA. Liquid phase dehydration of D-xylose in the presence of Keggin-type heteropolyacids. Appl. Catal. A. 2005;285(1-2) 126-131, DOI: 10.1016/j.apcata.2005. 02.016.

35. Sádaba I, Lima S, Valente AA, López Granados M. Catalytic dehydration of xylose to furfural: vanadyl pyrophosphate as source of active soluble species. Carbohydrate Research 2011;346(17) 2785-2791,

36. Win DT. Furfural-Gold from Garbage. AU J.T. 2005;8(4) 185-190, http://www.journal.au.edu/au_ techno/2005/apr05/vol8no4_abstract04.pdf.

37. Jinan ZZ International Trade Co., Ltd.

38. Zibo Pulisi Trading Co., Ltd.

39. Zhou Z, Ma Q, Zhang A, Wu M. Synthesis of water-soluble monotosylated ethylenediamines and their application in ruthenium and iridium-catalyzed transfer hydrogenation of aldehydes. Applied Organometallic Chemistry 2011;25(12)856-861,

40. Ajjou AN, Pinet JL. The biphasic transfer hydrogenation of aldehydes and ketones with isopropanol catalyzed by water-soluble rhodium complexes. Journal of Molecular Catalysis A: Chemical 2004;214(2)203-206, ISSN:1381-1169.

41. Li H, Luo H, Zhuang L, Dai W, Qiao M. Liquid phase hydrogenation of furfural to furfuryl alcohol over the Fe-promoted Ni-B amorphous alloy catalysts. Journal of Molecular Catalysis A: Chemical 2003;203(1-2)267-275,.

42. Inada K, Shibagaki M, Nakanishi Y, Matsushita H. The catalytic reduction of aldehydes and ketones with 2-propanol over silica-supported zirconium catalyst. ChemInform 1994;25(15) no pages, DOI: 10.1002/chin.199415090.

43. Wabnitz T, Breuninger D, Heimann J, Backes R, Pinkos R. Method for one-step synthesis of 2-methyltetrahydrofuran from furfural and catalyst 2009; PCT Int. Appl., WO/ 2009/003882 A1 20090108.

44. Wabnitz T, Breuninger D, Heimann J, Backes R, Pinkos R. Production of 2-methyltetrahydrofuran from furfural over two catalysts in a structured bed 2009; PCT Int. Appl., WO/2009/003881 A1 20090108.

45. Lange JP. Process for the hydrogenolysis of furfuryl derivatives PCT Int. Appl.,2009; PCT/EP2009/057899, WO/2009/156439 A1 20091230, (June 2009).

46. Zheng HY, Zhu YL, Teng BT, Bai ZQ, Zhang CH, Xiang HW, Li YW. Towards understanding the reaction pathway in vapor phase hydrogenation of furfural to 2-methylfuran. Journal of Molecular Catalysis A: Chemical 2006;246(1-2) 18-23,

47. TPFTZ Qingshan Int'l Business Technology Co., Ltd.

48. Zhang W, Zhu Y, Niu S, Li Y. A study of furfural decarbonylation on K-doped Pd/Al2O3 catalysts. Journal of Molecular Catalysis A: Chemical 2011;335(1-2) 71-81, ISSN:1873-314X.

49. Stevens JG, Bourne RA, Twigg MV, Poliakoff M. Real-Time Product Switching Using a Twin Catalyst System for the Hydrogenation of Furfural in Supercritical CO2. Angewandte Chemie International Edition 2010; 49(47), ISSN:1433-7851.

50. Popoff T, Theander O. Reactions of D-xylose and D-glucose in alkaline aqueous solutions.Carbohydrate Research1976;48(1) 13-21,

51. Johansson MH, Samuelson O. Reducing end groups in brich xylan and their alkaline degradation. Wood Science and Technology 1977; 11(4) 251-163, ISSN:0043-7719.

52. El Khadem HS, Ennifar S, Isbell HS. Contribution of the reaction pathways involved in the isomerization of monosaccharides by alkali. Carbohydrate Research 1987;169 13-21, ISSN: 0008-6215.

53. Jackson JE, Miller DJ, Marincean S. Process for the preparation of lactic acid and glyceric acid from aqueous solutions of hexoses and pentoses in the presence of an excess of a strongly basic anionic exchange resin 2007; U.S. Pat. Appl. Publ., 20070066844.

54. Epane G, Laguerre JC, Wadouachi A, Marek D. Microwave-assisted conversion of D-glucose into lactic acid under solvent-free conditions. Green Chemistry 2010;12(3) 502-506,

55. Rahubadda A, Montoya A, Haynes BS. C5 sugar decomposition products under hot compressed water conditions [online]. In:CHEMECA 2011: Engineering a Better World: Sydney Hilton Hotel, NSW, Australia, 2011;18-21 Barton, A.C.T.: Engineers Australia, 2011: 698-706. Conference paper. Availability: http:// search. informit. com. au/documentSummary;dn=168748429919269;res=IELENGISBN: 9780858259676.

56. Onda A, Ochi T, Kajiyoshi K, Yanagisawa K. A new chemical process for catalytic conversion of D-glucose into lactic acid and gluconic acid. Applied Catalysis, A: General 2008;343(1-2) 49-54,

57. Ma C, Jin F,Cao J,Wu B. Hydrothermal Conversion of Carbohydrates into Lactic Acid with Alkaline Catalysts: 4th International Conference onBioinformatics and Biomedical Engineering (iCBBE), June 18-20, 2010, Chengdu, China, 1-4,

58. Aspinall GO,Greenwood CT,Sturgeon RJ. The degradation of xylans by alkali. J. Chem. Soc. 1961;3667-3674,

59. Yang BY. Montgomery R. Alkaline degradation of fructofuranosides. Carbohydrate Research 1996;280(1) 47-57,

60. Raharja S, Rigal L, Barre L, VidalPF. Design of a continuous process for the alkaline treatment of xylose into lactic acid. The Canadian Journal of Chemical Engineering 1997;75(5)913-920,

61. Okano K, Tanaka T, Ogino C, Fukuda H, Kondo A. Biotechnological production of enantiomeric pure lactic acid from renewable resources: Recent achievements, perspectives, and limits. Applied Microbiology and Biotechnology 2010;85(3)413-423,

62. Yadav AK, Chaudhari AB, Kothari RM. Bioconversion of renewable resources into lactic acid: An industrial view. Critical reviews in biotechnology 2011;31(1) 1-19, PMID: 20476870.

63. Abdel-Rahman MA, Tashiro Y, Sonomoto K. Lactic acid production from lignocellulose-derived sugars using lactic acid bacteria: Overview and limits. Journal of Biotechnology 2011;156(4) 286-301,

64. Datta R, Henry M. Lactic acid: Recent advances in products, processes and technologies-A review. Journal of Chemical Technology and Biotechnology 2006;81(7) 1119-1129, DOI: 10.1002/jctb.1486.

65. Garvie EI. Bacterial lactate dehydrogenases. Microbiological reviews 1980;44(1) 106-139. PMCID: PMC373236

66. John RP, Nampoothiri KM, Pandey A. Fermentative production of lactic acid from biomass: An overview on process developments and future perspectives. Applied Microbiology and Biotechnology 2007;74(3)524-534,

67. Wee Y, Kim J, Ryu H. Biotechnological production of lactic acid and its recent applications. Food Technology and Biotechnology 2006;44(2) 163-172, ISSN: 1330-9862.

68. Zhang ZY, Jin B, Kelly JM. Production of lactic acid from renewable materials by Rhizopus fungi. Biochemical engineering journal 2007;35(3) 251-263, ISSN:1369-703X.

69. Zhao B, Wang L, Li F, Hua D, Ma C, Ma Y, Xu P. Kinetics of D-lactic acid production by Sporolactobacillus sp. strain CASD using repeated batch fermentation. Bioresource technology 2010;101(16) 6499-6505,

70. Zhao B, Wang L, Ma C, Yang C, Xu P, Ma Y. Repeated open fermentative production of optically pure L-lactic acid using a thermophilic Bacillus sp. strain. Bioresource technology 2010;101(16) 6494-6498, DOI: 10.1016/j.biortech.2010.03.051.

71. Salminen S, Ramos P, Fonden R. Substrates and lactic acid bacteria. Food Science and Technology (New York, NY, United States) 1993;58295-306. ISBN:0-8247-8907-5.

72. Hang YD. Direct fermentation of corn to L(+)-lactic acid by Rhizopus oryzae. Biotechnology Letters 1989;11(4) 299-300,

73. Krištofíková L, Rosenberg M, Vlnová A, Šajbidor J, Čertík M. Selection of Rhizopus strains for L(+)-lactic acid and gamma-linolenic acid production. Folia Microbiologica (Prague, Czech Republic) 1991;36(5) 451-455,

74. Miura S, Arimura T, Itoda N, Dwiarti L, Feng JB, Bin CH, Okabe M. Production of L-lactic acid from corncob. Journal of Bioscience and Bioengineering 2004;97(3) 153-157, ISSN: 1389-1723.

75. Ding S, Tan T. L-lactic acid production by Lactobacillus casei fermentation using different fed-batch feeding strategies. Process Biochemistry (Amsterdam, Netherlands) 2006;41(6) 1451-1454, ISSN: 1359-5113.

76. Chang HN, Kim N, Kang J, Jeong CM, Choi J, Fei Q, Kim BJ, Kwon S, Lee SY, Kim J. Multi-stage high cell continuous fermentation for high productivity and titer. Bioprocess and Biosystems Engineering 2011;34(4) 419-431, ISSN 16157591.

77. Dumbrepatil A, Adsul M, Chaudhari S, Khire J, Gokhale D. Utilization of molasses sugar for lactic acid production by Lactobacillus delbrueckii subsp. delbrueckii mutant Uc-3 in batch fermentation. Applied and Environmental Microbiology 2008;74(1) 333-335,

78. Bustos G, de la Torre N, Moldes AB, Cruz JM, Dominguez JM. Revalorization of hemicellulosic trimming vine shoots hydrolyzates trough continuous production of lactic acid and biosurfactants by L. pentosus. Journal of Food Engineering 2006;78(2) 405-412, ISBN:0260-8774.

79. Zhu Y, Lee YY, Elander RT. Conversion of aqueous ammonia-treated corn stover to lactic acid by simultaneous saccharification and cofermentation. Applied Biochemistry and Biotechnology 2007;137-140(1-2) 721-738, ISSN:0273-2289.

80. Givry S, Prevot V, Duchiron F. Lactic acid production from hemicellulosic hydrolyzate by cells of Lactobacillus bifermentans immobilized in Ca-alginate using response surface methodology. World Journal of Microbiology & Biotechnology 2008;24(6) 745-752,

81. Chaillou S, Bor Y, Batt CA, Postma PW, Pouwels PH. Molecular cloning and functional expression in Lactobacillus plantarum 80 of xylT, encoding the D-xylose-H+ symporter of Lactobacillus brevis. Applied and Environmental Microbiology 1998;64(12) 4720-4728, PMCID:PMC90914.

82. Helanto M, Kiviharju K, Leisola M, Nyyssölä A. Metabolic engineering of Lactobacillus plantarum for production of L-ribulose. Applied and Environmental Microbiology 2007;73(21) 7083-7091,.

83. Ohara H, Owaki M, Sonomoto K. Xylooligosaccharide fermentation with Leuconostoc lactis. Journal of Bioscience and Bioengineering 2006;101(5) 415-420, ISSN:1389-1723.

84. Tanaka K, Komiyama A, Sonomoto K, Ishizaki A, Hall SJ, Stanbury PF. Two different pathways for D-xylose metabolism and the effect of xylose concentration on the yield coefficient of L-lactate in mixed-acid fermentation by the lactic acid bacterium Lactococcus lactis IO-1. Applied Microbiology and Biotechnology 2002;60(1-2) 160-167, ISSN:0175-7598.

85. Abdel-Rahman M, Tashiro Y. Zendo T, Sonomoto K. Effective (+)-Lactic Acid Production by Co-fermentation of Mixed Sugars. Journal of Biotechnology 2010;150(Supplement) 347-348,

86. Abdel-Rahman MA, Tashiro Y, Zendo T, Hanada K, Shibata K, Sonomoto K. Efficient homofermentative L-(+)-lactic acid production from xylose by a novel lactic acid bacterium, Enterococcus mundtii QU 25. Applied and Environmental Microbiology 2011;77(5) 1892-1895,.

87. Okano K, Yoshida S, Yamada R, Tanaka T, Ogino C, Fukuda H, Kondo A. Improved production of homo-D-lactic acid via xylose fermentation by introduction of xylose assimilation genes and redirection of the phosphoketolase pathway to the pentose phosphate pathway in L-lactate dehydrogenase gene-deficient Lactobacillus plantarum. Applied and Environmental Microbiology 2009;75(24) 7858-7861, DOI: 10.1128/AEM.01692-09.

88. Tamakawa H, Ikushima S, Yoshida S. Efficient production of L-lactic acid from xylose by a recombinant Candida utilis strain. Journal of Bioscience and Bioengineering 2012;113(1) 73-75, ISSN:1389-1723.

89. Okano K, Yoshida S, Tanaka T, Ogino C, Fukuda H, Kondo A. Homo-D-lactic acid fermentation from arabinose by redirection of the phosphoketolase pathway to the pentose phosphate pathway in L-lactate dehydrogenase gene-deficient Lactobacillus plantarum. Applied and Environmental Microbiology 2009;75(15)5175-5178, ISSN: 0099-2240.

90. Ilmen M, Koivuranta K, Ruohonen L, Suominen P, Penttila M. Efficient production of L-lactic acid from xylose by Pichia stipitis. Applied and Environmental Microbiology 2007;73(1)117-123,

91. Wee Y, Ryu H. Lactic acid production by Lactobacillus sp. RKY2 in a cell-recycle continuous fermentation using lignocellulosic hydrolyzates as inexpensive raw materials. Bioresource technology 2009; 100(18)4262-4270, ISSN:0960-8524.

92. John RP, Nampoothiri KM, Pandey A. Simultaneous saccharification and fermentation of cassava bagasse for L-(+)-lactic acid production using Lactobacilli. Applied Biochemistry and Biotechnology 2006;134(3)263-272, ISSN:0273-2289.

93. Adsul M, Khire J, Bastawde K, Gokhale D. Production of lactic acid from cellobiose and cellotriose by Lactobacillus delbrueckii mutant Uc-3. Applied and Environmental Microbiology 2007;73(15) 5055-5057,

94. Singhvi M, Joshi D, Adsul M, Varma A, Gokhale D. D-(-)-Lactic acid production from cellobiose and cellulose by Lactobacillus lactis mutant RM2-24. Green Chemistry 2010;12(6)1106-1109,

95. Marques S, Santos JA. L Gírio, FM, Roseiro JC. Lactic acid production from recycled paper sludge by simultaneous saccharification and fermentation. Biochemical engineering journal 2008;41(3)210-216,

96. Gullon B, Yáñez R, Alonso JL, Parajó JC. L-Lactic acid production from apple pomace by sequential hydrolysis and fermentation. Bioresource technology 2008;99(2) 308-319, ISSN:0960-8524.

97. Ishida N, Saitoh S, Onishi T, Tokuhiro K, Nagamori E, Kitamoto K, Takahashi H. The effect of pyruvate decarboxylase gene knockout in Saccharomyces cerevisiae on L-lactic acid production. Bioscience, biotechnology, and biochemistry 2006;70(5)1148-1153, PMID: 16717415.

98. Tokuhiro K, Ishida N, Nagamori E, Saitoh S, Onishi T, Kondo A, Takahashi H. Double mutation of the PDC1 and ADH1 genes improves lactate production in the yeast Saccharomyces cerevisiae expressing the

bovine lactate dehydrogenase gene. Applied Microbiology and Biotechnology 2009;82(5)883-890, ISSN:0175-7598.

99. Bianchi MM, Brambilla L, Protani F, Liu C, Lievense J, Porro D. Efficient homolactic fermentation by Kluyveromyces lactis strains defective in pyruvate utilization and transformed with the heterologous LDH gene. Applied and Environmental Microbiology 2001;67(12)5621-5625, ISSN: 0099-2240.

100. Tianjin Tiger International Trade Co., Ltd.

101. Qingdao Abel Technology Co., Ltd.

102. PRWeb Online visibility on Vocus, Global Lactic Acid Market to Reach 367.3 Thousand Metric Tons by 2017, According to New Report by Global Industry Analysts, Inc.

103. Wisniak J, Hershkowitz M, Stein S. Hydrogenation of Xylose to Xylitol, Ind. Eng Chem., Prod. Res. Develop. 1974;13(1)75,

104. Wisniak J, Hershkowitz M, Stein S. Hydrogenation of Xylose over Platinum Group Catalysts. Ind. Eng. Chem. Prod. Res. Dev. 1974;13(4)232-236,

105. Mikkola JP, Salmi T. Three-phase catalytic hydrogenation of xylose to xylitol - prolonging the catalyst activity by means of on-line ultrasonic treatment. Catalysis Today 2001;64(3-4)271-277,

106. Mikkola JP, Kubicka D, Kuusisto J, Granholm N, Salmi T, Holmbom B. Non-traditional three-phase reactor setup for simultaneous acoustic irradiation and hydrogenation. Journal of Chemical Technology and Biotechnology 2003;78(2-3)203-207,

107. Heikkila H, Ojamo H, Tylli M, Ravanko V, Nurmi J, Haimi P, Alen R, Koivikko H. Preparation of L-xylose and its use for the production of xylitol. U.S. Pat. Appl. Publ., 20030097029, (May 2003).

108. Kuusisto J, Heikkilae H, Tylli M, Golde M, Riihimaeki T. Catalytic hydrogenation of sugar to sugar alcohol, using increasing reaction temperature and/or addition of monocarboxylic acid to maintain catalyst activity.Brit. UK Pat. Appl. GB 2437517(October2007).

109. Vanoppen D, Maas-Brunner M, Kammel U, Arndt JD. Preparation and use of silicon dioxide-supported ruthenium catalysts for saccharide hydrogenation, Ger. Offen. DE10128205(December 2002).

110. Guha SK, Kobayashi H, Hara K, Kikuchi H, Aritsuka T, Fukuoka A. Hydrogenolysis of sugar beet fiber by supported metal catalyst. Catalysis Communications 2011;12(11) 980-983,

111. Kwak BS, Lee BI, Kim TY, Kim JW, Lee SI. Process for preparation of sugar alcohols by ruthenium-catalyzed hydrogenation of sugars. PCT Int. Appl., WO 2005021475 A1 20050310 (2005).

112. Kwak BS, Lee BI, Kim TY, Kim JW, Lee SI. Method for preparing sugar alcohols using ruthenium- and zirconia- catalyzed hydrogenation of monosaccharides. PCT Int. Appl., WO 2006093364 A1 20060908 (2006).

113. Crabtree SP, Tyers DV. Hydrogenolysis of sugar feedstock. PCT Int. Appl., PCT/GB2004.004391, WO/2005/051874, (June 2005).

114. Sun J, Liu H. Selective hydrogenolysis of biomass-derived xylitol to ethylene glycol and propylene glycol on supported Ru catalysts. Green Chemistry 2011;13(1)135-142,

115. Chopade SP, Miller DJ, Jackson JE, Werpy TA, Frye Jr JG, Zacher AH. Catalysts and process for hydrogenolysis of sugar alcohols to polyols. US Patent 6291725, (January 2001).

116. Dubeck M, Knapp GG. Two stage hydrogenolysis of carbohydrate to glycols using sulfide modified ruthenium catalyst in second stage. US Patent 4,476,331, (September 1984).

117. Holcomb DE. Catalyst and method for production of polyols by hydrogenolysis of carbohydrates. US Patent 7,692,001, (April 2010).

118. Kasehagen L. Hydrogenolysis of reducible sugars to obtaini a high percentage of glycerol. US Patent 3,396,199, (August 1968).

119. Wright LW. Hydrogenation and hydrogenolysis of carbohydrates with tungsten oxide promoted supported nickel catalyst. US Patent 3,965,199, (June 1976).

120. Werpy TA, Frye Jr JG, Zacher AH, Miller DJ. Hydrogenolysis of 6-carbon sugars and other organic compounds. US Patent 6,841,085, (January 2005).

121. Suzuki S, Sugiyama M, Mihara Y, Hashiguchi K, Yokozeki K. Novel enzymatic method for the production of xylitol from D-arabitol by Gluconobacter oxydans. Bioscience, biotechnology, and biochemistry 2002;66(12)2614-2620,

122. Rangaswamy S, Agblevor FA. Screening of facultative anaerobic bacteria utilizing D-xylose for xylitol production. Applied Microbiology and Biotechnology 2002;60(1-2)88-93, ISSN:0175-7598.

123. Yoshitake J, Ohiwa H, Shimamura M, Imai T. Production of polyalcohol by a Corynebacterium species. I. Production of pentitol

from aldopentose. Agricultural and Biological Chemistry 1971;35(6) 905-911.

124. Yoshitake J, Ishizaki H, Shimamura M, Imai T. Xylitol production by an Enterobacter species. Agricultural and Biological Chemistry 1973;37(10) 2261-2266.

125. Cirino PC, Chin JW, Ingram LO. Engineering Escherichia coli for xylitol production from glucose-xylose mixtures. Biotechnology and bioengineering 2006;95(6)1167-1176, ISSN:0006-3592.

126. Parajó JC, Domínguez H, Domínguez JM. Biotechnological production of xylitol. Part 1: Interest of xylitol and fundamentals of its biosynthesis. Bioresource technology 1998;65(3)191-201,

127. Izumori K, Tuzaki K. Production of xylitol from D-xylulose by Mycobacterium smegmatis. Journal of Fermentation Technology 1988;66(1) 33-36, DOI:10.1016/0385-6380(88)90126-4.

128. Monedero V, Pérez-Martínez G, Yebra MJ. Perspectives of engineering lactic acid bacteria for biotechnological polyol production. Applied Microbiology and Biotechnology 2010;86(4)1003-1015, ISSN:0175-7598.

129. Nyyssoelae A, Pihlajaniemi A, Palva A, von Weymarn N, Leisola M. Production of xylitol from D-xylose by recombinant Lactococcus lactis. Journal of Biotechnology 2005;118(1)55-66, ISSN:0168-1656.

130. Fridkin SK, Jarvis WR. Epidemiology of nosocomial fungal infections. Clinical microbiology reviews 1996;9(4)499-511, ISSN: 0893-8512.

131. Meinander NQ, Hahn-Haegerdal B. Fed-batch xylitol production with two recombinant Saccharomyces cerevisiae strains expressing XYL1 at different levels, using glucose as a cosubstrate: a comparison of production parameters and strain stability. Biotechnology and bioengineering 1997;54(4)391-399, ISSN:0006-3592.

132. Parajó JC, Domínguez H, Domínguez JM. Biotechnological production of xylitol. Part 2: Operation in culture media made with commercial sugars. Bioresource technology 1998;65(3)203-212, DOI:10.1016/S0960-8524(98)00036-4.

133. Lee WJ, Ryu YW, Seo JH. Characterization of two-substrate fermentation processes for xylitol production using recombinant Saccharomyces cerevisiae containing xylose reductase gene. Process Biochemistry (Oxford) 2000;35(10)1199-1203, DOI:10.1016/S0032-9592(00)00165-5.

134. Govinden R, Pillay B, Van Zyl WH, Pillay D. Xylitol production by recombinant Saccharomyces cerevisiae expressing the Pichia stipitis and Candida shehatae XYL1 genes. Applied Microbiology and Biotechnology 2001;55(1)76-80, ISSN:0175-7598.

135. Sasaki M, Jojima T, Inui M, Yukawa H. Xylitol production by recombinant Corynebacterium glutamicum under oxygen deprivation. Applied Microbiology and Biotechnology 2010;86(4)1057-1066, ISSN:0175-7598.

136. Akinterinwa O, Khankal R, Cirino PC. Metabolic engineering for bioproduction of sugar alcohols. Current opinion in biotechnology 2008;19(5)461-467,

137. Nigam P, Singh D. Processes for fermentative production of xylitol - a sugar substitute. Process Biochemistry (Oxford) 1995;30(2)117-124,

138. Dien BS, Cotta MA, Jeffries TW. Bacteria engineered for fuel ethanol production: current status. Applied Microbiology and Biotechnology 2003;63(3)258-266,

139. Hahn-Hagerdal B, Karhumaa K, Fonseca C, Spencer-Martins I, Gorwa-Grauslund M. Towards industrial pentose-fermenting yeast strains. Applied Microbiology and Biotechnology 2007;74(5)937-953, ISSN:0175-7598.

140. Khankal R, Chin JW, Cirino PC. Role of xylose transporters in xylitol production from engineered Escherichia coli. Journal of Biotechnology 2008;134(3-4)246-252, ISSN:0168-1656.

141. Sasaki M, Jojima T, Kawaguchi H, Inui M, Yukawa H. Engineering of pentose transport in Corynebacterium glutamicum to improve simultaneous utilization of mixed sugars. Applied Microbiology and Biotechnology 2009;85(1)105-115, ISSN:0175-7598.

142. Kinoshita S. Glutamic acid bacteria. In: Demain AL, Solomon NA (eds) Biology of industrial microorganisms. Benjamin/Cummins, London 1985;6 115-142.

143. Terasawa M, Yukawa H. Industrial production of biochemicals by native immobilization. Bioprocess technology 1993;16, Ind. Appl. Immobilized Biocatal.37-52.

144. Zhang J, Geng A, Yao C, Lu Y, Li Q. Xylitol production from D-xylose and horticultural waste hemicellulosic hydrolysate by a new isolate of Candida athensensis SB18. Bioresource technology 2012;105134-141, ISSN: 0960-8524.

145. Lopez F, Delgado OD, Martinez MA, Spencer JFT, Figueroa LIC. Characterization of a new xylitol-producer Candida tropicalis strain. Antonie van Leeuwenhoek 2004;85(4)281-286, ISSN:0003-6072.

146. Ikeuchi T, Azuma M, Kato J, Ooshima H. Screening of microorganisms for xylitol production and fermentation behavior in high concentrations of xylose. Biomass and Bioenergy 1999;16(5)333-339,

147. Kwon S, Park S, Oh D. Increase of xylitol productivity by cell-recycle fermentation of Candida tropicalis using submerged membrane bioreactor. Journal of Bioscience and Bioengineering 2006;101(1)13-18, ISSN:1389-1723.

148. Oh DK, Kim SY. Increase of xylitol yield by feeding xylose and glucose in Candida tropicalis. Applied Microbiology and Biotechnology 1998;50(4)419-425, ISSN:0175-7598.

149. Ojamo H. Yeast xylose metabolism and xylitol production. VTT Publications 1994;176, ISBN:9513844145.

150. Choi J, Moon K, Ryu Y, Seo J. Production of xylitol in cell recycle fermentations of Candida tropicalis. Biotechnology Letters 2000;22(20)1625-1628, DOI:10.1023/A:1005693427389.

151. Granström TB. Biotechnological production of xylitol with Candida yeasts. Aalto University, Library Publication Archive, 160, ISBN:951-22-5997-4.

152. Hallborn J, Walfridsson M, Airaksinen U, Ojamo H, Hahn-Hagerdal B, Penttila M, Kerasnen S. Xylitol production by recombinant Saccharomyces cerevisiae. Biotechnology (Nature Publishing Company) 1991;9(11)1090-1095, PMID 1367625

153. Salgado JM, Rodríguez N, Cortés S, Domínguez JM. Effect of nutrient supplementation of crude or detoxified concentrated distilled grape marc hemicellulosic hydrolysates on the xylitol production by Debaryomyces hansenii. Preparative biochemistry & biotechnology 2012;42(1) 1-14, ISSN: 1082-6068.

154. Oh E, Bae Y, Kim K, Park Y, Seo J. Effects of overexpression of acetaldehyde dehydrogenase 6 and acetyl-CoA synthetase 1 on xylitol production in recombinant Saccharomyces cerevisiae. Biocatalysis and Agricultural Biotechnology 2012;1(1)15-19,

155. Li M, Meng X, Diao E, Du F. Xylitol production by Candida tropicalis from corn cob hemicellulose hydrolysate in a two-stage fed-batch fermentation process. Journal of Chemical Technology and Biotechnology 2012;87(3)387-392, DOI:10.1002/jctb.2732.

156. Jeon YJ, Shin HS, Rogers PL. Xylitol production from a mutant strain of Candida tropicalis. Letters in applied microbiology 2011;53(1)106-113, ISSN:0266-8254.

157. Huang C, Jiang Y, Guo G, Hwang W. Development of a yeast strain for xylitol production without hydrolysate detoxification as part of the integration of co-product generation within the lignocellulosic ethanol process. Bioresource technology 2012;102(3)3322-3329, ISSN:0960-8524.

158. Barbosa MFS, de Medeiros MB, de Mancilha IM, Schneider H, Lee H. Screening of yeasts for production of xylitol from D-xylose and some factors which affect xylitol yield in Candida guilliermondii. Journal of industrial microbiology 1988;3(4)241-251,

159. Roseiro JC, Peito MA, Gírio FM, Amaral-Collaco MT. The effects of the oxygen transfer coefficient and substrate concentration on the xylose fermentation by Debaryomyces hansenii. Archives of Microbiology 1991;156(6) 484-490,

160. Ligthelm ME, Prior BA, du Preez JC. The oxygen requirements of yeasts for the fermentation of D-xylose and D-glucose to ethanol. Applied Microbiology and Biotechnology 1988;28(1)63-68, ISSN:0175-7598.

161. Debus D, Methner H, Schulze D, Dellweg H. Fermentation of xylose with the yeast Pachysolen tannophilus. European Journal of Applied Microbiology and Biotechnology 1983;17(5) 287-291,

162. Kötter P, Ciriacy M. Xylose fermentation by Saccharomyces cerevisiae. Applied Microbiology and Biotechnology 1993;38(6)776-783,

163. Hallborn J, Gorwa MF, Meinander N, Penttilä M, Keränen S, Hahn-Haegerdal B. The influence of cosubstrate and aeration on xylitol formation by recombinant Saccharomyces cerevisiae expressing the XYL1 gene. Applied Microbiology and Biotechnology 1994;42(2-3)326-333, ISSN: 0175-7598.

164. Sonderegger M, Jeppsson M, Hahn-Hagerdal B, Sauer U. Molecular Basis for Anaerobic Growth of Saccharomyces cerevisiae on Xylose, Investigated by Global Gene Expression and Metabolic Flux Analysis. Applied and Environmental Microbiology 2004;70(4)2307-2317,

165. Ahmad I, Shim WY, Jeon WY, Yoon BH, Kim J. Enhancement of xylitol production in Candida tropicalis by co-expression of two genes involved in pentose phosphate pathway. Bioprocess and Biosystems Engineering 2012;35(1-2)199-204, ISSN:1615-7591.

166. Shanghai Yanda Biotechnology Co., Ltd.

167. Wuhan Xuyadi Chemicals Co., Ltd.

168. Tianjin Yuanlong Chemical Industry Co., Ltd.

169. http://english.jl.gov.cn/Investment/Opportunities/Majorproject/201104/t20110401_971857.html

170. http://www.azom.com/news.aspx?newsID=24757

171. Bonrath W, Fischesser J. Novel reaction with a gold catalyst. WO2008148549 (December 2008).

172. Pruesse U, Heidkamp K, Decker N, Herrmann M, Vorlop KD. Processing of raw materials from renewable resources by selective oxidation with gold catalysts. Chemie Ingenieur Technik 2010;82(8) 1231-1237.

173. Van der Weijden RD, Mahabir J, Abbadi A. Reuter, MA. Copper recovery from copper (II) sulfate solutions by reduction with carbohydrates. Hydrometallurgy 2002;64(2)131-146,

174. Ericsson T, Samuelson O. Treatment of Birch Xylan with Chlorine in Aqueous Solution. Acta Chemica Seandinavica B,Summary citation from AGRICOLA, the online catalog of the National Agricultural Library (NAL)1975;29(3) 309-314, DOI: 10.3891/acta.chem.scand.29b-0309.

175. Jokic A, Ristic N, Jaksic MM, Spasojevic M, Krstajic N. Simultaneous electrolytic production of xylitol and xylonic acid from xylose. Journal of applied Electrochemistry 1991;21(4) 321-326, ISSN: 0021-891X.

176. Kiely DE, Hash KR. Process for the oxidization of monosaccharides to glycosyl carboxylic acids using oxygen and nitric acid as co-oxidants. U.S. Pat. Appl. Publ, US 20080033205(February 2008).

177. Venema FR, Peters JA, van Bekkum H. Platinum-catalyzed oxidation of aldopentoses to aldaric acids. Journal of Molecular Catalysis 1992;77(1)75-85;

178. Boussie TR, Dias EL, Fresco ZM, Murphy VJ. Production of glutaric acid and derivatives from carbohydrate-containing materials. PCT Int. Appl., AU-A-2010259935, WO 2010/144871, (December 2010).

179. Fleche G. Process for the manufacture of xylaric acid and uses thereof US Patent 5731467, (March 1998).

180. Isbell HS, Frush HL, Martin ET. Reactions of carbohydrates with hydroperoxides: Part I. Oxidation of aldoses with sodium peroxide. Carbohydrate Research 1973;26(2)287-295, DOI:10.1016/S0008-6215(00)84515-2.

181. Pullanikat P, Jung SJ, Yoo KS, Jung KW. Oxidative degradation of reducing carbohydrates to ammonium formate with H2O2 and NH4OH. Tetrahedron Letters 2010;51(47)6192-6194, ISSN:0040-4039.

182. Wölfel R,Taccardi N,Bösmann A,Wasserscheid P. Selective catalytic conversion of biobased carbohydrates to formic acid using molecular oxygen. Green Chemistry 2011;13(10)2759-2763,

183. Bernhauer K, Riedl-Tumova E. Oxidations by acetic acid bacteria. X. Oxidation of enantiomorphic forms of xylose and arabinose. Biochemische Zeitschrift 1950;321 26-30.

184. Ishizaki H, Ihara T, Yoshitake J, Shimamura M, Imai T. D-Xylonic acid production by Enterobacter cloacae. Nippon Nogei Kagaku Kaishi 1973;47(12) 755-761, DOI: 10.1271/nogeikagaku1924.47.755.

185. Buchert J, Viikari L, Linko M, Markkanen P. Production of xylonic acid by Pseudomonas fragi. Biotechnology Letters 1986;8(8) 541-546,

186. Uchida K, Suzuki Y. Physiology of plant pathogens. 3. Production of D-galactonic acid, 2-keto-D-galactonic acid, L-arabonic acid, and D-xylonic acid by Erwinia milletiae. Nippon Nogei Kagaku Kaishi 1975;49(5) 257-262.

187. Hayasida A. Demonstration of pentose oxidase and catalase in fusaria. Biochemische Zeitschrift 1938;298169-178.

188. Ohsugi M, Tochikura T, Ogata K. Production of D-xylonic acid by Micrococcus species. Agricultural and Biological Chemistry 1970;34(3)357-363.

189. Lockwood LB, Nelson GEN. The oxidation of pentoses by Pseudomonas. Journal of Bacteriology1946;52(5)581-586.

190. Weimberg R. Pentose oxidation by Pseudomonas fragi. Journal of Biological Chemistry 1961;236(3)629-635, PMID 13783864.

191. Kiessling H, Lindberg B, McKay J. Some products of the metabolism of D-xylose by Pullularia pullulans. Acta chemica Scandinavica 1962;16 1858-1862,

192. Buchert J, Puls J, Poutanen K. Comparison of Pseudomonas fragi and Gluconobacter oxydans for production of xylonic acid from

hemicellulose hydrolyzates. Applied Microbiology and Biotechnology 1988;28(4-5)367-372,

193. Stephens C, Christen B, Fuchs T, Sundaram V, Watanabe K, Jenal U. Genetic analysis of a novel pathway for D-xylose metabolism in Caulobacter crescentus. Journal of Bacteriology 2007;189(5)2181-2185, ISSN: 0021-9193.

194. Yamanaka K, Gino M, Kaneda R. A specific NAD-D-xylose dehydrogenase from Arthrobacter sp. Agricultural and Biological Chemistry 1977;41(8)1493-1499, ISSN:0002-1369.

195. Nygard Y, Toivari MH, Penttila M, Ruohonen L, Wiebe MG. Bioconversion of d-xylose to d-xylonate with Kluyveromyces lactis. Metabolic engineering 2011;13(4)383-391, ISSN:1096-7176.

196. Meijnen J, de Winde JH, Ruijssenaars HJ. Establishment of oxidative D-xylose metabolism in Pseudomonas putida S12. Applied and Environmental Microbiology 2009;75(9)2784-2791, ISSN: 0099-2240.

197. Novick NJ, Tyler ME. (1982). L-Arabinose metabolism in Azospirillum brasiliense. Journal of Bacteriology 1982;149(1)364-367, PMCID PMC216631.

198. Toivari MH, Ruohonen L, Richard P, Penttilae M, Wiebe MG. Saccharomyces cerevisiae engineered to produce D-xylonate. Applied Microbiology and Biotechnology 2010;88(3)751-760, ISSN:0175-7598.

199. Liu H, Valdehuesa KNG, Nisola GM, Ramos KRM, Chung W. High yield production of D-xylonic acid from D-xylose using engineered Escherichia coli. Bioresource technology 2012;115244-248, ISSN:0960-8524.

200. Henan Premtec Enterprise Corporation

201. http://www.prweb.com/releases/formic_acid/leather_textiles/prweb 1569064.htm

CHAPTER 2

An overview of key pretreatment processes employed for bioconversion of lignocellulosic biomass into biofuels and value added products

Venkatesh Chaturvedi , Pradeep Verma

School of Biotechnology, Banaras Hindu University, Department of Biotechnology, Guru Ghasidas Vishwavidyalaya

Department of Microbiology, Central University of Rajasthan

ABSTRACT

The hunt for alternative sources of energy generation that are inexpensive, ecofriendly, renewable and can replace fossil fuels is on, owing to the increasing demands of energy. One approach in this direction is the conversion of plant residues into biofuels wherein lignocellulose, which forms the structural framework of plants consisting of cellulose, hemicellulose and lignin, is first broken down and hydrolyzed into simple fermentable sugars, which upon fermentation form biofuels such as ethanol. A major bottleneck is to disarray lignin which is present as a protective covering and makes cellulose and hemicellulose recalcitrant to enzymatic hydrolysis. A number of biomass deconstruction or pretreatment processes (physical, chemical and biological) have been used to break the structural framework of plants and depolymerize lignin. This review surveys and discusses some major pretreatment processes pertaining to the pretreatment of plant biomass, which are used for the production of biofuels and other value added products. The emphasis is given on processes that provide maximum amount of sugars, which are subsequently used for the production of biofuels.

KEYWORDS

Pretreatment, Biomass, Lignin, Cellulose, Biofuel

1. INTRODUCTION

In recent years, due to the world's energy demand and with the uncertainty in the supply, ever increasing prices of fossil fuels, causing increased environmental pollution, it has become imperative to scout alternative sources of energy which can replace conventional fossil fuels (Wan and Li 2011). In this context, the conversion of plant biomass/products into biofuels and biochemicals has gained impetus due to the feasibility of an alternative processes available to convert the complex biomass into biofuels and biomaterials (Adsul et al. 2011). As a result, the lignocellulosic materials become again an important resource to provide and support a sustainable and environmentally friendly system (Itziar et al. 2012). Figure 1 summarizes the conversion of plant biomass into biofuels and other value added products.

Biofuels can be broadly categorized as first- and second-generation biofuels. Biofuels which are derived from starch, sugars, and plant oils are termed as first-generation biofuels. These include bioethanol derived from food crops such as sugar beet, cereals, sugarcane and biodiesel obtained from rape, sunflower, soya, and palm, respectively (Ghose et al. 1983). Biofuels derived from lignocellulosic biomass such as hard/soft wood, agricultural wastes are called as second-generation biofuels (Sun and Cheng 2002) (Fig. 2).

Among these, the production of the researches in the direction of second-generation biofuels is in stride due to easy availability of lignocellulosic biomass (Sánchez and Cardona 2008). According to National Renewable Energy Laboratory (NREL), in 2010, production of ethanol in the USA was around 5 billion gallons/annum and was contributed by food grains. With the advent of second-generation biofuels, it is expected that in 2015, the production will rise to 8 billion gallons/annum and a major portion will be contributed by lignocellulosic biomass.

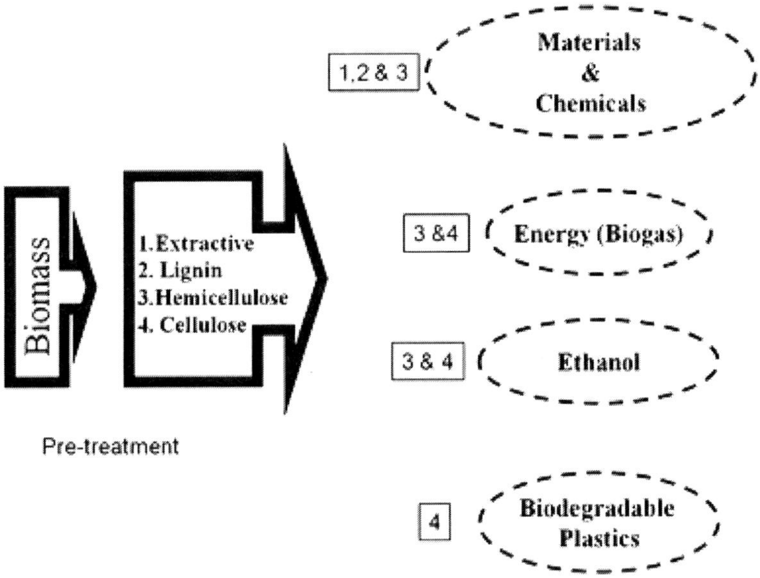

Figure. 1 Schematic representation of steps involved in the conversion of plant biomass into biofuels and other value added products (Alvira et al. 2010)

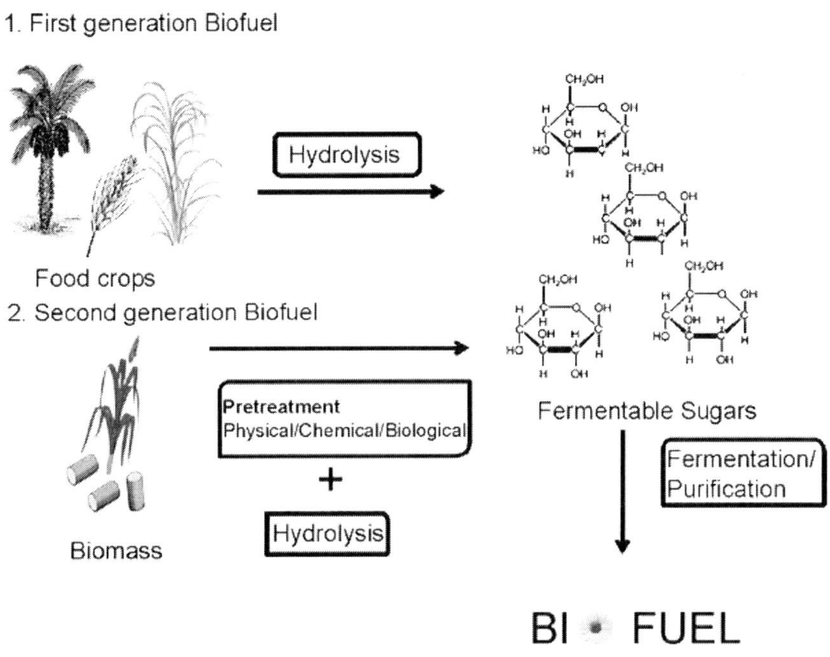

Figure. 2 Steps involved in the synthesis of the first- and second-generation biofuels (Sun and Cheng 2002)

A major portion of this biomass comprises cellulose, hemicelluloses, lignin and pectin. Among these constituents, cellulose and hemicelluloses are polymers of sugars and can be hydrolyzed to form fermentable sugars. Lignin forms a protective covering, which protects cellulose and hemicelluloses from degradation by various fungi and bacteria. Presence of lignin impedes enzymatic hydrolysis of cellulose and hemicellulose. Therefore, a pretreatment processing step is required for depolymerization of lignin before conversion of cellulose and hemicellulose into simple sugars (Balan et al. 2009; Adsul et al. 2011).

A number of pretreatment technologies based on numerous physical, chemical and biological methods have been developed, which alter/damage the structure of lignocelluloses and remove lignin (Alvira et al. 2010; Geddes et al. 2011). The exposed complex carbohydrates such as cellulose and hemicelluloses are then hydrolyzed to form fermentable sugars (Fig. 3). An effective pretreatment process should have following advantages: it should preserve and decrystallize the celluloses, depolymerize hemicelluloses; it should restrict the formation of inhibitors which resist the hydrolysis of carbohydrates; it should have low energy input, recovery of value added products such as lignin; and at last it should be cost effective (Agbor et al.2011).

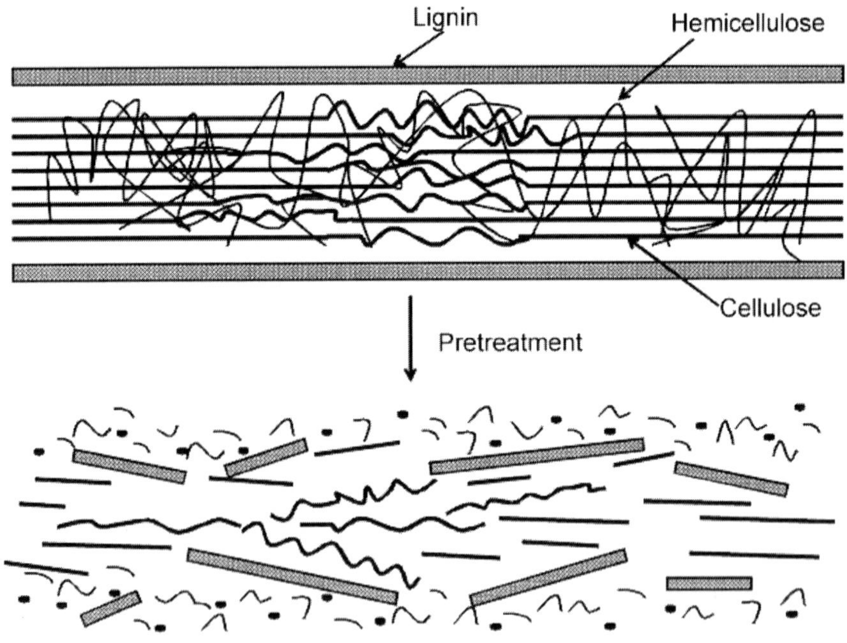

Figure. 3 Effects of various pretreatment processes on structure of lignocellulose (adapted from Hsu et al. 1980)

In this review, we have emphasized on some of the major and widely used chemical, and biological pretreatment processes which are employed for treatment of different lignocellulosic biomass aiming at removal of lignin and conversion of cellulose and hemicellulose into reducing sugars, which can be further employed for the production of bioethanol or other value added products. We have discussed the main principles behind these pretreatment processes, their mechanisms, merits/demerits and maximum yield of sugars which is obtained. The processes are as follows.

2. PRETREATMENT OF BIOMASS UNDER ACIDIC CONDITIONS

This process involves the treatment of lignocellulosic biomass with different acids such as sulfuric acid, oxalic acid, peracetic acid, respectively. Dilute acid treatment is considered as a cheap and effective pretreatment method due to low cost and easy availability of acids (Kim et al. 2005). Acid treatment is carried out in the presence of high and low concentrations of acids, and at high and low temperatures. At high temperatures, pretreatment of biomass is performed at temperature greater than 160 °C in continuous process containing a low concentration of biomass (5–10 % substrate concentration). In general, dilute sulfuric acid is sprayed on raw biomass and incubated at 160–220 °C for few minutes (González et al. 1986). At high temperature (160–220 °C), hemicellulose gets hydrolyzed, releasing monomeric sugars and soluble oligomers from the cell wall matrix. Hemicellulose removal increases the porosity and improves enzymatic digestibility of cellulose (Kim et al. 2005). During dilute acid treatment, hemicellulose is hydrolyzed, whereas cellulose and lignin are unaffected. The major pit hole of such process is the synthesis of furfural and hydroxymethyl furfural, which are produced by dehydration of hexoses and pentoses, respectively (Liu and Song 2009). These are considered as a strong inhibitor in microbial fermentation. While at low temperatures, the process is carried out at temperature below 160 °C in batch process at high biomass concentration (10–40 % of substrate). In this process, concentrated strong acids such as H_2SO_4 and HCl are widely used for treating lignocellulosic materials because they are powerful agents for cellulose hydrolysis (Sun and Cheng2002). In the presence of concentrated acids, hemicellulose and cellulose get hydrolyzed completely, whereas lignin is unaffected. The advantage of using concentrated acids in the pretreatment process is the high yield of monomeric sugar, but the major drawback of concentrated acid is their corrosive nature and the need to recycle acids to lower the cost of pretreatment.

Hog fuel and pin chips are byproducts produced by sawmills. A mixture of Hemlock hog fuel/pin chips (85:15 ratio) was pretreated by a two-stage dilute sulfuric acid treatment (Kim et al. 2005). The optimum conditions in this two stage treatment regime were; in the first stage biomass was treated with 1.1 % acid at 190 °C for 150 s and in the second stage biomass was treated with 2.5 % acid at 210 °C for 115 s respectively. Results indicated that at these pretreatment conditions, total sugar yield of 26.9 % was obtained. Similarly, olive tree biomass consisting of thin branches and leaves was pretreated with dilute sulfuric acid (Cara et al. 2008). Pretreatment was performed in the presence of varying acid concentration (0.2–1.4 %) and at different temperatures (170–210 °C). Highest sugar yield (48.6 %) was observed at 170 °C and 1.0 % acid concentration. Biomass (5–20 wt%), sulfuric acid (0.5–1.5 vol. %) and temperature at 120 and 140 °C were varied using dilute sulfuric acid treatment of corn fibers by Noureddini and Byun (2010). They observed that maximum yield of reducing sugars (56.8 %) was obtained at 0.5 % sulfuric acid, 5 % biomass and at 140 °C. Furfural formation was low as compared to other conditions. Similarly, sugarcane tops were treated with dilute acid treatment under optimized conditions for ethanol production by Sindhu et al. (2011). Maximum (68.5 %) sugar yield was obtained in the presence of 25 % w/w biomass, 3 % sulfuric acid for 60 min. Apart from sulfuric acid, many workers have employed other acids such as Peracetic acid (PAA) and oxalic acid for the treatment of biomass. Corn cobs were pretreated with oxalic acid (Lee et al. 2011). In their study, Corncob pellets were impregnated with oxalic acid. The pellets were rapidly heated to 168 °C and held for 26 min. Results indicated that in oxalic acid treatment, a total sugar yield of 13.1 % was obtained. The treated biomass also contained a low level of inhibitors. Further, Lee and Jeffries (2011) compared the effect of maleic, oxalic and sulfuric acid on the treatment of corn cob. Their results clearly indicated that in the presence of maleic acid, highest yield of reducing sugars 10 % was obtained. The treatment resulted in higher levels of furfural and hydroxymethyl furfural formation. Many workers have evaluated the combination of two pretreatment processes for biomass. Kim et al. (2011) studied pretreatment of rice straw in a two-stage pretreatment process using aqueous ammonia and dilute sulfuric acid in percolation mode. Under optimum conditions, the yield of reducing sugars in both the stages was 96.9 and 90.8 %, respectively. Their study clearly indicated that combination of these two processes resulted in better removal of lignin and hemicellulose from rice straw. Similar to this study, pretreatment of *Eulaliopsis binata*, a perennial grass commonly found in India and China, was performed using dilute sulfuric acid (Tang et al. 2013). The optimum conditions for the pretreatment were 0.5 % dilute sulfuric acid, 160 °C for 30 min, and at a solid-to-liquor ratio of 1:5. The results showed that the pretreatment liquor contained 21.02 % total sugars, 3.22 % lignin and 3.34 % acetic acid and low levels of inhibitors. Steam pretreatment of corn stover in combination with sulfuric acid was studied by Bondesson et al. (2013). The highest yield using sulfuric acid (78 %) was achieved using pretreatment at 200 °C for 10 min. In order to obtain maximum sugar yield, the synergetic effect of mild acid and alkali with Electron Beam Irradiation (EBI) on the enzymatic hydrolysis of grass was evaluated by Karthika et al. (2012). In their experimental setup, grass samples were soaked with 1 % H_2SO_4/1 % NaOH, exposed to 75 and 150 kGy of EBI. Water-presoaked biomass was used as control. Hydrolysis of pretreated samples was carried out using cellulase (15 FPU/g biomass) for 120 h. Reducing sugar and glucose yields from enzymatic hydrolysis were 32 % in NaOH treated and exposed to 75 kGy, and 43 % in NaOH treated and 150 kGy treated, respectively. The yields were significantly higher when compared with untreated control (19 %). Removal of hemicellulose, decreased crystallinity and structural changes were major factors for the combined treatment effect favoring the hydrolysis. This survey clearly indicates that only acid treatment is insufficient for higher sugar yields and thus combination of two different pretreatment processes, such as the study conducted by Kim et al. (2011), where rice husk was pretreated in two stages with two pretreatment processes and very high yield of reducing sugars (90.8 %) was obtained.

Acid pretreatment of biomass is inexpensive because acids such as sulfuric acid and hydrochloric acid employed are cheap. The process is carried out at high temperatures, and therefore, it requires high energy input, which is costly. Secondly, presence of acids at high temperatures can be corrosive, and thus, the process requires specific reaction vessels which must be resistant to these conditions. In addition, acid treatment generates inhibitors which have to be removed and as a result downstream processing costs are high.

3. PRETREATMENT OF BIOMASS UNDER ALKALINE CONDITIONS

In the alkaline treatment, biomass is treated with alkali such as sodium, potassium, calcium and ammonium hydroxides at normal temperature and pressures. The main advantage of the process is efficient removal of lignin from the biomass. Many reports have demonstrated that the process removes acetyl and uronic acid groups present on hemicellulose and thus enhances the accessibility of the enzyme that degrades hemicellulose (Chang and Holtzapple 2000). In the alkali treatment process, ester linkages that join xylan and hemicelluloses residues are also hydrolyzed (Sun and Cheng 2002). Alkaline treatment is also divided into two types on the basis of alkali employed. These include:

4. PRETREATMENT WITH CALCIUM, SODIUM AND POTASSIUM HYDROXIDE

Lime (calcium hydroxide) or sodium hydroxide is most commonly employed. The reaction conditions are generally mild, which prevent condensation of lignin leading to its high solubility and greater removal. In these conditions, degradation of sugars is minimal (Sharma et al. 2012). Such process is usually applied to biomass having low lignin content such as grasses and softwood. Addition of air or oxygen has shown to significantly enhance the delignification process (Chang and Holtzapple 2000). Lime pretreatment of corn stover was studied by Kim and Holtzapple (2005). Corn stover was pretreated with an excess of calcium hydroxide (0.5 g $Ca(OH)_2$/g raw biomass) in non-oxidative and oxidative conditions at 55 °C for 4 weeks with aeration. Under these conditions, the total yield of glucose and xylose was 91.3 and 51.8 % at 15 FPU/g cellulose, and 95.5 and 53.5 % at 60 FPU/g cellulose, respectively. Similarly, lime pretreatment of rice hulls for the production of ethanol was studied by Saha and Cotta (2008). In their study, milled rice hulls (15.0 % w/v) and lime at different concentrations were mixed in water, and autoclaved at 121 °C for 1 h. Results indicated that the maximum yield of total sugars (12.6 %) was achieved at 100 mg lime/g hulls and 1 h pretreatment time. Also, lime pretreatment did not generate furfural and hydroxymethyl furfural in the hydrolysate. Many workers have evaluated the effect of sodium hydroxide on different biomasses. Zhao et al. (2008) studied alkaline pretreatment using sodium hydroxide of spruce at low temperature in both presence and absence of urea. It was observed that the enzymatic hydrolysis rate and efficiency were significantly improved by treatment process. Results showed that after the treatment process, lignin was slightly removed; however, the structure was altered that cellulose was accessible to hydrolytic enzymes and thus enzymatic hydrolysis greatly improved after the treatment process. Optimization of the process leads to 70 % glucose yield at low temperature (−15 °C) using 7 % NaOH/12 % urea solution. The optimum conditions for pretreatment was found to be 3 % NaOH/12 % urea and −15 °C and around 60 % glucose conversion was achieved. Similar to this study, effects of potassium hydroxide (KOH) treatment on switch grass was studied by Sharma et al. (2012). Results indicated that about 99.26 % sugar retention was observed at 0.5 % KOH, at 21 °C and after 12 h treatment. Further, mentioned treatment caused production of 58.2 % of reducing sugar. Many workers have used combination of two pretreatment processes. Combination of alkaline treatment (lime) with oxidative delignification process for treatment of Alamo switch grass (*Panicum virgatum*) was studied by Falls and Holtzapple (2011). Optimum conditions were found to be 110 °C, 6.89 bar O_2, 240 min, 0.248 g $Ca(OH)_2$/g biomass. Results indicated that 2.6 % glucan and significant levels of reducing sugars (26.0 %) in raw biomass were recovered from the pretreatment liquor. Pretreatment of *Miscanthus* sp. biomass using two-stage dilute sulfuric acid and lime pretreatment was studied by Guo et al. (2013). The optimized conditions using response surface methodology were, 0.73 wt% H_2SO_4, 150 °C, 6.1 min in the first stage, and 0.024 g lime/g biomass, 202 °C, 30 min in the second stage showed more than 80 % recovery of glucose and more than 70 % recovery of xylose.

The present survey clearly suggests that alkaline pretreatment process involving calcium, sodium and potassium hydroxides gave higher yields of reducing sugars with biomass having low lignin content, such as rice hulls and grasses. Although, lime and other hydroxides are inexpensive, down stream processing costs are high, thus making it a costly process. The process utilizes a huge amount of water for washing salts of calcium and sodium, which are incorporated into the biomass and are difficult to remove (González et al.1986). In addition, the process also generates inhibitors during depolymerization of lignin which have to be removed prior to hydrolysis step.

5. PRETREATMENT WITH AMMONIA

The process involves aqueous ammonia treatment at elevated temperatures. The process sufficiently reduces lignin content and removes some hemicellulose, while cellulose is decrystallized. Ammonia pretreatment techniques can be divided into three broad types, viz., the Ammonia Recycle Percolation (ARP) (Kim et al. 2003), Ammonia Fiber Explosion method (AFEX) (Teymouri et al. 2005) and Soaking in Aqueous Ammonia (SAA) treatments (Kim et al. 2008), respectively. In ARP treatment, the biomass is pretreated with aqueous ammonia in a flow-through column reactor. The liquid flows at high temperature through the reactor column, which is packed with biomass (Kim et al. 2003; Kim and Lee 2005). Incubation in the presence of ammonia leads to hydrolysis of hemicellulose and destruction of lignin by ammonolysis. In this process, cellulose forms a complex with ammonia resulting in breaking of hydrogen bonds in cellulose. This leads to the loss of crystallinity of cellulose due to, which it becomes accessible to enzymatic hydrolysis (Schuerch 1963; Mittal et al. 2011). After the reaction, the solid fraction, rich in cellulose and hemicellulose, is separated from the liquid. The liquid fraction is sent into a steam-heated evaporator for ammonia recovery and lignin and other sugar separation. Ammonia is then recycled to the reactor inlet, whereas the separated fraction is sent into a crystallizer. After crystallization, a washing step is carried out to extract the sugars that have been retained in the solid matrix. Pretreatment of corn stover by ARP was evaluated by Kim and Lee (2005), at optimized conditions consisting of 0.47 g of NH_3 and 2.7 g of water at 170 °C for pretreatment of 1 g of corn stover. Results indicated that 73.4 %

delignification and 88.5 % digestibility with enzyme loading of 15 FPU/g of glucan were observed.

Soaking in Aqueous Ammonia (SAA) at low temperature removes lignin efficiently in the raw material by minimizing the interaction with hemicellulose. As a result an increase of surface area, pore size increases and retained hemicellulose/cellulose can be hydrolyzed to fermentable sugars by commercial enzymes. Pretreatment of corn stover in a two-stage percolation process consisting of hot water treatment followed by treatment with aqueous ammonia was evaluated by Kim and Lee (2006). The optimum operating conditions of the two-stage process in the first stage were: 190 °C, 5.0 mL/min, 30 min for hot water treatment and 170 °C, 5.0 mL/min, 60 min for aqueous ammonia treatment. Under these conditions, 92 % of xylan was hydrolyzed, of which 83 % was recovered, and 75 % of delignification was achieved. The treated biomass contained 78 % glucan, 3.6 % xylan and 9.8 % Klason lignin. The digestibility of hot water–ARP treated sample was 94 % with 60 FPU/g of glucan and 85 % with 15 FPU/g of glucan. Kim et al. (2008) evaluated SAA as a pretreatment method for barley hull. In their study, destarched barley hull was treated with 15–30 % aqueous ammonia, at 30–75 °C for 12 h to 77 days with no agitation and a solid-to-liquid ratio of 1:12. After soaking, the solids were recovered by filtrating, washed and analyzed. About 66 % of lignin was solubilized, and 83 % of glucan, 63 % of xylan were saccharified.

6. PRETREATMENT WITH AMMONIA FIBER EXPLOSION PROCESS

AFEX is considered a potential technique for pretreatment of lignocellulosic material (Holtzapple et al. 1992). In the AFEX process, biomass is treated with liquid anhydrous ammonia at temperatures (60–100 °C) and high pressure (250–300 psi) for 5 min. Then the pressure was released rapidly. The combined effect of ammonia and high pressure leads to swelling of lignocellulose biomass, disruption of lignocellulose architecture leading to hemicellulose hydrolysis, and decrystallization of cellulose. However, lignin remains unaffected during the process. Results have demonstrated that cellulose and hemicelluloses obtained by the process can be completely converted to fermentable sugars (Mosier et al. 2005). The biomass recovered using this process is dry as water is not used. The biomass is stable for long periods and no degradation of cellulose and hemicelluloses is observed. The main advantage of the process is that nearly all the ammonia which is employed can be recovered and reused. AFEX can be employed for various lignocellulosic materials, including rice straw, municipal solid waste, newspaper, sugar beet pulp, sugar cane bagasse, corn stover, switch grass, etc. Herbaceous crops and agricultural residues are well suited for AFEX.

Only drawback of this process is that it cannot be employed for treatment of biomass having high lignin content such as hardwood and newspaper.

Pretreatment of growth and seed stage of reed canary grass by Ammonia Fiber Expansion (AFEX) followed by enzymatic hydrolysis using 15 Filter Paper Units (FPU) cellulase/g of glucan was studied by Bradshaw et al. (2007). Optimum conditions for pretreatment were: for vegetative growth stage of reed canary grass: 100 °C, 60 % moisture content, 1.2:1 kg ammonia/kg of dry matter and for seed stage of reed canary grass: 100 °C, 60 % moisture content, 0.8:1 kg ammonia/kg of dry matter. Maximum yield of reducing sugars after 168 h hydrolysis using 15 FPU Spezyme CP cellulase/g of glucan were 86 % glucose and 78 % xylose for vegetative growth stage of reed canary grass, and 89 % glucose and 81 % xylose for seed stage of reed canary grass. In addition, pretreatment of Empty Palm Fruit Bunch Fiber (EPFBF), generated from palm processing industry was investigated by Lau et al. (2010). Pretreatment was carried out using AFEX and enzymatic hydrolysis. The optimum conditions were at 135 °C, 45 min retention time, water to dry biomass loading of 1:1 (weight ratio), and ammonia to dry biomass loading of 1:1 (weight ratio). The particle size of the pretreated biomass was reduced post-AFEX. The optimized enzyme formulation consisting of Accellerase (84 μL/g biomass), Multifect Xylanase (31 μL/g biomass), and Multifect Pectinase (24 μL/g biomass) gave yield of 90 % of the total reducing sugars after 72 h of enzymatic hydrolysis. Shao et al. (2010) have demonstrated that AFEX-pretreated corn grain gave 1.5- to 3-fold higher enzymatic hydrolysis yield compared with the untreated substrates. Sequential addition of cellulases after hydrolysis of starch resulted in 15–20 % higher hydrolysis yield compared with simultaneous addition of hydrolytic enzymes. AFEX-pretreated corn stover gave 70 % glucan conversion after 72 h of hydrolysis for 6 % glucan loading (at 8 mg total enzyme loading per gram of glucan). Microbial inoculation of corn stover before ensilation yielded a 10–15 % lower glucose hydrolysis yield for the pretreated substrate, due to the loss in starch content. Ethanol fermentation of AFEX-treated (at 6 % w/w glucan loading) corn stover gave 93 % ethanol yield.

Pretreatment with ammonia has been proved to be an efficient method for biomasses having low lignin content. Among various alkaline pretreatment processes discussed, the highest yield of reducing sugars (80–90 %) is obtained in this process (Bradshaw et al. 2007). This process has shown high sugar yields and high percent removal of lignin from biomass. Pretreatment of biomass by employing ammonia in processes such as SAA, ARP, AFEX has shown promising results in terms of depolymerization of lignin. A major drawback of using ammonia is the cost involved in pretreatment process, which is very high. Recycling of ammonia during pretreatment has shown to decrease the operating cost. Another important concern while using ammonia is the environmental concerns that are related with ammonia usage. Since ammonia is toxic to the environment, suitable strategy has to be maintained to prevent its leakage in the environment. This additional feature also increases the cost of the process.

7. PRETREATMENT OF BIOMASS USING ORGANOSOLV PROCESS

Organosolv pretreatment process involves extraction of lignin from plant biomass by employing organic solvent or mixture of solvents in combination with water. Solvents which are most commonly employed are ethanol, methanol, acetone, and ethylene glycol (Ichwan and Son 2011). Temperatures used in the process can be as high as 200 °C, but lower temperatures can be sufficient depending on the type of biomass and the use of a catalyst (Ghose et al. 1983). Possible catalysts include inorganic or organic acids (Sun and Cheng 2002). Organosolv extraction process causes hydrolysis of the internal bonds in lignin and also between lignin and hemicellulose. The organic solvents also cause hydrolysis of the glycosidic bonds in hemicelluloses and to a smaller extent in cellulose. The presence of acids as a catalyst causes acid-catalyzed degradation of the monosaccharides into furfural and 5-hydroxymethyl furfural followed by condensation reactions between lignin and these reactive aldehydes (Chum et al. 1985). After removal of lignin, the biomass (rich in cellulose) is used for enzymatic hydrolysis (Zhao et al. 2009).

Pretreatment of wheat straw by glycerol-based autocatalytic organosolv pretreatment was studied by Sun and Chen (2008). At optimized conditions corresponding to liquid–solid ratio of 20 g/g at 220 °C for 3 h, 70 % hemicelluloses and 65 % lignin were removed from the biomass. Also, 98 % cellulose retention was achieved. The pretreated fibers showed 90 % of the enzymatic hydrolysis yield after 48 h of incubation. Organosolv pretreatment of Horticultural Waste (HW) for bioethanol production was studied by Geng et al. (2012). A modified organosolv method using ethanol under mild conditions followed by H_2O_2 post-treatment was investigated for HW. Result showed that enzymatic hydrolysis of the organosolv pretreated HW with 17.5 % solid content, enzyme loading of 20 FPU/g HW of filter paper cellulase, and 80 CBU/g HW of β-glycosidase resulted in a HW hydrolysate containing 15.4 % reducing sugar after 72 h. Fermentation of the above hydrolysate medium produced 11.69 g/L ethanol at 8 h using *Saccharomyces cerevisiae*. Hideno et al. (2013) have shown that application of alcohol-based organosolv treatment in combination with Ball Milling (BM) to Japanese cypress (*Chamaecyparis obtusa*) greatly improved the enzymatic digestibility and decreased the required severity of organosolv treatment. Moreover, alcohol-based organosolv treatment increased the efficiency and reduced the time required for BM despite small quantity of removed lignin. It was found that the combination of alcohol-based organosolv treatment in mild condition and short-time BM had a synergistic effect on the enzymatic digestibility of Japanese cypress.

Ichwan and Son (2011) studied organosolv extraction of oil palm pulp using various solvents such as ethylene glycol–water, ethanol–water and acetic acid–water mixture to extract cellulose. The yield of organosolv pulping with ethylene glycol–water, ethanol–water and acetic acid–water mixture was 50.1; 48.1 and 41.7 %, respectively, while the kappa number was 74.7; 72.7; and 67. Fourier Transform Infrared Spectroscopy (FTIR) spectra showed the degradation of cellulose pulp occurred during acid pulping. The X-ray Diffraction (XRD) measurement showed that the ethanol–water mixture pulping resulted in higher crystallinity of pulp (68.67), followed by ethylene glycol–water mixture as (58.14), acid pulping (54.21). Further Panagiotopoulos et al. (2012) performed pretreatment approach, employing steam followed by organosolv treatment to fractionate hemicellulose, lignin and cellulose components from poplar wood chips. The results indicated that lignin extraction was enhanced as more than 66 % of lignin was extracted. More than 98 % of the original cellulose was recovered after the two-stage pretreatment and 88 % of the cellulose could be hydrolyzed to glucose at enzyme loading of 5 FPU/g of cellulose after 72 h. Organosolv extraction of sugarcane bagasse was evaluated by Mesa et al. (2011). Under optimized conditions, 30 % (v/v) ethanol at 195 °C for 60 min produced 29.1 % of reducing sugars. The scale-up of the process, by performing the acid pretreatment in a 10-L semi-pilot reactor fed with direct steam, was successfully performed. Recently, Pan and Saddler (2013) employed extracted lignin obtained from ethanol organosolv and used it as a replacement of petroleum-based polyol for the production of Rigid Polyurethane Foams (RPFs). The results suggested that the prepared foams contained 12–36 % (w/w) extracted lignin. The density, compressive strength, and cellular structure of the prepared foams were investigated and compared and it was found that the lignin-containing foams had comparable structure and strength with polyol containing foams.

Organosolv process has been extensively utilized for extraction of high-quality lignin, which is a value added product, and once the lignin is removed from the biomass, the cellulose fibers become accessible to enzymes for hydrolysis and absorption of cellulolytic enzymes to lignin is minimized, which leads to higher conversion of biomass. According to the present survey, this process has shown high amounts of enzymatic hydrolysis of treated biomass (around 90 %) due to efficient removal of lignin. The main drawback of the process is that the solvent and the catalysts that are employed are costly. Removal and recovery of the solvent can considerably reduce the operational cost (Sun and Cheng 2002). Another important aspect is safety measures which have to be implemented because organic solvents are inflammable and uncontrolled use can cause fires and explosions. This additional requirement increases the cost of the process. Organic solvents are inhibitors of enzymatic hydrolysis, so their removal is necessary for proper enzymatic hydrolysis of biomass (Mosier et al. 2005; Agbor et al. 2011). Removal of organic solvents also poses an additional cost on the process.

8. PRETREATMENT OF BIOMASS BY OXIDATIVE DELIGNIFICATION

Delignification of lignocellulose can also be achieved by treatment with an oxidizing agent such as hydrogen peroxide, ozone, oxygen or air (Hammel et al. 2002; Nakamura et al. 2004). In such process, lignin is converted to acids. However, these acids can act as inhibitors during fermentation process. Therefore, acids have to be removed (Alvira et al. 2010). In addition, oxidative treatment also damages the hemicellulose fraction of the lignocellulose complex and a major portion of the hemicellulose gets degraded and becomes unavailable for fermentation (Lucas et al. 2012). The major types of oxidative delignification process are as follows.

9. PRETREATMENT BY HYDROGEN PEROXIDE

Hydrogen peroxide is the most commonly employed oxidizing agent. Studies have shown that dissolution of about 50 % of lignin and most of the hemicellulose has been achieved in a solution of 2 % H_2O_2 at 30 °C. Studies have shown that hydrolysis of hydrogen peroxide leads to the generation of hydroxyl radicals, which degrade lignin and produce low molecular weight products. Removal of lignin from lignocellulose leads to the exposure of cellulose and hemicellulose causing increased enzymatic hydrolysis (Hammel et al. 2002). The yield of enzymatic hydrolysis followed can be as high as 95 %. Combined chemical (H_2O_2) and biological treatment of rice hull was performed by Yu et al. (2009). Their study consisted of chemical treatment (H_2O_2) followed by biological treatment with *Pleurotus ostreatus*. The combined pretreatments led to significant increases of the lignin degradation. At optimum conditions, H_2O_2 (2 %, 48 h) and *P. ostreatus* (18 days) led to 39.8 and 49.6 % of net yields of total sugar and glucose, respectively. It was about 5.8 times and 6.5 times more than that of the sole fungal pretreatment (18 days), and it was also slightly higher than that of the sole fungal pretreatment for 60 days. A study aimed at increasing the enzymatic digestibility of sweet sorghum bagasse for bioethanol production was performed by Cao et al. (2012). Among different chemical pretreatment methods employed, the best yields were obtained when sweet sorghum bagasse was treated with dilute NaOH solution followed by autoclaving and H_2O_2 immersing pretreatment. The highest cellulose hydrolysis yield, total sugar yield and ethanol concentration were 74.3, 90.9 % and 6.1 g/L, respectively, which were 5.9, 9.5 and 19.1 times higher than the control. Lucas et al. (2012) studied the effect of manganese acetate as a catalyst in oxidative delignification of wood with hydrogen peroxide at room temperature. When exposed to a mixture of hydrogen peroxide and manganese acetate in aqueous solution, poplar wood sections were converted into a fine powder-like material. Optical and Raman microscopy showed oxidation of lignin containing middle lamellae. Raman spectra from the solid residue revealed a delignified and cellulose-rich material. Glucose yields following enzymatic hydrolysis were 20–40 % higher in poplar sawdust pretreated with hydrogen peroxide and manganese acetate for 2, 4, and 7 days at room temperature than those in sawdust exposed to water only for identical durations, suggesting the viability of the mild, inexpensive method for pretreatment of lignocellulosic biomass. Alkaline Hydrogen Peroxide (AHP) pretreatment of Cashew Apple Bagasse (CAB) was evaluated by Correia et al. (2013). In their study, the effects of the concentration of hydrogen peroxide at pH 11.5, the biomass loading and the pretreatment duration performed at 35 °C and 250 rpm were evaluated after the subsequent enzymatic saccharification of the pretreated biomass using a commercial cellulase enzyme. At optimized conditions, consisting of solid loading of 5 % (w/v) at 4.3 % AHP, 6 h, 35 °C, a total reducing sugar yield of 42.9 % was obtained.

Pretreatment with hydrogen peroxide is considered a very harsh treatment which leads to lignin removal and high yields of reducing sugars. Reducing sugar yields up to 90 % have been obtained (Cao et al. 2012). A major drawback of this process is that it is costly owing to high cost of hydrogen peroxide and requirement of reactions vessels that can withstand such conditions. Another drawback of the process is the generation of inhibitors, which are needed to be removed prior to enzymatic hydrolysis. This additional step also increases the cost significantly.

10. PRETREATMENT BY OZONOLYSIS

In the ozonolysis process, the biomass is treated with ozone, which causes degradation of lignin by attacking aromatic rings structures, while hemicellulose and cellulose are not damaged (Nakamura et al.2004). It can be used to disrupt the structure of many different lignocellulosic materials, such as wheat straw, bagasse, pine, peanut, cotton straw and poplar sawdust (Sun and Cheng 2002). García-Cubero et al. (2009) investigated the pretreatment of wheat and rye straws with ozone to increase the enzymatic hydrolysis extent of potentially fermentable sugars. Enzymatic hydrolysis yields of up to 88.6 and 57 % were obtained compared to 29 and 16 % in non-ozonated wheat and rye straw, respectively. Moisture content and type of biomass showed the most significant effects on ozonolysis. In addition, ozonolysis experiments in basic medium with sodium hydroxide showed a reduction in solubilization of lignin and reliable cellulose and hemicellulose degradation. Miura et al. (2012) also studied the effect of ozonolysis and Wet Disk Milling (WDM) on Japanese cedar (*Cryptomeria japonica*) to improve sugar production by enzymatic saccharification. When the moisture content reached more than 40 %, ozone consumption decreased, resulting in less delignification. Ozone treatment removed mainly lignin, but also small amounts of polysaccharides. The application of WDM following the ozone treatment further increased glucose and xylose yields of 68.8 and 43.2 %, respectively, but had no significant effect on mannose yield. Similar to this study, ozonolysis in combination with WDM was used for pretreatment of sugarcane bagasse and straw (Barros Rda et al. 2013). Under optimized conditions, for sugar cane bagasse, yields of glucose (89.7 %) and xylose (48.8 %) were obtained after 60 min ozonolysis followed by WDM (4 cycles and 1.2 min/g). In case of straw, glucose and xylose saccharification yields of 92.4 and 52.3 % were obtained after WDM (4 cycles and 0.2 min/g) followed by 60 min ozonolysis, indicating that bagasse and straw responded differently to the same process conditions.

Ozonolysis alone has been shown to be ineffective for removal of lignin and yields of reducing sugars. However, combination of ozonolysis with other pretreatment process such as wet disk milling has shown promising results. Reports have demonstrated that sugar yields up to 90 % can be obtained by combination of these processes (Barros Rda et al. 2013). A major disadvantage of ozonolysis is the high costs that are involved, which inhibits its use for large-scale applications.

11. PRETREATMENT BY WET OXIDATION

Wet oxidation involves oxygen or air in combination with water at elevated temperature and pressure (Varga et al. 2003). As the name implies, in this process, oxidation of organic matter in the presence of oxygen takes place. When the process takes place at low temperatures, hydrolysis of lignocellulose occurs. At high temperatures, oxidation of lignocellulose occurs with liberation of carbon dioxide and water. Wet oxidation process solubilizes hemicellulose and lignin is degraded into carbon dioxide, water, and carboxylic acids such as succinic acid, glycolic acid, formic acid, acetic acid, phenolic compounds (Bjerre et al. 1996). It was presented as an alternative to steam explosion which had become the most widely used pretreatment method (Palonen et al. 2004). Industrially, wet air oxidation processes have been used for the treatment of wastes with a high organic matter by oxidation of soluble or suspended materials using either an oxidizing agent such as hydrogen peroxide or oxygen in aqueous phase at high temperatures (180–200 °C) (Jorgensen et al. 2007). The effect of wet oxidation to enhance anaerobic biodegradability and methane yields from different biowastes (food waste, yard waste, and digested biowaste) by employing thermal wet oxidation was studied by Lissens et al. (2004b). Measured methane yields for raw yard waste, wet oxidized yard waste, raw food waste, and wet oxidized food waste were 345, 685, 536, and 571 mL of CH_4/g of volatile suspended solids, respectively. Higher oxygen pressure during wet oxidation of digested biowaste considerably increased the total methane yield and digestion kinetics and permitted lignin utilization during a subsequent second digestion. The increase of the specific methane yield for the full-scale biogas plant by applying thermal wet oxidation was 35–40 %, showing that there is still a considerable amount of methane that can be harvested from anaerobic digested biowaste. Fox and Noike (2004) evaluated wet oxidation method for increasing methane fermentation of newspaper waste. Wet oxidation was carried out at 170, 190, and 210 °C, with a retention time of 1 h. The highest removal efficiencies of newspaper waste in terms of change in total Chemical Oxygen Demand (TCOD) and cellulose were achieved at 210 °C during which approximately 40 % TCOD and 69 % Cellulose were reduced, respectively. On the other hand, highest lignin removal efficiency was achieved at 190 °C in which approximately 65 % was removed. Batch methane fermentation tests were performed in 2-L glass bottles filled with the wet oxidized newspaper samples. Methane fermentation of newspaper pretreated at 190 °C gave the highest methane conversion efficiency (59 % of the initial TCOD was recovered as methane gas). Similarly, pretreatment of organic municipal solid waste enriched with wheat straw by wet oxidation for subsequent enzymatic conversion and fermentation into bioethanol was investigated by Lissens et al. (2004a). The effect of temperature (185–195 °C), oxygen pressure (3–12 bar) and sodium carbonate (0–2 g) addition on enzymatic cellulose and hemicellulose convertibility was studied at a constant wet oxidation retention time of 10 min. An enzyme convertibility assay at high enzyme loading showed that up to 78 % of the cellulose and up to 68 % of the hemicellulose in the treated waste could be converted into, respectively, hexose and pentose sugars compared to 46 % for cellulose and 36 % for hemicellulose in the raw waste. For all wet oxidation conditions tested, total carbohydrate recoveries were high (>89 %) and 44–66 % of the original lignin could be converted into non-toxic carboxylic acids. Simultaneous Saccharification and Fermentation (SSF) of the treated waste by*Saccharomyces cerevisiae* yielded average ethanol concentrations of 16.5–22 g/L for enzyme loadings of 5 and 25 FPU/g, respectively. The cellulose to ethanol conversion efficiency during SSF was 50, 62, 65 and 70 % for a total enzyme loading of 5, 10, 15 and 25 FPU/g of biomass, respectively. Pretreatment of rice husk by wet air oxidation for the production of ethanol was studied by Banerjee et al. (2009). Optimum conditions were found to be of 185 °C, 0.5 MPa, and 15 min which yielded 67 % (w/w) cellulose content in the solid fraction, along with 89 % lignin removal, and 70 % hemicellulose solubilization.

Szijártó et al. (2009) studied the treatment of a common reed (*Phragmites australis*) by employing wet oxidation to enhance the enzymatic digestibility of reed cellulose to soluble sugars, thus improving the convertibility of reed to ethanol. The most effective treatment increased the digestibility of reed cellulose by cellulases more than three times compared to the untreated control. During this wet oxidation, 51.7 % of the hemicellulose and 58.3 % of the lignin was solubilized, whereas 87.1 % of the cellulose remained in the solids. After enzymatic hydrolysis of pretreated fibers from the same treatment, the conversion of cellulose to glucose was 82.4 %. Simultaneous saccharification and fermentation of pretreated solids resulted in a final ethanol concentration as high as 8.7 g/L, yielding 73 % of the theoretical yield. Banerjee et al. (2011) investigated pretreatment of rice husk by Alkaline Peroxide-Assisted Wet Air Oxidation (APAWAO) to increase the enzymatic convertibility of cellulose in pretreated rice husk. Rice husk was presoaked overnight in 1 % (w/v) H_2O_2solution at room temperature, followed by Wet Air Oxidation (WAO). APAWAO pretreatment resulted in solubilization of 67 wt% of hemicellulose and 88 wt% of lignin initially present in raw rice husk. APAWAO pretreatment resulted in 13-fold increase in the amount of glucose that could be obtained from otherwise untreated rice husk. Up to 86 % of cellulose in the pretreated rice husk could be converted into glucose within 24 h, yielding over 21 % glucose. Wet oxidation is considered a suitable process for pretreatment of biomass having high lignin content.

This survey indicates that with a variety of biomasses, this process has shown promising results. Reducing sugar yields up to 70 % have been obtained by utilizing this pretreatment process (Banerjee et al. 2009). The main drawback of the process is that it requires the maintenance of high temperature, pressure and the presence of strong oxidizing agents such as hydrogen peroxide. These requirements lead to high costs to maintain such conditions and also require large-scale reaction vessels to withstand such harsh conditions. Therefore, application of this process in large-scale pretreatment of biomass is limited.

12. PRETREATMENT OF BIOMASS BY BIOLOGICAL METHODS

During conversion of lignocellulosic biomass, the pretreatment step is considered as rate limiting step and governs the final cost of products. Conventional physicochemical methods for lignin degradation require large inputs of energy and also cause pollution. Therefore, biological pretreatment of lignocellulosic biomass is considered as an efficient, ecofriendly and cheap alternative (Wan and Li 2012). The biological pretreatment of lignocellulosic biomass is usually performed by employing cellulolytic microorganisms which synthesize potent cellulolytic enzymes during hydrolysis. The common candidates are filamentous fungi, which are ubiquitous in the environment such as soil, living plants and lignocellulosic waste material. Studies have shown that white-rot fungi are the most effective microorganisms for the pretreatment of lignocelluloses such as wood chips, wheat straw, Bermuda grass and softwood (Akin et al. 1995). These fungi are capable of degrading cellulose, hemicellulose and lignin and are considered as major degraders of woods in forest ecosystems. Another class of fungi, commonly known as brown-rot fungi can selectively degrade cellulose and hemicellulose, without affecting lignin (Rasmussen et al. 2010). The mechanism of degradation of lignocellulose by fungi is fully understood and can be broadly divided into two main categories: oxidative and hydrolytic. In oxidative mechanism, lignin is degraded via production of Reactive Oxygen Species (ROS) mainly hydroxyl radicals by the fungi (Hammel et al. 2002). Many fungi are known to produce hydrogen peroxide by the action of enzymes such as glyoxal oxidase, pyranose-2 oxidase, and aryl-alcohol oxidase, respectively (Martinez et al. 2009). Hydrogen peroxide then reacts with iron via the Fenton reaction to produce hydroxyl radicals. These radicals degrade lignin and produce low molecular weight products (Hammel et al. 2002). Another group of enzymes that degrade lignin are laccases and manganese peroxidase (MnP). Laccase is a multicopper oxidase and can catalyze free radicals-mediated reactions that degrade lignin (Eggert et al. 1997). Manganese peroxidase on the other hand oxidizes Mn^{2+} into Mn^{3+} by hydrolysis of hydrogen peroxide (Hofrichter 2002). Mn^{3+} is a powerful oxidant that degrades lignin. In hydrolytic mechanism, the fungi produce hydrolytic enzymes which degrade glycosidic linkages in cellulose and hemicellulose releasing monomeric sugars (Feijoo et al. 2008). Cellulose degradation is achieved by synergistic action of three classes of hydrolytic enzymes: endo-(1,4)-β-glucanase (endocellulase), cellobiohydrolase (exocellulase), and β-glucosidase (Baldrian and Valaskova 2008). Whereas, hemicellulose degradation is achieved by action of hydrolytic enzymes such as endo-xylanases, endo-α-L-arabinase, endo-mannanase, β-galactosidase, and β-glucosidases, respectively (Shallom and Shoham 2003). By synergistic action of both these pathways, lignocellulose is degraded by fungi. Biological pretreatment of wheat straw by white-rot fungus Euc-1 was performed by Dias et al. (2010). The effect of fungal pretreatment on the cellulose accessibility toward commercial cellulolytic enzyme preparation (Onozuka R-10) has been evaluated. Results indicated that after 23 days of treatment following enzyme treatment, maximum reducing sugar yield of 23 % was observed. Also 7 % decrease in total lignin content was also recorded at these treatment conditions. Biological pretreatment of different feed stocks such as corn stover, wheat straw, soybean straw, switch grass, and hardwood by *Ceriporiopsis subvermispora* was evaluated by (Wan and Li2011). Results of their study indicated that after 18-day pretreatment, corn stover, switch grass, and hardwood were effectively delignified by the fungus by the combined action of manganese peroxidase and laccase. Glucose yields during enzymatic hydrolysis reached 56.5, 37.2, and 24.2 %, respectively, which were a 2- to 3-fold increase over those of the raw materials. A further 10–30 % increase in glucose yields was observed when pretreatment time extended to 35 days. Biological pretreatment of rice husks by a white-rot fungus *Phanerochete chrysosporium* is performed in one step consisting of biological and enzymatic treatment, respectively (Potumarthi et al. 2013). Results demonstrated that fungal pretreated rice husk produced highest 44.7 % reducing sugars on the 18th day of fungal treatment. Similarly, Yu et al. (2010) evaluated the effects of biological treatment using *Ganoderma lucidum*, *Trametes versicolor* and*Echinodontium taxodii* prior to alkaline/oxidative (A/O) pretreatment. Results showed that *Echinodontium taxodii* significantly enhanced the efficiency of chemical pretreatment. Subsequent to treatment of corn straw with *Echinodontium taxodii* for 15 days, the straw was subjected to digestion by 0.0016 % NaOH and 3 % H_2O_2 at room temperature for 24 h, which increased the reducing sugar yield by 50.7 %. Biological treatment of wheat straw by solid state and submerged fermentations in the presence of white-rot basidiomycetes such as *Bjerkandera adusta*, *Fomes fomentarius*, *Ganoderma resinaceum*, *Irpex lacteus*,*Phanerochaete chrysosporium*, *Trametes versicolor*, Euc-1, and *Lepista nuda* were evaluated by Pinto et al. (2012). Results of their study indicated that enzymatic hydrolysis of holocellulose after solid-state pretreatment showed that treatment with *T. versicolor* leads to a significant increase of saccharification process. The total reducing sugar yield was 7 %, which was 91 % higher as compared to untreated straw. López-Abelairas et al. (2012) studied the biological pretreatment of lignocellulosic biomass for the production of bioethanol using ligninolytic fungi (*Pleurotus eryngii* and *Irpex lacteus*) in a solid-state fermentation of sterilized wheat straw complemented with a mild alkali treatment. The largest digestibility was achieved with fermentation with *I. lacteus* under optimized conditions, under which cellulose and hemicellulose digestibility increased after 21 days of pretreatment from 16 to 100 % and 12 to 87 %, respectively. The maximum glucose yield (84 %) of cellulose available in raw material was obtained after only 14 days of pretreatment with an overall ethanol yield of 74 % of the theoretical value, which is similar to that reached with SE. Saritha et al. (2012) studied the treatment of hardwood and softwood by employing *Streptomyces griseus*isolated from leaf litter and showed that it enhanced the mild alkaline solubilization of lignin and also produced high levels of the cellulase when growing on wood substrates. Lignin loss (Klason lignin) observed was 10.5 and 23.5 % in case of soft wood and hard wood, respectively. Thus, biological pretreatment process for lignocellulosic substrate using lignolytic organisms such as actinomycetes and white-rot fungi was found to be efficient and cost effective. Performing saccharification and fermentation processes at high-substrate concentration has shown that it increases the concentration of inhibitors such as furan derivatives and phenolic compounds, which eventually inhibit the fermentation process. To inhibit production of inhibitors, treatment with enzymes such as laccases has been suggested. Alvira et al. (2012) studied the treatment of laccase on concentration of phenolic compounds. The results demonstrated that supplementation of laccase reduced the phenolic content and increased the performance of *Saccharomyces cerevisiae* during fermentation. This survey clearly suggests that total reducing sugar in yield in biological treatment of biomass ranges from 30 to 60 %, after 10–30 days of incubation (Wan and Li 2011). Degradation of lignin and extraction of phenolic compounds from the leaves of *Larrea tridentata* by a combination of biological treatment with basidiomycete *Phanerochaete chrysosporium* followed by organosolv extraction by methanol were evaluated by Martins et al. (2013). Results clearly indicated that recovery of phenolic compounds was 33 % more with combination of biological treatment followed by methanol extraction as compared to methanol extraction alone.

Pretreatment of biomass with fungal and bacterial strains for cellulose hydrolysis is inexpensive and easy to operate. Biological pretreatment of biomass is considered to be an inexpensive process when compared to other pretreatment processes such as organosolv or AFEX, where the cost of chemicals and infrastructure is high (Hasunuma et al. 2013). However, large-scale operation of the process leads to higher operational costs. To carry out biological pretreatment of biomass on a large scale, a large, sterile area is required and it is necessary to maintain the sterile conditions during the treatment process. This important requirement increases the cost of biological pretreatment process. The second key disadvantage of this process is that a major portion of carbohydrates such as cellulose and hemicellulose are consumed by the microorganisms, i.e., fungi. This leads to lower outputs and increases the cost of treatment process. Furthermore, the process is too slow and thus it is not recommended for industrial purposes.

13. PRETREATMENT OF BIOMASS BY MICROWAVE IRRADIATION

Microwaves are radio waves with wavelengths ranging from 1 m to 1 mm, or equivalently, with frequencies between 300 MHz (0.3 GHz) and 300 GHz. When these waves interact with organic matter, they get absorbed by water, fats and sugars and their energy gets transferred to organic molecules generating enormous amount of heat. In this way, microwaves exhibit heating effect (Banik et al. 2003). Microwaves are also being employed in the treatment of lignocellulosic biomass (Ooshima et al. 1984). Studies have shown that microwaves cause localized heating of biomass leading to disruption of lignocellulose architecture. Thus, cellulose and hemicellulose get accessible to enzymatic hydrolysis (Sarkar et al. 2012). Microwave pretreatment of rice straw was studied by Maa et al. (2009). In their study, Box–Behnken design and response surface methodology were used to optimize the microwave pretreatment of rice straw. Under optimized conditions consisting of microwave intensity 680 W, irradiation time 24 min and substrate concentration 75 g/L, maximal efficiencies of cellulose, hemicellulose and total saccharification were, respectively, increased by 30.6, 43.3 and 30.3 %. In a similar study, Beech wood powder was treated with H_2O_2 and nine different ammonium salts (sulfate, chloride, hydrogen carbonate, nitrate, hydrogen phosphate, tartrate, carbonate, molybdate and acetate) at 80 °C for 5 h in an autoclave, and enzymatic saccharification of the resultant pulp was carried out using cellulolytic enzymes from *T. viride* (Verma et al. 2011). Ammonium molybdate (50 mM) gave the highest sugar yield of 50.1 %. The same process was repeated with microwave treatment. Under optimized conditions, the maximum sugar yield of 59.5 % was obtained by microwave irradiation at 140 °C for 30 min with ammonium molybdate and H_2O_2. Su et al. (2010) studied the effects of microwave treatment on sorghum liquor waste for bioethanol production. Their results indicated that reducing sugar yield following microwave treatment was tremendously high as compared to untreated waste. Microwave-assisted pretreatment of recalcitrant softwood in the presence of aqueous glycerol and different organic and inorganic acids was studied by Liu et al. (2010). Their study indicated that pulp obtained by organosolv lysis with 0.1 % hydrochloric acid (pK_a −6) at 180 °C for 6 min gave the highest sugar yield, 53.1 %. With other acids such as phosphoric acid, malonic acid having lower pK_a value, a lower sugar yield was obtained. These results clearly indicate that microwave-assisted glycerolysis is a suitable process for treatment of a variety of soft woods.

Microwave pretreatment of wheat straw for ethanol production was studied by Xu et al. (2011). In their study, an orthogonal design was used to optimize the microwave pretreatment on wheat straw for ethanol production. According to the orthogonal analysis, pretreatment with the ratio of biomass to liquid at 80 g/kg, the NaOH concentration of 10 kg^3, the microwave power of 1,000 W for 15 min was confirmed to be the optimal condition. The ethanol yield was 14.8 and was much higher from untreated material which was only 2.6 %. Boonmanumsin et al. (2012) studied the microwave-assisted ammonium hydroxide treatment of*Miscanthus sinensis*. The results indicated that yield of monomeric sugars was substantially increased in the presence of microwave treatment. In a new approach, microwave pretreatment of oil palm empty fruit bunch fiber in the presence of alkaline conditions was studied by Nomanbhay et al. (2013). Their study showed that at optimum conditions of pretreatment, 3 % (w/v) NaOH at 180 W for 12 min led to the loss of 74 % lignin and 24.5 % holocellulose. The yield of total reducing sugars was 41 %.

Microwave treatment of biomass is considered a harsh process which leads to high lignin removal and increased sugar yields. This survey indicates that the yield of reducing sugars ranges from 40 to 60 % (Verma et al. 2011).

These results have clearly indicated that microwave pretreatment of biomass gives promising results. But when efficacy of this process in terms of cost is considered, the process is very costly (Feng and Chen 2008). For large-scale pretreatment of biomass, a large microwave irradiator is required, which is costly, energy consuming and limits its use in large-scale operations. Another major disadvantage of the process is generation of high temperature and non-uniform heating of biomass, which leads to the formation of inhibitors and thus, the yields are generally lower as expected. This drawback also increases the operational cost (Jackowiak et al. 2011).

A summary of major treatment processes and a comparison between the yields of reducing sugars obtained are summarized in Table 1. It is clear from the table that none of the process discussed are efficient enough to give 100 % yield of reducing sugars from all types of biomasses. The efficiency of the process depends mainly on the type of biomass used as raw material, its structure and lignin content (Mckendry 2002; Cao et al. 2012). One major point that emerges from this discussion is that combination of two or more pretreatment processes is proven to be efficient when compared with single pretreatment process alone in terms of reducing sugar yield and lignin removal from different biomasses. Therefore, combinations of pretreatment processes are being tested for increasing the yield of reducing sugars (Cara et al. 2006; Yu et al. 2009; Miura et al. 2012).

14. CONCLUSION

The impact of green house gases on the climate change has been recognized as a serious environmental threat. Efforts are being made for the search of more efficient, sustainable and environmental friendly technology to prevent such emission. Due to immense potential for conversion of renewable biomaterials into biofuels, production of ethanol from lignocellulosics has received much attention since the last decade. A major bottleneck in this technology is the presence of lignin, which is a major inhibitor of hydrolysis of cellulose and hemicellulose. This has led to extensive research in the development of various pretreatment processes for the treatment of lignocellulosic biomass. These processes were based on physical, chemical and biological principles. We have discussed various pretreatment processes that are commonly employed in the treatment of biomass. The major advantages, disadvantages and yield of pretreatment processes have been discussed. One important point that emerges is that no treatment technology offers 100 % conversion of biomass into fermentable sugars. There is always a loss of biomass, which affects the final yield and increases the cost of finished product, i.e., biofuel. Although pretreatment of lignocellulosic biomass with combination of two or more pretreatment processes have shown promising results, we still feel that there is a need for extensive research in this area, so that either a new efficient treatment process is developed or an existing process is upgraded to give promising results.

Table 1 Summary of various pretreatment processes, their advantages/disadvantages, pretreatment conditions and percent yield

Acid treatment	Dilute/ concentrated acid/few minutes/high and low temperature	Increase in porosity/ increased enzymatic hydrolysis	Synthesis of furfural/ hydroxymethyl furfural/need for recycling/costly	Olive tree biomass/170 °C and 1.0 % acid	48.6 % Reducing sugar	Cara et al. (2008)
				Corn fibers/0.5 % sulfuric acid, 5 % biomass and at 140 °C	56.8 % Reducing sugar	Noureddini and Byun (2010)
				Sugarcane tops/25 % w/w biomass, 3 % sulfuric acid for 60 min	68.5 % Reducing sugar	Sindhu et al. (2011)
Alkaline treatment	Alkali treatment at NTP	Removal of lignin/ hemicellulose hydrolysis	Formation of salts of calcium and magnesium	Corn stover/0.5 g $Ca(OH)_2$/g, 55 °C for 4 weeks with aeration	91.3 % Glucose, 51.85 % xylose	Kim and Holtzapple (2005)
				Spruce/3 % NaOH/12 % urea and −15 °C	60 % Glucose	Zhao et al. (2008)
				Switch grass/0.5 % KOH, at 21 °C and after 12 h treatment	58.2 % Reducing sugar	Sharma et al. (2012)
Ammonia treatment	Ammonia at elevated temperatures	Removal of lignin/ decrystallizing cellulose	Removal of ammonia/costly	Corn stover/first stage 190 °C, 5.0 mL/min, 30 min for hot water treatment and 170 °C, 5.0 mL/min, 60 min for aqueous ammonia treatment	78 % Glucan, 3.6 % xylan	Kim and Lee (2006)
				Barley hull/15–30 % aqueous ammonia, at 30–75 °C for 12 h to 77 days	83 % Glucan, 63 % xylan	Kim et al. (2008)
AFEX	Liquid ammonia at high temperature and pressure	Lignin removal/ hydrolysis of hemicellulose/ decrystallization of cellulose	Costly/not employed for high lignin content	Canary grass/100 °C, 60 % moisture content, 1.2:1 kg ammonia/kg of dry matter	86 % Glucose, 78 % xylose	Bradshaw et al. (2007)
				Empty palm fruit bunch fiber/ 135 °C, 45 min retention time, water to dry biomass loading of 1:1 (weight ratio), and ammonia to dry biomass loading of 1:1 (weight ratio)	90 % Reducing sugar	Lau et al. (2010)
Organosolv	Organic solvents at both high and low temperatures	Pure lignin obtained and used as value added product	Solvents inhibit enzymatic hydrolysis/ costly	Wheat straw/glycerol, 20 g/g at 220 °C for 3 h	90 % Enzymatic hydrolysis	Sun and Chen (2008)
				Sugarcane bagasse/30 % (v/v) ethanol at 195 °C, for 60 min	29.1 % Reducing sugars	Mesa et al. (2011)
				Oil palm pulp/ethylene glycol–water	50.1 % Reducing sugars	Ichwan and Son (2011)
Oxidative delignification	Treatment with oxidizing agents	Fast process	Formation of acids which act as enzyme inhibitors/ costly/ hemicellulose degradation	Rice hull/H_2O_2 (2 %, 48 h) and *P. ostreatus* (18 days)	39.8 % Reducing sugar	Yu et al. (2009)
				Sweet sorghum bagasse/dilute NaOH, autoclaving and H_2O_2	90.9 % Reducing sugar	Cao et al. (2012)
				Cashew apple bagasse/5 % (w/ v) at 4.3 % AHP, 6 h, 35 °C	42.9 % Reducing sugar	Correia et al. (2013)

ACKNOWLEDGMENTS

PV is thankful to DBT for providing the financial support (Grant No. BT/304/NE/TBP/2012).

REFERENCES

1. Adsul MG, Singhvi MS, Gaikaiwari SA, Gokhale DV (2011) Development of biocatalysts for production of commodity chemicals from lignocellulosic biomass. Bioresour Technol 102(6):4304–4312

2. Agbor VB, Cicek N, Sparling R, Berlin A, Levin DB (2011) Biomass pretreatment: fundamentals toward application. Biotechnol Adv 29(6):675–685

3. Akin DE, Rigsby LL, Sethuraman A, Morrison WH, Gamble GR, Eriksson KEL (1995) Alterations in structure, chemistry, and biodegradability of grass lignocellulose treated with the white rot fungi *Ceriporiopsis* sub*Vermispora* and *Cyathus stercoreus*. Appl Environ Microbiol 61:1591–1598

4. Alvira P, Tomás-Pejó E, Ballesteros M, Negro MJ (2010) Pretreatment technologies for an efficient bioethanol production process based on enzymatic hydrolysis: a review. Bioresour Technol 101(13):4851–4861

5. Alvira P, Moreno AD, Ibarra D, Sáez F, Ballesteros M (2012) Improving the fermentation performance of*Saccharomyces cerevisiae* by laccase during ethanol production from steam-exploded wheat straw at high-substrate loadings. Biotechnol Prog 29(1):74–82

6. Balan V, Bals B, Chundawat SP, Marshall D, Dale BE (2009) Lignocellulosic biomass pretreatment using AFEX. Methods Mol Biol 581:61–77

7. Baldrian P, Valaskova V (2008) Degradation of cellulose by basidiomycetous fungi. FEMS Microbiol Rev 32(3):501–521

8. Banerjee S, Sen R, Pandey RA, Chakrabarti T, Satpute D, Giri BS, Mudliar S (2009) Evaluation of wet air oxidation as a pretreatment strategy for bioethanol production from rice husk and process optimization. Biomass Bioener 33:1680–1686

9. Banerjee S, Sen R, Mudliar S, Pandey RA, Chakrabarti T, Satpute D (2011) Alkaline peroxide assisted wet air oxidation pretreatment approach to enhance enzymatic convertibility of rice husk. Biotechnol Prog 27(3):691–697

10. Banik S, Bandyopadhyay S, Ganguly S (2003) Bioeffects of microwave—a brief review. Bioresour Technol 87:155–159

11. Barros Rda R, Paredes Rde S, Endo T, Bon EP, Lee SH (2013) Association of wet disk milling and ozonolysis as pretreatment for enzymatic saccharification of sugarcane bagasse and straw. Bioresour Technol 136:288–294

12. Bjerre AB, Olesen AB, Fernqvist T, Plöger A, Schmidt AS (1996) Pretreatment of wheat straw using combined wet oxidation and alkaline hydrolysis resulting in convertible cellulose and hemicellulose. Biotechnol Bioeng 49(5):568–577

13. Bondesson PM, Galbe M, Zacchi G (2013) Ethanol and biogas production after steam pretreatment of corn stover with or without the addition of sulphuric acid. Biotechnol Biofuels 6(1):11.

14. Boonmanumsin P, Treeboobpha S, Jeamjumnunja K, Luengnaruemitchai A, Chaisuwan T, Wongkasemjit S (2012) Release of monomeric sugars from *Miscanthus sinensis* by microwave-assisted ammonia and phosphoric acid treatments. Bioresour Technol 103(1):425–431CrossRef

15. Bradshaw TC, Alizadeh H, Teymouri F, Balan V, Dale BE (2007) Ammonia fiber expansion pretreatment and enzymatic hydrolysis on two different growth stages of reed canary grass. Appl Biochem Biotechnol 140:395–405

16. Cao W, Sun C, Liu R, Yin R, Wu X (2012) Comparison of the effects of five pretreatment methods on enhancing the enzymatic digestibility and ethanol production from sweet sorghum bagasse. Bioresour Technol 111:215–221

17. Cara C, Ruiz E, Ballesteros I, Negro MJ, Castro E (2006) Enhanced enzymatic hydrolysis of olive tree wood by steam explosion and alkaline peroxide delignification. Proc Biochem 41(2):423–429

18. Cara C, Ruiz E, Oliva J, Sáez F, Castro E (2008) Conversion of olive tree biomass into fermentable sugars by dilute acid pretreatment and enzymatic saccharification. Bioresour Technol 99:1869–1876

19. Chang VS, Holtzapple MT (2000) Fundamental factors affecting biomass enzymatic reactivity. Appl Biochem Biotechnol Part A Enzyme Eng Biotechnol 84–86:5–37

20. Chum HL, Douglas LJ, Feinberg DA, Schroeder HA (1985) Evaluation of pretreatments for enzymatic hydrolysis of cellulose.

21. Correia JA, Júnior JE, Gonçalves LR, Rocha MV (2013) Alkaline hydrogen peroxide pretreatment of cashew apple bagasse for ethanol production: study of parameters. Bioresour Technol 139:249–256

22. Dias AA, Freitas GS, Marque GSM, Sampaio A, Fraga IS, Rodrigue MAM, Evtuguin DV, Bezerra RMF (2010) Enzymatic saccharification of biologically pre-treated wheat straw with white-rot fungi. Bioresour Technol 101:6045–6050

23. Eggert C, Temp U, Eriksson KE (1997) Laccase is essential for lignin degradation by the white-rot fungus*Pycnoporus cinnabarinus*. FEBS Lett 407(1):89–92

24. Falls M, Holtzapple MT (2011) Oxidative lime pretreatment of Alamo switchgrass. Appl Biochem Biotechnol 165(2):506–522

25. Feijoo G, Moreira MT, Alvarez P, Lú-Chau TA, Lema JM (2008) Evaluation of the enzyme manganese peroxidase in an industrial sequence for the lignin oxidation and bleaching or eucalyptus kraft pulp. J Appl Polym Sci 109(2):1319–1327

26. Feng L, Chen Z (2008) Research progress on dissolution and functional modification of cellulose in ionic liquids. J Mol Liq 142:1–5

27. Fox M, Noike T (2004) Wet oxidation pretreatment for the increase in anaerobic biodegradability of newspaper waste. Bioresour Technol 91(3):273–281

28. García-Cubero MA, González-Benito G, Indacoechea I, Coca M, Bolado S (2009) Effect of ozonolysis pretreatment on enzymatic digestibility of wheat and rye straw. Bioresour Technol 100(4):1608–1613

29. Geddes CC, Nieves IU, Ingram LO (2011) Advances in ethanol production. Curr Opin Biotechnol 22(3):312–319

30. Geng A, Xin F, Ip J (2012) Ethanol production from horticultural waste treated by a modified organosolv method. Bioresour Technol 104:715–721

31. Ghose TK, Pannir Selvam PV, Ghosh P (1983) Catalytic solvent delignification of agricultural residues: organic catalysts. Biotechnol Bioeng 25(11):2577–2590

32. González G, López-Santín J, Caminal G, Solà C (1986) Dilute acid hydrolysis of wheat straw hemicellulose at moderate temperature: a simplified kinetic model. Biotechnol Bioeng 28(2):288–293

33. Guo B, Zhang Y, Yu G, Lee WH, Jin YS, Morgenroth E (2013) Two-stage acidic-alkaline hydrothermal pretreatment of lignocellulose for the high recovery of cellulose and hemicellulose sugars. Appl Biochem Biotechnol 169(4):1069–1087

34. Hammel KE, Kapich AN, Jensen KA Jr, Ryan ZC (2002) Reactive oxygen species as agents of wood decay by fungi. Enz Microb Technol 30(4):445–453

35. Hasunuma T, Okazaki F, Okai N, Hara KY, Ishii J, Kondo A (2013) A review of enzymes and microbes for lignocellulosic biorefinery and the possibility of their application to consolidated bioprocessing technology. Bioresour Technol.

36. Hideno A, Kawashima A, Endo T, Honda K, Morita M (2013) Ethanol-based organosolv treatment with trace hydrochloric acid improves the enzymatic digestibility of Japanese cypress (*Chamaecyparis obtusa*) by exposing nanofibers on the surface. Bioresour Technol 18:64–70

37. Hofrichter M (2002) Review: lignin conversion by manganese peroxidase (MnP). Enz Microb Technol 30(4):454–466

38. Holtzapple MT, Lundeen JE, Sturgis R, Lewis JE, Dale BE (1992) Pretreatment of lignocellulosic municipal solid-waste by ammonia fiber explosion (AFEX). Appl Biochem Biotechnol 34(5):5-21

39. Hsu TA, Ladisch MR, Tsao GT (1980) Alcohol from cellulose. Chem Technol 10(5):315–319

40. Ichwan M, Son TW (2011) Study on organosolv pulping methods of oil palm biomass. In: International Seminar on Chemistry, pp 364–370

41. Itziar E, Cristina S, Iñaki M, Jalel L (2012) Effect of alkaline and autohydrolysis processes on the purity of obtained hemicelluloses from corn stalks. Bioresour Technol 103:239–248

42. Jackowiak D, Bassard D, Pauss A, Ribeiro T (2011) Optimisation of a microwave pretreatment of wheat straw for methane production. Bioresour Technol 102(12):6750–6756

43. Jorgensen H, Kristensen JB, Felby C (2007) Enzymatic conversion of lignocellulose into fermentable sugars: challenges and opportunities. Biofuels Bioprod Bioref 1:119–134

44. Karthika K, Arun AB, Melo JS, Mittal KC, Kumar M, Rekha PD (2012) Hydrolysis of acid and alkali presoaked lignocellulosic biomass exposed to electron beam irradiation. Bioresour Technol 129:646–649

45. Kim S, Holtzapple MT (2005) Lime pretreatment and enzymatic hydrolysis of corn stover. Bioresour Technol 96:1994–2006

46. Kim TH, Lee YY (2005) Pretreatment of corn stover by soaking in aqueous ammonia. Appl Biochem Biotechnol Part A 124(1–3):1119–1131

47. Kim TH, Lee YY (2006) Fractionation of corn stover by hot-water and aqueous ammonia treatment. Bioresour Technol 97:224–232

48. Kim JW, Kim KS, Lee JS, Park SM, Cho HY, Park JC, Kim JS (2011) Two-stage pretreatment of rice straw using aqueous ammonia and dilute acid. Bioresour Technol 102(19):8992–8999

49. Kim TH, Kim JS, Sunwoo C, Lee YY (2003) Pretreatment of corn stover by aqueous ammonia. Bioresour Technol 90(1):39–47

50. Kim KH, Tucker M, Nguyen Q (2005) Conversion of bark-rich biomass mixture into fermentable sugar by two-stage dilute acid-catalyzed hydrolysis. Bioresour Technol 96:1249–1255

51. Kim TH, Taylor F, Hicks KB (2008) Bioethanol production from barley hull using SAA (soaking in aqueous ammonia) pretreatment. Bioresour Technol 99(13):5694–5702

52. Lau MJ, Lau MW, Gunawan C, Dale BE (2010) Ammonia Fiber Expansion (AFEX) pretreatment, enzymatic hydrolysis, and fermentation on Empty Palm Fruit Bunch Fiber (EPFBF) for cellulosic ethanol production. Appl Biochem Biotechnol 162:1847–1857

53. Lee J, Jeffries TW (2011) Efficiencies of acid catalysts in the hydrolysis of lignocellulosic biomass over a range of combined severity factors. Bioresour Technol 102:5884–5890

54. Lee J, Houtman CJ, Kim H, Choi I, Jeffries TW (2011) Scale-up study of oxalic acid pretreatment of agricultural lignocellulosic biomass for the production of bioethanol. Bioresour Technol 102:7451–7456

55. Lissens G, Klinke H, Verstraete W, Ahring B, Thomsen AB (2004a) Wet oxidation treatment of organic household waste enriched with wheat straw for simultaneous saccharification and fermentation into ethanol. Environ Technol 25(6):647–655

56. Lissens G, Thomsen AB, De Baere L, Verstraete W, Ahring BK (2004b) Thermal wet oxidation improves anaerobic biodegradability of raw and digested biowaste. Environ Sci Technol 38(12):3418–3424

57. Liu ZLL, Song MZ (2009) Genomic adaptation of *Saccharomyces cerevisiae* to inhibitors for lignocellulosic biomass conversion to ethanol. Appl Mycol 136

58. Liu J, Takada R, Karita S, Watanabe T, Honda Y, Watanabe T (2010) Microwave-assisted pretreatment of recalcitrant softwood in aqueous glycerol. Bioresour Technol 23:9355–9360

59. López-Abelairas M, Alvarez Pallín M, Salvachúa D, Lú-Chau T, Martínez MJ, Lema JM (2012) Optimisation of the biological pretreatment of wheat straw with white-rot fungi for ethanol production. Bioprocess Biosyst Eng.

60. Lucas M, Hanson SK, Wagner GL, Kimball DB, Rector KD (2012) Evidence for room temperature delignification of wood using hydrogen peroxide and manganese acetate as a catalyst. Bioresour Technol 119:174–180

61. Maa H, Liu W, Chen X, Wua Y, Yu Z (2009) Enhanced enzymatic saccharification of rice straw by microwave pretreatment. Bioresour Technol 100:1279–1284

62. Martinez AT, Ruiz-Duenas FJ, Martinez MJ, Del Rio JC, Gutierrez A (2009) Enzymatic delignification of plant cell wall: from nature to mill. Curr Opin Biotechnol 20(3):348–357

63. Martins S, Teixeira JA, Mussatto SI (2013) Solid-state fermentation as a strategy to improve the bioactive compounds recovery from *Larrea tridentata* leaves. Appl Biochem Biotechnol.

64. Mckendry P (2002) Energy production from biomass (part 1) overview of biomass. Bioresour Technol 83:37–46

65. Mesa L, González E, Cara C, González M, Castro E, Mussatto SI (2011) The effect of organosolv pretreatment variables on enzymatic hydrolysis of sugarcane bagasse. Chem Eng J 168:1157–1162

66. Mittal A, Katahira R, Himmel ME, Johnson DK (2011) Effects of alkaline or liquid-ammonia treatment on crystalline cellulose: changes in crystalline structure and effects on enzymatic digestibility. Biotechnol Biofuels 4:41

67. Miura T, Lee SH, Inoue S, Endo T (2012) Combined pretreatment using ozonolysis and wet-disk milling to improve enzymatic saccharification of Japanese cedar. Bioresour Technol 126:182–186

68. Mosier N, Wyman C, Dale B, Elander R, Lee YY, Holtzapple M, Ladisch M (2005) Features of promising technologies for pretreatment of lignocellulosic biomass. Bioresour Technol 96(6):673–686

69. Nakamura Y, Daidai M, Kobayashi F (2004) Ozonolysis mechanism of lignin model compounds and microbial treatment of organic acids produced. Water Sci Technol 50(3):167–172

70. Nomanbhay SM, Hussain R, Palanisamy K (2013) Microwave assisted enzymatic saccharification of oil palm empty fruit bunch fiber for enhanced fermentable sugar yield. J Sustain Bioenergy Syst 3:7–17

71. Noureddini H, Byun J (2010) Dilute-acid pretreatment of distillers' grains and corn fiber. Bioresour Technol 101:1060–1067

72. Ooshima H, Aso K, Harano Y, Yamamoto T (1984) Microwave treatment of cellulosic materials for their enzymatic-hydrolysis. Biotechnol Lett 6:289–294

73. Palonen H, Thomsen AB, Tenkanen M, Schmidt AS, Viikari L (2004) Evaluation of wet oxidation pretreatment for enzymatic hydrolysis of softwood. Appl Biochem Biotechnol 117(1):1–17

74. Pan X, Saddler JN (2013) Effect of replacing polyol by organosolv and kraft lignin on the property and structure of rigid polyurethane foam. Biotechnol Biofuels 6(1):12

75. Panagiotopoulos IA, Chandra RP, Saddler JN (2012) A two-stage pretreatment approach to maximise sugar yield and enhance reactive lignin recovery from poplar wood chips. Bioresour Technol 130:570–577

76. Pinto PA, Dias AA, Fraga I, Marques G, Rodrigues MAM, Colaco J, Sampaio A, Bezerra RMF (2012) Influence of ligninolytic enzymes on straw saccharification during fungal pretreatment. Bioresour Technol 111:261–267

77. Potumarthi R, Baadhe RR, Nayak P, Jetty A (2013) Simultaneous pretreatment and saccharification of rice husk by *Phanerochete chrysosporium* for improved production of reducing sugars. Bioresour Technol 128:113–117

78. Rasmussen ML, Shrestha P, Khanal SK, Pometto Iii AL, van Leeuwen J (2010) Sequential saccharification of corn fiber and ethanol production by the brown rot fungus *Gloeophyllum trabeum*. Bioresour Technol 101(10):3526–3533

79. Saha BC, Cotta MA (2008) Lime pretreatment, enzymatic saccharification and fermentation of rice hulls to ethanol. Biomass Bioenergy 32:971–977

80. Sánchez OJ, Cardona CA (2008) Trends in biotechnological production of fuel ethanol from different feedstocks. Bioresour Technol 99(13):5270–5295

81. Saritha M, Arora A, Lata (2012) Biological pretreatment of lignocellulosic substrates for enhanced delignification and enzymatic digestibility. Indian J Microbiol 52(2):122–130

82. Sarkar N, Ghosh SK, Bannerjee S, Aikat K (2012) Bioethanol production from agricultural wastes: an overview. Renew Energy 37(1):19–27

83. Schuerch C (1963) Plasticizing wood with liquid ammonia. J Ind Eng Chem 55:39

84. Shallom D, Shoham Y (2003) Microbial hemicellulases. Curr Opin Microbiol 6(3):219–228

85. Shao Q, Chundawat SP, Krishnan C, Bals B, Sousa Lda C, Thelen KD, Dale BE, Balan V (2010) Enzymatic digestibility and ethanol fermentability of AFEX-treated starch-rich lignocellulosics such as corn silage and whole corn plant. Biotechnol Biofuels 9:3–12

86. Sharma R, Palled V, Sharma-Shivappa RR, Osborne J (2012) Potential of potassium hydroxide pretreatment of Switch grass for fermentable sugar production. Appl Biochem Biotechnol 169(3):761–772

87. Sindhu R, Kuttiraja M, Binod P, Janu KU, Sukumaran RK, Pandey A (2011) Dilute acid pretreatment and enzymatic saccharification of sugarcane tops for bioethanol production. Bioresour Technol 102:10915–10921

88. Su MY, Tzeng WS, Shyu YT (2010) An analysis of feasibility of bioethanol production from Taiwan sorghum liquor waste. Bioresour Technol 101(17):6669–6675

89. Sun F, Chen H (2008) Organosolv pretreatment by crude glycerol from oleochemicals industry for enzymatic hydrolysis of wheat straw. Bioresour Technol 99:5474–5479

90. Sun Y, Cheng J (2002) Hydrolysis of lignocellulosic materials for ethanol production: a review. Bioresour Technol 83:1–11

91. Szijártó N, Kádár Z, Varga E, Thomsen AB, Costa-Ferreira M, Réczey K (2009) Pretreatment of reed by wet oxidation and subsequent utilization of the pretreated fibers for ethanol production. Appl Biochem Biotechnol 155(1–3):386–396

92. Tang J, Chen K, Huang F, Xu J, Li J (2013) Characterization of the pretreatment liquor of biomass from the perennial grass, *Eulaliopsis binata*, for the production of dissolving pulp. Bioresour Technol 129:548–552

93. Teymouri F, Laureano-Perez L, Alizadeh H, Dale BE (2005) Optimization of the ammonia fiber explosion (AFEX) treatment parameters for enzymatic hydrolysis of corn stover. Bioresour Technol 96(18):2014–2018

94. Varga E, Schmidt AS, Réczey K, Thomsen AB (2003) Pretreatment of corn stover using wet oxidation to enhance enzymatic digestibility. Appl Biochem Biotechnol 104(1):37–50

95. Verma P, Watanabe T, Honda Y, Watanabe T (2011) Microwave-assisted pretreatment of woody biomass with ammonium molybdate activated by H_2O_2. Bioresour Technol 102:3941–3945

96. Wan C, Li Y (2011) Effectiveness of microbial pretreatment by *Ceriporiopsis subvermispora* on different biomass feedstocks. Bioresour Technol 102:7507–7512

97. Wan C, Li Y (2012) Fungal pretreatment of lignocellulosic biomass. Biotechnol Adv 30(6):1447–1457

98. Xu J, Chen H, Kádár Z, Thomsen AB, Schmidt JE, Peng H (2011) Optimization of microwave pretreatment on wheat straw for ethanol production. Biomass Bioenergy 35:385–386

99. Yu J, Zhang J, He J, Liu Z, Yu Z (2009) Combinations of mild physical or chemical pretreatment with biological pretreatment for enzymatic hydrolysis of rice hull. Bioresour Technol 100:903–908

100. Yu H, Zhang X, Song L, Ke J, Xu C, Du W, Zhang J (2010) Evaluation of white-rot fungi-assisted alkaline/oxidative pretreatment of corn straw undergoing enzymatic hydrolysis by cellulase. J Biosci Bioeng 110(6):660–664

101. Zhao Y, Wang Y, Zhu JY, Ragauskas A, Deng Y (2008) Enhanced enzymatic hydrolysis of spruce by alkaline pretreatment at low temperature. Biotechnol Bioeng 99(6):1320–1328

102. Zhao X, Cheng K, Liu D (2009) Organosolv pretreatment of lignocellulosic biomass for enzymatic hydrolysis. Appl Microbiol Biotechnol 82(5):815–827.

CHAPTER 3

Ligninolytic enzymes: A biotechnological alternative for bioethanol production

Jersson Plácido and Sergio Capareda

Department of Biological and Agricultural Engineering, Texas A&M University, 201 Scoates hall, 2117 TAMU, College Station, TX 77843, USA

ABSTRACT

Ligninolytic fungi and enzymes (i.e., laccase, manganese peroxidase, and lignin peroxidase) have been applied recently in the production of second-generation biofuels. This review contains the analysis of ligninolytic enzymes and their applications in second-generation biofuels. In here, each of the ligninolytic enzymes was described analyzing their structures, catalysis, and reaction mechanism. Additionally, delignification and detoxification, the two most important applications of ligninolytic enzymes, were reviewed and analyzed. The analysis includes an evaluation of the biochemical process, feedstocks, and the ethanol production. This review describes the current situation of the ligninolytic enzymes technology and its future applications in bioethanol industry.

KEYWORDS

Ligninolytic fungi, Bioethanol, Laccases, Manganese peroxidase, Lignin peroxidase, Detoxification, Delignification

1. INTRODUCTION

The United States regulation requires 36 billion gallons of biofuels, and 21 billion should be produced from lignocellulosic material or other new advanced fuels by 2020 [1]. Lignocellulose, the most abundant source of organic material in the world, is a polysaccharide combination of cellulose, hemicellulose, and lignin [2]. Bioethanol is one of the most essential biofuels produced from lignocellulosic material. In general, bioethanol production from lignocellulosic material includes three principal steps: 1) pretreatment, 2) saccharification, and 3) fermentation. The pretreatment process modifies the lignocellulose structure by removing the lignin and altering the cellulose and hemicellulose structure [3]. Saccharification is the enzymatic transformation of cellulose and hemicellulose into monosaccharides as glucose and xylose. Whereas, fermentation is the transformation of these monosaccharides into ethanol. The current pretreatment methodologies utilize energy-intensive processes (high pressures and temperatures) and harsh chemical compounds (NaOH, H_2SO_4). This combination generates undesirable compounds and process inefficiencies. To overcome these issues, biomass pretreatment has explored more sustainable technologies such ligninolytic enzymes.

In nature, different organisms degrade lignin; however, the most effective ones are the fungi belonging to the white rot fungi class [4]. These microorganisms possess an enzymatic system to effectively degrade lignin. This system contains three principal enzymes laccase, manganese peroxidase (MnP), and lignin peroxidase (LiP) [5]. These microorganisms have been used to delignify different substrates such paper, animal feed, and biofuels [3,6–8]. In biofuel production, the ligninolytic enzymes have two principal purposes, delignification and detoxification. Delignification methods apply ligninolytic fungus and their enzymes to reduce the lignin content in several feedstocks [3,5–8]. Detoxification utilizes the ligninolytic enzymes to reduce the toxic compounds present in the biomass hydrolysates after chemical or physicochemical pretreatments [9–14].

The aim of this review is to describe the relevant aspects of the ligninolytic enzymes laccase, MnP, and LiP and to analyze the present and future application of this technology in the production of biofuels.

2. LIGNINOLYTIC ENZYMES

2.1 Laccases

Laccases (E.C. 1.10.3.2) also known as benzenediol: oxygen oxidoreductase or p-diphenol oxidase belong to the oxidoreductase class. Yoshida, at the end of the nineteenth century, was the first to extract laccases from exudates of the Japanese tree *Rhus vernicifera.*Laccases are widely distributed in microorganisms, insects, and plants, showing a specific function in each of them. From this group, white rot fungi are the most studied laccases. These fungi use their enzymes to break the plant's lignocellulosic wall and obtain the host's nutrients. These fungal enzymes degrade the complex polyphenol structure that constitutes lignin, the principal recalcitrant component in the lignocellulosic wall. Besides their catalytic characteristics, these enzymes are attractive for biotechnological process, because they are extracellular and inducible, do not need a cofactor, and have low specificity. Laccase production is induced and increased by the addition of molecules as copper, dyes, or other recalcitrant compounds [15]. Laccase employs oxygen as an oxidizing agent and cofactor instead of other expensive cofactors such as pantothenic acid, thiamine, or biotin [16].

Laccases have low substrate specificity; this characteristic allows the degradation of several compounds with a phenolic structure [17]. Therefore, laccases have been employed in several areas such as bioremediation of aromatic recalcitrant compounds [18], treatment of effluents polluted with lignin [19], chemical synthesis [20,21], degradation of a wide number of textile dyes [22–25], and biomass pretreatment for biofuel production [26–28].

2.2 Laccase structure and catalytic mechanism

Laccases have a primary structure of approximately 500 amino acid residues organized in three consecutive domains, with a Greek key β barrel topology. The three domains are distributed in a first domain with 150 initial amino acids, a second domain between the 150 and 300 residue, and a third domain from the 300 to 500 amino acid. The structure is stabilized by two disulfide bridges localized between domains I and II and between domains I and III [16,29,30]. However, some laccases present three disulfide bridges. *Melanocarpus albomyces* has disulfide bridges inside domain I, another between domain I and domain III, and the last one between domain II and III [31].

Laccase belongs to multicopper oxidases (MCOs) and blue multicopper oxidases [21]. Laccases have four copper (Cu) molecules in their active site which participate in oxygen reduction and water production (Figure 1) [32]. The laccases' four copper atoms are disseminated in three types of cores or places: type 1 Cu (T1), type 2 Cu (T2), and type 3 Cu (T3). These cores are in two metallic active sites: the mononuclear location T1 and the trinuclear location T2/T3 [16,21,29,30]. Type 1 Cu (T1) or blue Cu is a paramagnetic copper with a strong absorption at 600 nm (blue coloration). This signal is generated by the covalent union between Cu-Cys in the mononuclear location [16,33]. In addition, T1 has the highest redox potential in the enzyme and participates in the enzymatic catalysis by oxidizing the substrate [16]. Laccase's oxidative power is affected by the amino acids surrounding T1 [16]. Similar to T1, T2 is a paramagnetic cooper; however, it lacks absorption in the visible region of the spectrum. This Cu exhibits a strong absorption under electron paramagnetic resonance (EPR); technique employed to identify this spot [16]. In addition, the T2/T3 core participates in the inhibition of the enzymatic activity by its interaction with anions such as fluoride or cyanide [34]. On the other hand, T3 Cu is a binuclear core formed by a Cu (II) dimer. This site presents absorption at 330 nm and a lack of EPR signal. Type 3 coppers are diamagnetic and have a Cu-Cu union which participates in the compound oxidation as an electron acceptor.

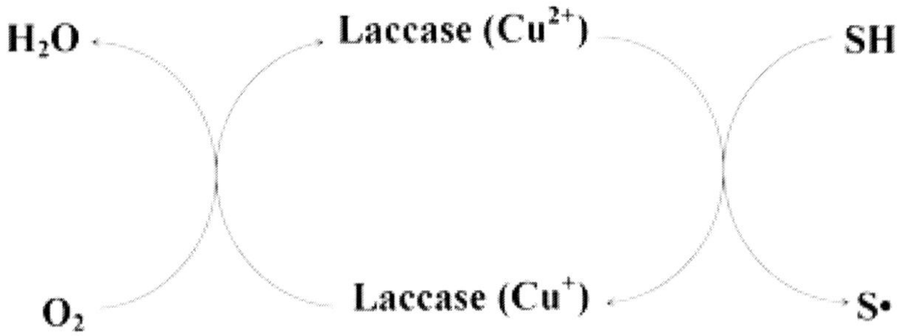

Figure 1. General diagram of laccase redox mechanism (SH reduced substrate, S• oxidized substrate).

In laccases, the different types of coppers have variations associated with the coordination between them and the amino acids in the active site. T1 Cu exhibits a different triangular planar coordination similar to other MCOs. Laccases display two histidines (His) and one cysteine (Cys) as equatorial ligands. Meanwhile, the other MCOs have an additional axial ligand with a methionine. In contrast, laccases did not have this axial extra bond, they have a leucine (Leu) or a phenylalanine (Phe) substituting methionine [16,29–31]. On the other hand, T2 Cu exhibits coordination with two His and a water molecule; whereas, T3 Cu interacts with a water molecule and six His. In T3 Cu, the two Cu molecules share the water molecule and split the six His in two groups of three [16,29,30]. The two T3 Cu interact with the oxygen atom; however, the interaction between water and the two Cu at the same time only happens when the enzyme is in its oxidized form [16,29–31,33].

Laccase crystalline structures evidenced that T2 Cu (EPR active) could be any of the three Cu in the trinuclear site. In this site, two of the Cu reduce and become silent to EPR. Meanwhile, the other one remains active to EPR

[35]. This interchange between T2 Cu and one of T3 coppers could be related to the reduction mechanism of the two oxygen molecules (cofactor), mechanism which is not clear for blue multicopper oxidases [35].

The enzymes' redox potential oscillates between 300 and 800 mV; these fluctuations depend on different factors such Cu ligands, distance among Cu, and the second sphere of amino acids in the active site. The last factor is associated with the degradation of recalcitrant compounds as lignin or industrial dyes [16].

Another structural characteristic in laccases is the presence of ligands: monosaccharides (mannose, N-acetyl glucosamine), ions like Ca^{2+}, and organic molecules like glycerol [16,29–31,35]. The presence of these ligands produce significant differences between laccases, an example of that is the glycosylations exhibited in the isoenzymes from *Trametes versicolor* [30].

2.3 Heme-peroxidases

After laccases, lignin peroxidase (LiP) and manganese peroxidase (MnP) are the most significant ligninolytic enzymes. These enzymes belong to the heme-proteins because they have the protoporphyrin IX as a prosthetic group. Similar to other heme-peroxidases such cytochrome c peroxidase and horseradish peroxidase, the catalytic cycle of LiP and MnP have three reactions [32,36–39].

Hydrogen peroxide causes the enzyme oxidation to produce the compound I and water:

$$\text{Reduced peroxidase} + H_2O_2 \rightarrow \text{Compound I} + H_2O$$

The modified enzyme (compound I) catalyzes the production of a free radical (S•) and a second modified form of the enzyme (compound II) by an electron transfer from the substrate (SH: reduced substrate):

$$\text{Compound I} + SH \rightarrow \text{Compound II} + S\bullet$$

Finally, compound II reacts with a second substrate molecule to produce another free radical and water; meanwhile, the enzyme reduces to its original form.

$$\text{Compound II} + SH \rightarrow \text{Reduced peroxidase} + S\bullet + H_2O$$

2.4 Lignin peroxidase (LiP)

LiPs (E.C. 1.11.1.14) were originally discovered in nitrogen- and carbon-limited cultures of *Phanerochaete chrysosporium* [32]. LiP possess high redox potential (700 to 1,400 mV), low optimum pH 3 to 4.5, and the ability to catalyze the degradation of a wide number of aromatic structures such veratryl alcohol (3,4-dimethoxybenzyl) and methoxybenzenes [32,36,38]. LiP oxidizes aromatic rings moderately activated by electron donating substitutes; in contrast, common peroxidases participate in the catalysis of aromatic substrates highly activated (ammine, hydroxyl, etc.). An explanation for this type of catalysis is the production of veratryl alcohol radicals. These radicals have higher redox potential than LiP's compounds I and II and can participate in the degradation of compounds with high redox potential [40].

LiPs are monomeric glycosylated enzymes of 40 kDa, with 343 amino acids residues, 370 water molecules, a heme group, four carbohydrates, and two calcium ions [39]. Their secondary structure is principally helicoidal. It contains eight major helixes, eight minor helixes, and two anti-parallel beta sheets. LiPs contain two domains at both sides of the heminic group. This group is inlaid in the protein but accessible to solvents via two small channels [39]. The heminic cavity includes 40 residues, and it bonds to the protein via hydrogen bridges. Additionally, the heminic iron (Fe) coordinates with a His and a molecule of water. This His is associated with the high redox potential of LiP. The enzyme's redox potential rises when the His has a reduced imidazol character [39]. In addition, a greater distance between the His and the heminic group increases the redox potential of the enzyme. This increment in the redox potential is a response to the electronic deficiency in the Fe of the porphyrin ring. In fact, this distance causes most differences among enzymes with similar porphyrin cores [39,41].

Another characteristic related with LiP's high redox potential is the invariant presence of a tryptophan residue (Trp171) in the enzymes' surface. Trp171 seems to facilitate electronic transference to the enzyme from substrates that cannot access into the heminic oxidative group [41]. Additionally, Trp171 participates with the catalysis of veratryl alcohol, a metabolite produced by some ligninolytic fungi. Veratryl alcohol participates in the oxidation of different aromatic molecules. Some researchers conceptualized that this alcohol protects the enzyme from the action of H_2O_2 and participate as a redox mediator between the enzyme and substrates which cannot get inside the heminic center [38,39].

2.5 Manganese peroxidase (MnP)

Kuwahara in 1984 found the first MnP (E.C. 1.11.1.13) in batch cultures of *Phanerochaete chrysosporium* [32]. They are glycoproteins with a molecular weight between 38 and 62.5 kDa [42], approximately 350 amino acid residues, and a 43% of identity with LiP sequence. MnP structure has two domains with the heminic group in the middle, ten major helixes, a minor helix, and five disulfide bridges. One of those bridges participates in the manganese (Mn) bonding site. This site is a characteristic that distinguishes MnP from other peroxidases [37].

The enzyme's catalytic cycle starts with the transference of two electrons from the heminic group to H_2O_2; it produces compound I and water. After that, compound I catalyzes the oxidation of one substrate molecule with the production of a free radical and compound II. Compound II oxidizes Mn^{2+} to produce Mn^{3+}, the cation responsible to oxidize aromatic compounds. It is important to keep in mind that compound II demands Mn^{2+} presence for its reaction; in contrast, compound I can oxidize Mn^{2+} or the substrate. After Mn^{3+} is stabilized by organic acids, it reacts non-specifically with organic molecules by removing an electron and a proton from the substrates [42]. Mn^{3+} is a small size compound with high redox potential, which diffuses easily in the lignified cell wall. Therefore, Mn^{3+} starts the attack inside the plant cell wall which facilitates the penetration and action of the other enzymes [38,41].

2.6 Sources of ligninolytic enzymes

The ligninolytic enzymes are a ubiquitous group of enzymes found in different types of organisms as plants, bacteria, insects, and fungi. In plants, laccases are the most documented ligninolytic enzyme; these are extracellular glycoproteins composed by a monomeric protein with a sugar component. Laccases have been found in different types of plants: the Japanese lacquer tree, mango, mung bean, peach, tobacco, zea mays, etc. [43]. In plants, laccases realize different kind of functions such as lignin synthesis [43], iron oxidation from Fe(II) to Fe(III) [44], and regeneration of injured tissue [43]. In plants, lignin synthesis is laccases' most significant function. In these organisms, they catalyze the monolignol dimerization, the basic molecules of lignin polymers and oligomers [43].

In bacteria, lignin breakdown is exhibited by three groups: actinomycetes, α-proteobacteria, and γ-proteobacteria [45]. These three groups of bacteria were described in insects degrading wood microflora. These types of bacteria showed delignification in both *in vitro* and *in vivo*analyses [45]. *Streptomyces viridosporus*, actinomycetes species, and *Thermobifida fusca* exhibited LiP activities [45–47]. In contrast, *Azospirillum* *lipoferum*, *Thermus thermophilus*, *Marinomonas mediterranea*, *Bacillus subtilis*, and *Streptomyces cyaneus* showed laccase activity [48–51]. Bacterial laccases are intracellular enzymes with a monomeric, multimeric, or homotrimeric structure without carbohydrate moiety [43]. In nature, bacteria use laccases in spore protection and pigmentation. This type of enzymes exhibited resistance to high pH and temperatures [43,45,52].

In insects, the delignification process is done by ligninolytic enzymes produced by the insects or by the insects' microflora. The ligninolytic enzymes have been identified in different insects and different parts of them. An example of that are *Nephotettix cincticeps* (salivary glands) [53], *Manduca sexta* (Malpighian tubules, midgut, fat body, and epidermis) [54], *Reticulitermes flavipes* (gut) [55], and *Tribolium castaneum*(cuticles) [56]. The principal functions of insect's ligninolytic enzymes are cuticle sclerotization and pigmentation, toxic compound oxidation, and polymerization reactions [43,53].

The fungal ligninolytic enzymes are the most well-known enzymes, and they occur in ascomycetes, basidiomycetes, and deuteromycetes [57]. Inside these groups are several fungal species; however, the most studied fungi are: *Trametes versicolor*, *Phanerochaete chrysosporium*, *Pleurotus ostreatus*, *Dichomitus squalens*, *Lentinula edodes*, *Irpex lacteus*, and *Cerrena maxima* [32,58]. The principal functions of fungal ligninolytic enzymes are lignin degradation, spore and fruiting body formation, and degradation of plant's toxic compounds [32,57].

2.7 Biofuel applications

In the last years, ligninolytic fungi and their enzymes have appeared as a new alternative for pretreatment processes. This technology has been useful principally in two processes, delignification and detoxification. In delignification, ligninolytic fungi or their enzymes have been used to reduce the lignin content in different types of feedstock [3,4,59]. Whereas, in detoxification, ligninolytic enzymes have been used in the removal of chemical compounds from sugar hydrolysates after traditional pretreatments.

2.8 Delignification

In delignification process, ligninolytic enzymes have been applied in four different methods: 1) fungal delignification, 2) enzymatic delignification, 3) laccase-mediator system (LMS), and 4) integrated fungal fermentation (IFF). Table 1 resumes the application of these four delignification technologies. Fungal delignification utilizes the complete ligninolytic microorganisms for biomass delignification. In those cases, the microorganisms grow with the target biomass in a submerged culture [8,60] or a solid state fermentation [3,7]. Both cultivations achieved good delignification percentages; however, these were not always related with high glucose yields. The work of Salvachúa et al. (2011) is an example of that; in that case, the fungi with higher glucose yield were different from the ones with the highest delignification percentage. However, the lack of relation between high glucose yield and high delignification percentage occurs in different pretreatments and not only in microbial delignification. In fact, the principal issue with fungal pretreatment is the large duration of the process compared with the other pretreatment technologies [5]. The length of time used by the microorganisms to obtain high delignification percentages is not less than 13 days and can be up to 40 or 50 days [4,8] However, this period will depend on the strain used. The effectiveness of microbial delignification has been improved by adding an alkali treatment previous the fungal pretreatment. The alkali pretreatment benefits were the reduction of the process duration and the increment in the glucose and ethanol yield [61,62].

Table 1. Delignification methods using ligninolytic fungi and ligninolytic enzymes

Feedstock	Fungal strain	Type of process	Mediator	Delignification %	Glucose yield	Ethanol yield	Process duration	Ref
Wood pulp	*Trametes villosa*	Enzymatic Laccase	HBT HPA	50% 39%	-	-	15 h	[81]
Wood pulp	*Aspergillus fumigatus*	Enzymatic Laccase	HBT	14%	-	-	2 h	[59]
Hardwood chips	*Trametes versicolor*	Enzymatic Laccase	VA	54.3%	-	-	3 h	[62]
Sugarcane bagasse	*Ceriporiopsis subvermispora*	Fungal MnP, LiP, Laccase	-	33%	-	-	30 d	[64]
Eucalyptus globulus	*Myceliophthora thermophila*	Enzymatic Laccase	MS	>20%	-	-	12 h	[19]
Paper pulp	*Pycnoporus cinnabarinus*	Enzymatic Laccase	AC SA CA HBT	21% 25% −9% 40%	-	-	4-24 h	[68]
Corn stover Switchgrass	*Ceriporiopsis subvermispora*	Fungal Laccase	-	>25% >25%	56% 37%	-	18 d	[4]
Pinus contorta	-	Enzymatic Laccase	-	85.69%	288.83 mg/g	-	4 h	[83]
Eucalyptus globulus	*Pycnoporus cinnabarinus*	Enzymatic Laccase	HBT	27%	-	-	3-4 h	[84]
Wheat straw	Isolate Euc, *Irpex lacteus*	Fungal Laccase	-	35% 35%	70% 70%	-	15-46 d	[2]
Wheat straw	*Ceriporiopsis subvermispora* *Irpex lacteus*	Fungal MnP, LiP, Laccase	-	34% 39%	63% 66%	90%	21 d	[3]
Corn stover	*Ceriporiopsis subvermispora*	Fungal MnP Laccase	-	39.2%	66.9%	-	42 d	[7]
Corn stover (cs) Wheat straw (ws)	*Pycnoporus sanguineus H275*	Fungal Laccase	- - VA	27% (cs) 25% (ws) 21% (cs) 22% (ws)	-	-	21 d	[8]
Elephant grass (eg) Eucalypt wood (ew)	*Trametes villosa*	LMS	HBT	36% (eg) 58% (ew)	20.7% (eg) 57.9 (ew)	16.2% (eg) 12.3% (ew)	24 h	[22]
Corn stover	*Trametes versicolor*	LMS	HBT	-	91.7%	-	48 h	[20]
Wheat straw	*Sclerotium sp.*	LMS	ABTS	-	80.29%	-	48 h	[21]
Cotton gin trash	*Trametes versicolor*	LMS	PrimaGreen®	15%	23%	31.6%	96 h	[27]
Cotton gin trash	*Trametes versicolor*	LMS	PrimaGreen®	27%	41%	64%	96 h	[28]

Sugarcane bagasse	*Pleurotus ostreatus IBL-02*	Enzymatic MnP, LiP Laccase	-	33.6%	72.9%	16.3 g/l	48 h	[67]
Wheat straw	*T. versicolor I. lacteus*	Fungal MnP LiP Laccase	-	-	68.33% 56.26%	-	21 d	[85]
Sugarcane bagasse	*Phlebia sp. MG-60*	IFF		30%	-	45%	4 weeks	[56]
Sugarcane bagasse	*Phanerochaete chrysosporium Lentinula edodes Pleurotus ostreatus*	Fungal MnP	-	64%	-	-	12 weeks	[87]
Oil palm empty fruit bunches	*Pleurotus floridanus*	Fungal		5%	15.4%	22.9%	4 weeks	[88]
Corn stover	*Coprinus comatus*	IFF		66.5%	82%		5-7 d	[76]
Wheat straw	*Irpex lacteus*	Fungal MnP Laccase	-	-	8.4%	24%	14-21 d	[62]
Corn stover	*Phanerochaete chrysosporium Coriolus versicolor*	Enzymatic MnP Laccase	-	-	323 mg/g	-	24 h	[66]
Bamboo culms	*Punctularia sp.*	Fungal		-	10%	-	12 weeks	[89]
Poplar wood	*Trametes velutina*	Fungal	-		43.6%	22.2%	8 weeks	[90]
Sugarcane bagasse	*Pleurotus florida*	Fungal MnP Laccase	-	7.91%	203.33 mg/g	-	25 d	[91]
Corn stover	*Irpex lacteus*	Fungal MnP		25.48%	306.98 mg/g	144.03 mg/g	28 d	[92]
Sugarcane Bagasse	*Phlebia sp. MG-60*	IFF	-	-	-	44%	4 weeks	[75]
Hardwood kraft pulp, Waste newspaper	*Phlebia sp. MG-60*	IFF	-	-	-	71.8% 51.1%	68 h	[23]
Sugarcane bagasse	*Phlebia sp. MG-60*	IFF	-	-	-	66.2%	20 d	[61]
Oak wood	*Phlebia sp. MG-60*	IFF	-	-	-	43.9%	56 d	[74]

LMS: Laccase-Mediator System. IFF: Integrated fungal fermentation. HBT: 1-Hydroxybenzotriazole, HPA: N-Hydroxyphthalimide, MS: Methyl syringate. VA: Violuric acid. AC: Acetosyringone. SA: Syringaldehyde, CA: p-coumaric acid.

Enzymatic delignification is the application of enzymatic extracts and purified or semi-purified ligninolytic enzymes to realize the lignin degradation. Enzymatic delignification employs commercial or native ligninolytic enzymes [63]. Laccase is the most utilized enzyme followed by MnP and LiP; nevertheless, mixtures of two or three ligninolytic enzymes have been used for some delignification pretreatments. In those cases, the synergetic relationship between the ligninolytic enzymes improved biomass delignification [64,65]. In wheat straw, ligninolytic enzymes generated a reduction in the cellulose conversion during the saccharification process. To avoid cellulases' inhibition, ligninolytic enzymes were deactivated before the saccharification process [66]. Similar to fungal delignification, alkali pretreatments have been used in enzymatic delignification as an initial pretreatment to facilitate the delignification process [66]. Enzymatic processes achieved similar delignification percentages than microbial pretreatment. However, enzymatic processes employ less amount of time (between 24 and 96 h) to produce the same delignification achieved by fungal pretreatment [67]. Although this methodology reduces the delignification process duration, the current development of enzymatic delignification cannot compete with the conventional pretreatment technologies in terms of timespan and costs [5].

LMS utilize fungal laccases and redox mediators to achieve the delignification process. Redox mediators are chemical compounds, which act as electron carriers between the enzyme and the final substrate (Figure 2) [19]. Mediators' redox potential increases when the enzyme oxidizes them; after that, oxidized mediators react with the final substrate to recuperate the lost electrons. The increment in the redox potential allows the degradation of recalcitrant compounds such as lignin and different kinds of aromatic compounds that the enzyme by itself cannot break [19,68]. The mediator compounds most used are 1-hydroxybenzotriazole (HBT), 2,2′-azino-bis(3-ethylbenzothiazoline-6-sulphonic acid) (ABTS), and some natural mediators as syringaldehyde or vanillin [69]. The principal applications of laccase-mediator system are bleaching and delignification in the paper industry. However, this system has been used for delignification of some biofuel feedstocks. In cotton gin trash, the delignification was performed by a commercial LMS from Genencor (Rochester, NY, USA). In that case, LMS was used after an ultrasonication and hot water pretreatments [27]. LMS was improved by the addition of alkali-ultrasonication process before the hot water and LMS pretreatment. Alkali-ultrasonication duplicated the ethanol and cellulose yield compared with the previous method [28].

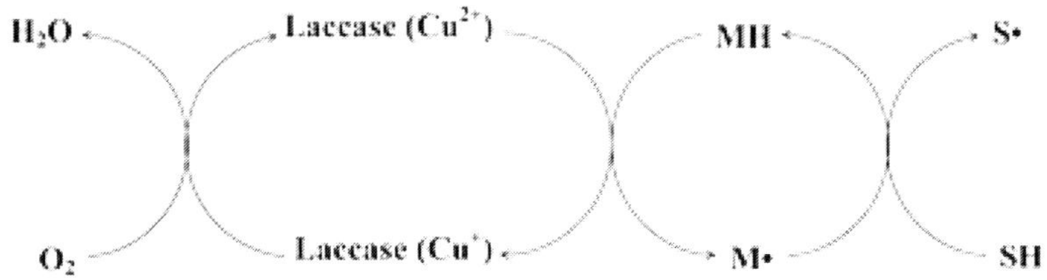

Figure 2. Laccase-mediator system's cycle.

LMS pretreatment was utilized over corn stover by using HBT as mediator [70]. In there, a correlation between laccase loading and enzymatic digestibility was found. Additionally, they found the importance of corn stover previous silage to increase LMS effectiveness [70]. In wheat straw, LMS pretreatment was used after the steam explosion pretreatment and was performed before and simultaneously with the enzymatic saccharification. The results in both processes were similar. Thus, LMS can participate in a semi-consolidate process for delignification and saccharification [71]. The application of LMS in elephant grass and eucalypt wood is the first report of LMS achieving significant delignification without a previous pretreatment. This improvement was achieved by using four steps of LMS pretreatment and NaOH washing [72]. LMS is less used in biofuels than the paper industry. This lack of applications is related with the small development of specific mediators for feedstock delignification.

Integrated fungal fermentation (IFF) is a consolidated process where a fungus or a group of fungi transform biomass into ethanol without the participation of other treatments or microorganisms [73]. *Phlebia* sp. MG-60 is a white rot fungus with the ability of selectively transforming lignin under solid state fermentation and produce ethanol from delignified biomasses under semi-aerobic submerged fermentation [74]. *Phlebia*sp. MG-60 was selected from a group of other 12 fungi. From this group, *Phlebia* sp. MG-60 was the only one to exhibit the uncommon characteristic of degrade lignin and produce ethanol from cellulose [73]. This fungal strain has been evaluated in the production of ethanol from different feedstocks such as hardwood kraft pulp, waste newspaper, sugarcane bagasse, and hard wood [73–75]. The addition of basal media, organic compounds, and minerals to the culture media increased the ethanol yield and the delignification percentage [75]. *Phlebia* sp. MG-60 produced an ethanol yield between 30% and 70% depending of the type of biomass utilized [61]. The process duration is between 6 and 29 days and is associated with the type of biomass and the presence of pretreatment. The addition of alkali pretreatment generated a reduction in the process timespan from 21 days to 240 h [61]. On the other hand, fungi co-culture is an alternative to generated and integrated process for ethanol production. In this case, ethanol is produced by mixing two types of fungi to develop delignification, saccharification, and fermentation. *Coprinus comatus* and *Trametes reesei* were cultivated to produce ethanol from corn stover. In this case, co-culture achieved greater delignification and cellulose conversion than monoculture [76].

Delignification using ligninolytic enzymes can be upgraded by improving the enzymatic catalysis using protein engineering. This area utilizes three types of approaches to modify ligninolytic enzymes: rational approaches, semi-rational approaches, and directed evolution [77]. Rational approaches are the molecular modification (site direct mutation) of the ligninolytic enzyme sequence using prior structural information as foundation for the specific sequence modification. Rational approaches have been used to improve the laccase ability to degrade non-phenolic substrates and to enhance the capacity to oxidize bulky phenolic molecules [78,79]. Semi-rational approaches employ saturation mutagenesis to modify hotspot residues in the enzymes. Saturation mutagenesis replaces the selected amino acids by all the codons that can generate the additional 19 amino acids [77]. This approach exhibited the production of enzymes with three- to eightfold higher catalytic efficiencies [80]. Directed molecular evolution is an approach that utilizes random mutation, gene recombination, and selection as fundaments. This molecular approach allows realizing the enzymatic design in absence of structural information. Direct molecular evolution has been used to improve catalytic activity and solvent tolerance [77].

The four delignification processes are compared in Table 2. In enzymatic and microbial process, the pros observed are: positive environmental impact, high delignification and detoxification, low sugar losses, and the possibility of develop a consolidate process. On the other hand, microbial and enzymatic have different cons. The enzymatic processes cons are: high costs, low commercial availability, and mediator necessity. Meanwhile, in microbial processes, the principal limitation is the process duration. Besides the limitations and assets, ligninolytic fungi and their enzymes represent an interesting option to the conventional chemical and physical processes used in biofuel production.

Table 2. Delignification technique comparison.

Delignification process	Delignification	Sugar losses	Process duration	Economic	Environmental impact
Enzymatic	High	Low	2 to 48 h	High costs	Positive
Fungal	High	Low	6 to 45 days	Low costs	Positive
LMS	High	Low	2 to 48 h	High costs	Depends of the mediator
IFF	Medium	Some	1 to 12 weeks	Low costs	Positive

2.9 Detoxification

Some of the conventional biomass pretreatment technologies produce toxic compounds after treat biomass. These toxic compounds are classified in four groups: 1) furan derivatives, 2) pentose and hexose degradation, 3) weak acids, and 4) phenolic compounds from lignin [81]. These compounds affect the fermentative microorganisms and the cellulolytic enzymes, which generate an overall ethanol yield reduction [26].

The detoxification process employs chemical, physical, and biological strategies. Chemical and physical techniques utilize technologies such as filtration, anion exchange chromatography, and NaOH precipitation; these techniques are expensive, produce sugar losses, and do not remove all the inhibitors in the hydrolysates [12]. The biological detoxification (Table 3) process employs ligninolytic fungi or their enzymes to reduce the concentration of toxic compounds in feedstock hydrolysates [14]. The detoxification process is performed after or before the saccharification process and has a duration between 1 to 12 h [11,13,14,93]. Detoxification after or before saccharification is a controversial point. Jurado et al. [93] recommended enzymatic detoxification after cellulose hydrolysis because enzymatic detoxification leads to a reduction in the saccharification process [93]. In contrast, Moreno et al. [26] suggested enzymatic detoxification before saccharification, because it enhances ethanol yield, reduces microbes lag phase, and generates an increment in substrate loading. The authors considered these positive factors more important than the reduction in glucose recovery [26].

From the four types of toxic compounds produced in biomass hydrolysates, phenolic compounds are the most degraded type by ligninolytic enzymes [14,93]. Phenolic compounds inhibited fermentation at concentration of 1 ppm [98]. On the other hand, the other types of toxic compounds exhibit a variable behavior. These compound detoxification is affected by feedstock nature, enzyme characteristics, and pretreatment harshness [26]. Enzymatic detoxification has been applied in the reduction of toxicity after different pretreatments such as steam explosion, strong acids, organosolv, and hot liquid water. In addition, ligninolytic enzyme detoxification increase fermentation rates and ethanol yield [26]. As an example, in sugarcane bagasse, enzymatic detoxification generated an ethanol yield five times greater than ion exchange detoxification [14]. Lee et al. [96] improved the detoxification of rice straw hydrolysates by optimizing the temperature, enzyme concentration, and mediator compounds. The optimization increased the detoxification of phenolic compounds up to 92% by using HBT as mediator [96]. Ligninolytic enzyme detoxification is a future option for biofuel industries. However, this enzymatic option needs to improve their ability to degrade other types of toxic compounds from hydrolysates.

Table 3. Feedstock hydrolysates detoxification

Feedstock	Microorganism	Enzyme	Pretreatment	Detoxification	Ethanol yield	Ref.
		Laccase		71%	0.55	
Wood	*Trametes versicolor*	LiP	Steam explosion, 205°C for 6 min.	53%	0.40	[9]
		Laccase+LiP		66%	0.47	
				80%	0.47	[10]
Spruce chips	*Trametes reesei*	Laccase	H_2SO_4 (0.5%) Steam explosion, 222°C, 23 bar 7 min.	80%	0.43	
Wheat straw	*Trametes versicolor*	Laccase	Hot liquid water at 180°C, 15psi, 30 min.	65%	-	[11]
Sugarcane bagasse	*Trametes versicolor*	Laccase	Steam explosion, 215°C during 10 min.	80%	0.43	[13]
Sugarcane bagasse	*Cyathus stercoreus*	Laccase	HCl (0.5–3.5% v/v),140°C for 30 min.	77.5%	0.374	[14]
Wheat straw	*Coriolopsis rigida*	Laccase	H_2SO_4 (1% w/w), 190°C, 12 bars, 10 min	75%	-	[93]
Wheat straw	*Trametes villosa*	Laccase	Steam explosion, 220°C, 2.5 min.	88-92%	-	[26]
Rice straw	*Coltricia perennis*	Laccase	H_2SO_4 (0.5% w/w), 121°C, 15 min	37-76%	-	[94]
Wheat straw	*Trametes versicolor*	Laccase	Ethanol organosolv process, 220°C 120 min	20-90%	0.374g/g	[95]
Rice straw	*Yarrowia lipolytica*	Laccase	H_2SO_4 (0.5%, w/w) 5 h, steam explosion 121°C 30 min	52%		[96]
		LMS		92%		
Wheat straw	*Pycnoporus cinnabarinus*	Laccase	Steam explosion 200°C, 10 min	-	-	[97]

In recent years, enzymatic detoxification of furan derivatives has been evaluated. This type of detoxification is performed by aryl-alcohol oxidase (AAO) which is a less investigated group of ligninolytic enzymes. AAO have been assessed in the degradation of 5-hydroxymethylfurfural (HMF) and polyunsaturated alcohols [99–101]. Carro et al. [101] evaluated the transformation of HMF into 2,5-furandicarboxylic acid (FDCA) by AAO catalysis. AAO (*Pleurotus eryngii*) transformed efficiently HMF into 2,5-formylfurancarboxylic acid (FFCA); however, it was not capable of transforming directly from HMF into FDCA. To complete the catalysis of HMF, a fungal heme peroxygenase (*Agrocybe aegerita*) was added to the reaction. This enzyme transforms FFCA into FDCA using the H_2O_2 previously generated by AAO [101]. Ferreira et al. [100] determined the ability of AAO to reduce polyunsaturated alcohols. They evaluated 12 different compounds; the best degradations were observed in benzyl alcohol, m-fluorobenzyl alcohol, and p-fluorobenzyl alcohol. The enzyme has the ability to catalyze 'non-activated' alcohols which is an important factor for future detoxification procedures [100]. Other enzymatic process to transform furfurals have been studied; however, these processes utilize non-ligninolytic enzymes from different types of bacteria. In the future, furfural detoxification using ligninolytic enzymes need to be tested in biomass hydrolysates to have a more accurate vision of their real efficiency and applicability.

3. CONCLUSIONS

Ligninolytic fungi and their enzymes exhibited several biophysics and biochemical characteristics that have been applied in the delignification and detoxification of biofuel feedstocks. Ligninolytic processes (enzymatic and fungal) produce ethanol yield, glucose conversion, and delignification and detoxification percentages similar or superior than conventional detoxification and delignification methodologies. Nevertheless, ligninolytic processes have low cellulose modification, the presence of mediators, and long duration process (fungal processes). The transformation of ligninolytic enzymes as viable option for biofuel industry requires the improvements in the process duration, cost reduction, and more specialized mediators.

ACKNOWLEDGEMENTS

The Colombian government and the Fulbright organization are acknowledged for the financial support via the granting of FULBRIGHT-COLCIENCIAS for the Ph.D. studies of Jersson Plácido in the USA.

AUTHORS' CONTRIBUTIONS

JP manuscript drafting and writing. SC was responsible of final manuscript approval. Both authors read and approved the final manuscript.

REFERENCES

1. Sissine F, Energy Independence and Security Act of 2007: a summary of major provisions. 2007; 2011.

2. Dias AA, Freitas GS, Marques GSM, Sampaio A, Fraga IS, Rodrigues MAM, Evtuguin DV, Bezerra RMF (2010) Enzymatic saccharification of biologically pre-treated wheat straw with white-rot fungi. Bioresour Technol 101:6045–6050.

3. Salvachúa D, Prieto A, López-Abelairas M, Lu-Chau T, Martínez ÁT, Martínez MJ (2011) Fungal pretreatment: an alternative in second-generation ethanol from wheat straw. Bioresour Technol 102:7500–7506.

4. Wan C, Li Y (2011) Effectiveness of microbial pretreatment by Ceriporiopsis subvermispora on different biomass feedstocks. Bioresour Technol 102:7507–7512.

5. Nigam P, Gupta N, Anthwal A. Pre-treatment of agro-industrial residues. In: Nigam P and Pandey A. Biotechnology for Agro-Industrial Residues Utilisation, Springer Netherlands; 2009, p. 13–33.

6. Li L, Li X, Tang W, Zhao J, Qu Y (2008) Screening of a fungus capable of powerful and selective delignification on wheat straw. Lett Appl Microbiol 47:415–420.

7. Wan C, Li Y (2010) Microbial delignification of corn stover by Ceriporiopsis subvermispora for improving cellulose digestibility. Enzyme Microb Technol 47:31–36.

8. Lu C, Wang H, Luo Y, Guo L (2010) An efficient system for pre-delignification of gramineous biofuel feedstock in vitro: application of a laccase from Pycnoporus sanguineus H275. Process Biochemistry 45:1141–1147.

9. Jönsson LJ, Palmqvist E, Nilvebrant N- and Hahn-Hägerdal B (1998) Detoxification of wood hydrolysates with laccase and peroxidase from the white-rot fungus *Trametes versicolor*. Appl Microbiol Biotechnol 49:691–697.

10. Larsson S, Reimann A, Nilvebrant N, Jönsson L (1999) Comparison of different methods for the detoxification of lignocellulose hydrolyzates of spruce. Appl Biochem Biotechnol 77:91–103.

11. Kolb M, Sieber V, Amann M, Faulstich M, Schieder D (2012) Removal of monomer delignification products by laccase from Trametes versicolor. Bioresour Technol 104:298–304.

12. Palmqvist E and Hahn-Hägerdal B (2000) Fermentation of lignocellulosic hydrolysates. I: inhibition and detoxification. Bioresour Technol 74:17–24.

13. Martín C, Galbe M, Wahlbom CF, Hahn-Hägerdal B and Jönsson LJ (2002) Ethanol production from enzymatic hydrolysates of sugarcane bagasse using recombinant xylose-utilising Saccharomyces cerevisiae. Enzyme Microb Technol 31:274–282.

14. Chandel AK, Kapoor RK, Singh A, Kuhad RC (2007) Detoxification of sugarcane bagasse hydrolysate improves ethanol production by Candida shehatae NCIM 3501. Bioresour Technol 98:1947–1950.

15. Minussi RC, Pastore GM, Durán N (2007) Laccase induction in fungi and laccase/N-OH mediator systems applied in paper mill effluent. Bioresour Technol 98:158–164

16. Matera I, Gullotto A, Tilli S, Ferraroni M, Scozzafava A, Briganti F (2008) Crystal structure of the blue multicopper oxidase from the white-rot fungus Trametes trogii complexed with p-toluate. Inorg Chim Acta 361:4129–4137.

17. Zouari-Mechichi H, Mechichi T, Dhouib A, Sayadi S, Martínez AT, Martínez MJ (2006) Laccase purification and characterization from Trametes trogii isolated in Tunisia: decolorization of textile dyes by the purified enzyme. Enzyme Microb Technol 39:141–148.

18. Corvini P, Schäffer A, Schlosser D (2006) Microbial degradation of nonylphenol and other alkylphenols - our evolving view. Appl Microbiol Biotechnol 72:223–243.

19. Babot ED, Rico A, Rencoret J, Kalum L, Lund H, Romero J, del Río JC, Martínez ÁT, Gutiérrez A (2011) Towards industrially-feasible delignification and pitch removal by treating paper pulp with Myceliophthora thermophila laccase and a phenolic mediator. Bioresour Technol 102:6717–6722.

20. Sergio R (2006) Laccases: blue enzymes for green chemistry. Trends Biotechnol 24:219–226

21. Couto S, Toca-Herrera J (2006) Lacasses in the textile industry. Biotechnol Mol Biol Rev 1:115-120

22. Kokol V, Doliška A, Eichlerová I, Baldrian P, Nerud F (2007) Decolorization of textile dyes by whole cultures of Ischnoderma resinosum and by purified laccase and Mn-peroxidase. Enzyme Microb Technol 40:1673–1677.

23. Michniewicz A, Ledakowicz S, Ullrich R, Hofrichter M (2008) Kinetics of the enzymatic decolorization of textile dyes by laccase from Cerrena unicolor. Dyes and Pigments 77:295–302.

24. Shanmugam S, Palvannan T, Sathish Kumar T, Michael A (2005) Biological decolourization of textile and paper effluents by *Pleurotus florida* and *Agaricus bisporus* (white-rot Basidiomycetes). World Journal of Microbiology and Biotechnology 21:1149–1151.

25. Chanagá Vera X, Plácido Escobar J, Marín Montoya M, Pérez Y, del Socorro M (2012) Native fungi with industrial dye degrading potential in the Aburrá Valley, Colombia. Revista Facultad Nacional de Agronomía, Medellín 65:6811–6821

26. Moreno AD, Ibarra D, Fernández JL, Ballesteros M (2012) Different laccase detoxification strategies for ethanol production from lignocellulosic biomass by the thermotolerant yeast Kluyveromyces marxianus CECT 10875. Bioresour Technol 106:101–109.

27. Plácido J, Imam T, Capareda S (2013) Evaluation of ligninolytic enzymes, ultrasonication and liquid hot water as pretreatments for bioethanol production from cotton gin trash. Bioresour Technol 139:203–208.

28. Plácido J, Capareda S (2014) Analysis of alkali ultrasonication pretreatment in bioethanol production from cotton gin trash using FT-IR spectroscopy and principal component analysis. Bioresources and Bioprocessing 1:1–9

29. Ferraroni M, Myasoedova N, Schmatchenko V, Leontievsky A, Golovleva L, Scozzafava A, Briganti F (2007) Crystal structure of a blue laccase from Lentinus tigrinus: evidences for intermediates in the molecular oxygen reductive splitting by multicopper oxidases. BMC Structural Biology 7:60

30. Bertrand T, Jolivalt C, Briozzo P, Caminade E, Joly N, Madzak C, Mougin C (2002) Crystal structure of a four-copper laccase complexed with an Arylamine: insights into substrate recognition and correlation with kinetics. Biochemistry (N Y) 41:7325–7333.

31. Hakulinen N, Kiiskinen L, Kruus K, Saloheimo M, Paananen A, Koivula A, Rouvinen J (2002) Crystal structure of a laccase from Melanocarpus albomyces with an intact trinuclear copper site. Nat Struct Mol Biol 9:601–605

32. Dias A, Sampaio A, Bezerra R. Environmental applications of fungal and plant systems: decolourisation of textile wastewater and related dyestuffs. In: Singh S and Tripathi R. Environmental Bioremediation Technologies, Springer Berlin Heidelberg; 2007, p. 445–463.

33. Holwerda RA, Wherland S, Gray HB (1976) Electron transfer reactions of copper proteins. Annu Rev Biophys Bioeng 5:363–396.

34. Alcalde M. Laccases: biological functions, molecular structure and industrial applications. In: Polaina J and MacCabe AP. Industrial Enzymes, Springer Netherlands; 2007, p. 461–476.

35. Garavaglia S, Teresa Cambria M, Miglio M, Ragusa S, Iacobazzi V, Palmieri F, D'Ambrosio C, Scaloni A, Rizzi M (2004) The structure of Rigidoporus lignosus laccase containing a full complement of copper ions, reveals an asymmetrical arrangement for the T3 copper pair. J Mol Biol 342:1519–1531.

36. Piontek K, Glumoff T, Winterhalter K (1993) Low pH crystal structure of glycosylated lignin peroxidase from Phanerochaete chrysosporium at 2.5 Å resolution. FEBS Lett 315:119–124.

37. Sundaramoorthy M, Kishi K, Gold MH, Poulos TL (1994) The crystal structure of manganese peroxidase from Phanerochaete chrysosporium at 2.06-A resolution. Journal of Biological Chemistry 269:32759–32767

38. Angel TM (2002) Molecular biology and structure-function of lignin-degrading heme peroxidases. Enzyme Microb Technol 30:425–444.

39. Choinowski T, Blodig W, Winterhalter KH, Piontek K (1999) The crystal structure of lignin peroxidase at 1.70 Å resolution reveals a hydroxy group on the Cβ of tryptophan 171: a novel radical site formed during the redox cycle. J Mol Biol 286:809–827.

40. Khindaria A, Yamazaki I, Aust SD (1996) Stabilization of the veratryl alcohol cation radical by lignin peroxidase. Biochemistry (N Y) 35:6418–6424.

41. Hammel KE, Cullen D (2008) Role of fungal peroxidases in biological ligninolysis. Curr Opin Plant Biol 11:349–355.

42. Martin H, (2002) Review: lignin conversion by manganese peroxidase (MnP). Enzyme Microb Technol 30:454–466.

43. Polak J, Jarosz-Wilkolazka A (2012) Fungal laccases as green catalysts for dye synthesis. Process Biochemistry 47:1295–1307.

44. Hoopes JT, Dean JFD (2004) Ferroxidase activity in a laccase-like multicopper oxidase from Liriodendron tulipifera. Plant Physiology and Biochemistry 42:27–33.

45. Bugg TD, Ahmad M, Hardiman EM, Singh R (2011) The emerging role for bacteria in lignin degradation and bio-product formation. Curr Opin Biotechnol 22:394–400.

46. Ramachandra M, Crawford DL, Hertel G (1988) Characterization of an extracellular lignin peroxidase of the lignocellulolytic actinomycete Streptomyces viridosporus. Appl Environ Microbiol 54:3057–3063

47. Adav SS, Ng CS, Arulmani M, Sze SK (2010) Quantitative iTRAQ secretome analysis of cellulolytic thermobifida fusca. Journal of Proteome Research 9:3016–3024

48. Gupta N, Farinas ET (2010) Directed evolution of CotA laccase for increased substrate specificity using Bacillus subtilis spores. Protein Engineering, Design and Selection 23:679–682

49. [49] Faure D, Bouillant M- and Bally R (1995) Comparative study of substrates and inhibitors of Azospirillum lipoferum and Pyricularia oryzae laccases. Appl Environ Microbiol 61:1144–1146

50. Miyazaki K (2005) A hyperthermophilic laccase from Thermus thermophilus HB27. Extremophiles 9:415–425

51. Arias ME, Arenas M, Rodríguez J, Soliveri J, Ball AS, Hernández M (2003) Kraft pulp biobleaching and mediated oxidation of a nonphenolic substrate by laccase from Streptomyces cyaneus CECT 3335. Appl Environ Microbiol 69:1953–1958

52. Dwivedi UN, Singh P, Pandey VP, Kumar A (2011) Structure-function relationship among bacterial, fungal and plant laccases. J Molec Catal B 68:117–128.

53. Hattori M, Tsuchihara K, Noda H, Konishi H, Tamura Y, Shinoda T, Nakamura M, Hasegawa T (2010) Molecular characterization and expression of laccase genes in the salivary glands of the green rice leafhopper, Nephotettix cincticeps (Hemiptera: Cicadellidae). Insect Biochem Mol Biol 40:331–338.

54. Dittmer NT, Suderman RJ, Jiang H, Zhu Y, Gorman MJ, Kramer KJ, Kanost MR (2004) Characterization of cDNAs encoding putative laccase-like multicopper oxidases and developmental expression in the tobacco hornworm, Manduca sexta, and the malaria mosquito, Anopheles gambiae. Insect Biochem Mol Biol 34:29–41

55. Coy MR, Salem TZ, Denton JS, Kovaleva ES, Liu Z, Barber DS, Campbell JH, Davis DC, Buchman GW, Boucias DG, Scharf ME (2010) Phenol-oxidizing laccases from the termite gut. Insect Biochem Mol Biol 40:723–732.

56. Arakane Y, Muthukrishnan S, Beeman RW, Kanost MR, Kramer KJ (2005) Laccase 2 is the phenoloxidase gene required for beetle cuticle tanning. Proc Natl Acad Sci U S A 102:11337–11342

57. dos Santos AB, Cervantes FJ, van Lier JB (2007) Review paper on current technologies for decolourisation of textile wastewaters: perspectives for anaerobic biotechnology. Bioresour Technol 98:2369–2385.

58. Martínez ÁT, Ruiz-Dueñas FJ, Martínez MJ, del Río JC, Gutiérrez A (2009) Enzymatic delignification of plant cell wall: from nature to mill. Curr Opin Biotechnol 20:348–357.

59. Vivekanand V, Dwivedi P, Sharma A, Sabharwal N, Singh R (2008) Enhanced delignification of mixed wood pulp by *Aspergillus fumigatus*laccase mediator system. World Journal of Microbiology and Biotechnology 24:2799–2804.

60. Martín-Sampedro R, Eugenio ME, Carbajo JM, Villar JC (2011) Combination of steam explosion and laccase-mediator treatments prior to Eucalyptus globulus kraft pulping. Bioresour Technol 102:7183–7189.

61. Khuong LD, Kondo R, De Leon R, Kim Anh T, Shimizu K and Kamei I (2014) Bioethanol production from alkaline-pretreated sugarcane bagasse by consolidated bioprocessing using Phlebia sp. MG-60. Int Biodeterior Biodegrad 88:62–68.

62. López-Abelairas M, Pallín MÁ, Salvachúa D, Lú-Chau T, Martínez M, Lema J (2013) Optimisation of the biological pretreatment of wheat straw with white-rot fungi for ethanol production. Bioprocess and biosystems engineering 36:1251–1260

63. Mattinen M, Maijala P, Nousiainen P, Smeds A, Kontro J, Sipilä J, Tamminen T, Willför S, Viikari L (2011) Oxidation of lignans and lignin model compounds by laccase in aqueous solvent systems. J Molec Catal B 72:122–129.

64. Costa SM, Goncalves AR, Esposito E. *Ceriporiopsis Subvermispota* used in delignification of sugarcane bagasse prior to soda/anthraquinone pulping. In: Davison BH, Evans BR, Finkelstein M and McMillan JD. Twenty-Sixth Symposium on Biotechnology for Fuels and Chemicals, Humana Press; 2005, p. 695–706.

65. Archibald FS, Bourbonnais R, Jurasek L, Paice MG and Reid ID (1997) Kraft pulp bleaching and delignification by Trametes versicolor. J Biotechnol 53:215–236.

66. Wang F, Xie H, Chen W, Wang E, Du F, Song A (2013) Biological pretreatment of corn stover with ligninolytic enzyme for high efficient enzymatic hydrolysis. Bioresour Technol 144:572–578

67. Asgher M, Ahmad Z and Iqbal HMN (2013) Alkali and enzymatic delignification of sugarcane bagasse to expose cellulose polymers for saccharification and bio-ethanol production. Industrial Crops and Products 44:488–495.

68. Fillat U, Blanca Roncero M (2009) Effect of process parameters in laccase-mediator system delignification of flax pulp: Part I. Pulp properties Chem Eng J 152:322–329.

69. Camarero S, García O, Vidal T, Colom J, del Río JC, Gutiérrez A, Gras JM, Monje R, Martínez MJ and Martínez ÁT (2004) Efficient bleaching of non-wood high-quality paper pulp using laccase-mediator system. Enzyme Microb Technol 35:113–120.

70. Chen Q, Marshall MN, Geib SM, Tien M, Richard TL (2012) Effects of laccase on lignin depolymerization and enzymatic hydrolysis of ensiled corn stover. Bioresour Technol 117:186–192.

71. Qiu W, Chen H (2012) Enhanced the enzymatic hydrolysis efficiency of wheat straw after combined steam explosion and laccase pretreatment. Bioresour Technol 118:8–12.

72. Gutiérrez A, Rencoret J, Cadena EM, Rico A, Barth D, del Río JC and Martínez ÁT (2012) Demonstration of laccase-based removal of lignin from wood and non-wood plant feedstocks. Bioresour Technol 119:114–122.

73. Kamei I, Hirota Y, Mori T, Hirai H, Meguro S and Kondo R (2012) Direct ethanol production from cellulosic materials by the hypersaline-tolerant white-rot fungus Phlebia sp. MG-60. Bioresour Technol 112:137–142.

74. Kamei I, Hirota Y and Meguro S (2012) Integrated delignification and simultaneous saccharification and fermentation of hard wood by a white-rot fungus, Phlebia sp. MG-60. Bioresour Technol 126:137–141.

75. Khuong LD, Kondo R, Leon RD, Anh TK, Meguro S, Shimizu K and Kamei I (2014) Effect of chemical factors on integrated fungal fermentation of sugarcane bagasse for ethanol production by a white-rot fungus, Phlebia sp. MG-60. Bioresour Technol 167:33–40.

76. Ma K and Ruan Z (2015) Production of a lignocellulolytic enzyme system for simultaneous bio-delignification and saccharification of corn stover employing co-culture of fungi. Bioresour Technol 175:586–593.

77. Mate DM and Alcalde M (2015) Laccase engineering: from rational design to directed evolution. Biotechnol Adv 33:25–40. doi:http://dx.doi.org.scopeesprx.elsevier.com/10.1016/j.biotechadv.2014.12.007.

78. Madzak C, Mimmi MC, Caminade E, Brault A, Baumberger S, Briozzo P, Mougin C, Jolivalt C (2006) Shifting the optimal pH of activity for a laccase from the fungus Trametes versicolor by structure-based mutagenesis. Protein Engineering, Design and Selection 19:77–84

79. Galli C, Gentili P, Jolivalt C, Madzak C, Vadalà R (2011) How is the reactivity of laccase affected by single-point mutations? Engineering laccase for improved activity towards sterically demanding substrates. Appl Microbiol Biotechnol 91:123–131

80. Andberg M, Hakulinen N, Auer S, Saloheimo M, Koivula A, Rouvinen J, Kruus K (2009) Essential role of the C-terminus in Melanocarpus albomyces laccase for enzyme production, catalytic properties and structure. FEBS Journal 276:6285–6300

81. Annunziatini C, Baiocco P, Gerini MF, Lanzalunga O, Sjögren B (2005) Aryl substituted N-hydroxyphthalimides as mediators in the laccase-catalysed oxidation of lignin model compounds and delignification of wood pulp. J Molec Catal B 32:89–96.

82. Oudia A, Queiroz J, Simões R (2008) The influence of operating parameters on the biodelignification of *Eucalyptus globulus* kraft pulps in a laccase-violuric acid system. Appl Biochem Biotechnol 149:23–32.

83. Mukhopadhyay M, Kuila A, Tuli DK, Banerjee R (2011) Enzymatic depolymerization of Ricinus communis, a potential lignocellulosic for improved saccharification. Biomass Bioenergy 35:3584–3591.

84. Ibarra D, Romero J, Martínez MJ, Martínez AT, Camarero S (2006) Exploring the enzymatic parameters for optimal delignification of eucalypt pulp by laccase-mediator. Enzyme Microb Technol 39:1319–1327.

85. Pinto PA, Dias AA, Fraga I, Marques G, Rodrigues MAM, Colaço J, Sampaio A, Bezerra RMF (2012) Influence of ligninolytic enzymes on straw saccharification during fungal pretreatment. Bioresour Technol 111:261–267.

86. Kondo R, De Leon R, Anh TK, Meguro S, Shimizu K, Kamei I (2014) Effect of chemical factors on integrated fungal fermentation of sugarcane bagasse for ethanol production by a white-rot fungus, Phlebia sp. MG-60. Bioresour Technol 167:33–40

87. Dong XQ, Yang JS, Zhu N, Wang ET, Yuan HL (2013) Sugarcane bagasse degradation and characterization of three white-rot fungi. Bioresour Technol 131:443–451

88. Ishola MM, Isroi and Taherzadeh MJ (2014) Effect of fungal and phosphoric acid pretreatment on ethanol production from oil palm empty fruit bunches (OPEFB). Bioresour Technol 165:9–12.

89. Suhara H, Kodama S, Kamei I, Maekawa N, Meguro S (2012) Screening of selective lignin-degrading basidiomycetes and biological

pretreatment for enzymatic hydrolysis of bamboo culms. Int Biodeterior Biodegrad 75:176–180

90. Wang W, Yuan T, Cui B, Dai Y (2012) Pretreatment of Populus tomentosa with Trametes velutina supplemented with inorganic salts enhances enzymatic hydrolysis for ethanol production. Biotechnol Lett 34:2241–2246
91. Deswal D, Gupta R, Nandal P and Kuhad RC (2014) Fungal pretreatment improves amenability of lignocellulosic material for its saccharification to sugars. Carbohydr Polym 99:264–269.
92. Song L, Ma F, Zeng Y, Zhang X and Yu H (2013) The promoting effects of manganese on biological pretreatment with Irpex lacteus and enzymatic hydrolysis of corn stover. Bioresour Technol 135:89–92.
93. Jurado M, Prieto A, Martínez-Alcalá Á, Martínez ÁT, Martínez MJ (2009) Laccase detoxification of steam-exploded wheat straw for second generation bioethanol. Bioresour Technol 100:6378–6384.
94. Kalyani D, Dhiman SS, Kim H, Jeya M, Kim I and Lee J (2012) Characterization of a novel laccase from the isolated Coltricia perennis and its application to detoxification of biomass. Process Biochemistry 47:671–678.
95. Ludwig D, Amann M, Hirth T, Rupp S and Zibek S (2013) Development and optimization of single and combined detoxification processes to improve the fermentability of lignocellulose hydrolyzates. Bioresour Technol 133:455–461..
96. Lee K, Kalyani D, Tiwari MK, Kim T, Dhiman SS, Lee J and Kim I (2012) Enhanced enzymatic hydrolysis of rice straw by removal of phenolic compounds using a novel laccase from yeast Yarrowia lipolytica. Bioresour Technol 123:636–645.
97. Oliva-Taravilla A, Moreno AD, Demuez M, Ibarra D, Tomás-Pejó E, González-Fernández C and Ballesteros M (2015) Unraveling the effects of laccase treatment on enzymatic hydrolysis of steam-exploded wheat straw. Bioresour Technol 175:209–215.
98. Delgenes JP, Moletta R and Navarro JM (1996) Effects of lignocellulose degradation products on ethanol fermentations of glucose and xylose by Saccharomyces cerevisiae, Zymomonas mobilis, Pichia stipitis, and Candida shehatae. Enzyme Microb Technol 19:220–225.
99. Dijkman WP, Groothuis DE, Fraaije MW (2014) Enzyme-catalyzed oxidation of 5-hydroxymethylfurfural to furan-2,5-dicarboxylic acid. Angewandte Chemie International Edition 53:6515–6518.

100. Ferreira P, Medina M, Guillén F, Martinez M, Van Berkel W, Martinez A (2005) Spectral and catalytic properties of aryl-alcohol oxidase, a fungal flavoenzyme acting on polyunsaturated alcohols. Biochem J 389:731–738

101. Carro J, Ferreira P, Rodríguez L, Prieto A, Serrano A, Balcells B, Ardá A, Jiménez-Barbero J, Gutiérrez A, Ullrich R, Hofrichter M, Martínez AT (2015) 5-hydroxymethylfurfural conversion by fungal aryl-alcohol oxidase and unspecific peroxygenase. FEBS Journal:n/a-n/a.

CHAPTER 4

Production of 2nd Generation of Liquid Biofuels

Leona Paulová[1], Petra Patáková[1], Mojmír Rychtera[1] and Karel Melzoch[1]

[1] Department of Biotechnology, Institute of Chemical Technology Prague, Prague, Czech Republic

1. INTRODUCTION

Fluctuations in the price of oil and projections on depletion of accessible oil deposits have led to national and international efforts to enhance the proportion of energy derived from renewable sources (bioenergy) with special emphasis on the transport sector (e.g. according to Directive 2009/28 EC, by 2020, 20% of energy in EU-27 should be met from renewable sources and 10% should be used in transportation). To fulfil the legal requirements, wider exploitation of biofuels made from renewable feedstocks, as a substitute for traditional liquid fuels, will be inevitable; e.g. the demand for bioethanol in the EU is expected to reach 28.5 billion litres by 2020 [1], while in America 36 billion gallons of ethanol must be produced by 2022 [2]. Bioethanol, which has a higher octane level then petrol but only contains 66% of the energy yield of petrol, can be used as blend or burned in its pure form in modified spark-ignition engines [2]. This will improve fuel combustion, and will contribute to a reduction in atmospheric carbon monoxide, unburned hydrocarbons, carcinogenic emissions and reduce emissions of oxides of nitrogen and sulphur, the main cause of acid rain [2]. Butanol-gasoline blends might outcompete ethanol-gasoline ones because they have better phase stability in the presence of water, better low-temperature properties, higher oxidation stability during long term storage, more favourable distillation characteristics and lower volatility with respect to possible air pollution. Recently performed ECE 83.03 emission tests [3] have shown negligible or no adverse effects on air pollution by burning butanol-gasoline blends (containing up to 30% v/v of butanol) in spark ignition engines of Skoda passenger cars.

Although most of the world's bioethanol is currently produced from starch or sugar raw materials, attention is increasingly turning to 2nd generation biofuels made from lignocellulose, e.g. agriculture and forest wastes, fast growing trees, herbaceous plants, industrial wastes or wastes from wood and paper processing. The concept of ethanol production from lignocellulose sugars is not new. Probably the first technical attempt to degrade polysaccharides in wood was carried out by the French scientist Henri Braconnot in 1819 using 90% sulfuric acid [4]. His findings were exploited much later, in 1898, with the opening of the first cellulosic ethanol plant in Germany, followed by another one in 1910 in the US [5, 6]. During World War II, several industrial plants were built to produce fuel ethanol from cellulose (e.g. in Germany, Russia, China, Korea, Switzerland, US), but since the end of the war, most of these have been closed due to their non-competitiveness with synthetically produced ethanol [7]. In spite of all the advantages of lignocellulosic as a raw material (e.g. low and stable price, renewability, versatility, local availability, high sugar content, noncompetitiveness with food chain, waste revaluation) and extensive efforts of many research groups to reduce bottlenecks in technology of lignocellulosic ethanol production (e.g. energy intensive pretreatment, costly enzymatic treatment, need for utilization of pentose/hexose mixtures, low sugar concentration, low ethanol concentration), large scale commercial production of 2nd generation bioethanol has not been reopened yet [8], although many pilot and demonstration plants operate worldwide [9]. Identically, only first generation biobutanol is produced in China (approx. annual amount 100 000 t) and Brazil (approx. annual amount 8 000 t) [10]. At the 2012 London Olympic Games, British Petrol introduced its three most advanced biofuels i.e. cellulosic ethanol, renewable diesel and biobutanol. At a demonstration plant at Hull UK, biobutanol, produced by Butamax (joint venture of BP and DuPont) was blended at 24 % v/v with standard gasoline and used in BMW-5 series hybrids without engine modifications [11]. As the final price of both ethanol and 1-butanol produced by fermentation is influenced mostly by the price of feedstock, the future success of industrial ABE fermentation is tightly linked with the cost of pre-treatment of lignocellulosic material into a fermentable substrate.

2. CHARACTERIZATION OF 2ND GENERATION FEEDSTOCK

Plant biomass can be used as a sustainable source of organic carbon to create bioenergy, either directly in the form of heat and electricity, or as liquid biofuels produced by thermo-chemical or biochemical methods or their combination [12]. In contrast to fossil energy sources, which are the result of long-term transformation of organic matter, plant biomass is created via photosynthesis using carbon dioxide as a source of carbon and sunlight as a source of energy and therefore is rapidly produced. The world annual production of biomass is estimated to be 146 billion metric tons [13], which could contribute 9-13% of the global energy supply yielding 45±10 EJ per year [14, 15].

Lignocellulose, which is stored in plant cell walls makes up a significant part of biomass representing 60-80% of woody tissue of stems, 15-30% of leaves or 30-60% of herbal stems [16]. Since it is not digestible for human beings, its use as a feedstock for bioprocesses does not compete with food production as in the case of sugar or starch raw materials.

All lignocellulose consist of three main polymeric components – cellulose, non-cellulosic carbohydrates (predominantly represented by hemicellulose) and lignin; its proportion and structure differs for different types of biomass (Table1) and it is also influenced by variety, climatic conditions, cultivation methods and location. Minor components of the cell wall are represented by proteoglycans, pectin, starch, minerals, terpenes, resins tannins and waxes.

Cellulose is a homopolymer of 500-1 000 000 D-glucose units (e.g. 10 000 units in wood, 15 000 in native cotton) linked by β-1,4-glycosidic bonds [19, 26, 29]; the cellulose chains (200-300) are grouped together to form cellulose fibres. The strong inter-chain hydrogen bonds between hydroxyl groups of glucose residues in radial orientation and the aliphatic hydrogen atoms in axial positions creates a semi-crystalline structure resistant to enzymatic hydrolysis; weaker hydrophobic interactions between cellulose sheets promote the formation of a water layer near the cellulose surface, which protects cellulose from acid hydrolysis [30]. Cellulose originating from different plants has the same chemical structure, but it differs in crystalline structure and inter-connections between other biomass components. Microfibrils made of cellulose are surrounded by covalently or non-covalently bound hemicellulose, which is a highly branched heteropolymer made from 70-300 monomers units of pentoses (xylose, arabinose), hexoses (galactose, glucose, mannose) and acetylated sugars (e.g. glucuronic, galacturonic acids). Unlike cellulose, hemicellulose is not chemically homogenous and its composition depends on the type of material - hardwood contains predominantly xylans while softwood consists mainly of glucomannans [17, 23, 29,31]. Lignin, an amorphous heteropolymer of three phenolic monomers of phenyl propionic alcohols, namely p-coumaryl, coniferyl and sinapylalcohol, creates a hydrophobic filler, which is synthesized as a matrix displacing water in the late phase of plant fibre synthesis, and forms a layer encasing the cellulose fibres. Its covalent crosslinking with hemicellulose and cellulose forms a strong matrix, which protects polysaccharides from microbial degradation, makes it resistant to oxidative stress, and prevents its extraction by neutral aqueous solvents [31]. Forest biomass has the highest content of lignin (30-60% and 30-55 % for softwoods and hardwoods, respectively), while grasses and agricultural residues contain less lignin (10-30% and 3-15% respectively) [17].

There are several groups of lignocellulosic plant biomasses that can be exploited as a feedstock for bioprocessing. Woody biomass is represented mainly by hardwoods (angiosperm trees, e.g. poplar, willow, oak, cottonwood, aspen) and softwoods (conifers and gymnosperm trees e.g. pine, cedar, spruce, cypress, fir, redwood) together with forest wastes such as sawdust, wood chips or pruning residues. Nowadays the trend in this area is to use fast growing trees (poplar, willow) with genetically changed wood structures e.g. lower lignin content [32]. The advantage of forest biomass is its flexible harvesting time, thus avoiding long storage periods, and its high density, contributing to cost-effective transportation. Agricultural residues are represented mainly by corn stover or stalks, rice and wheat straw or sugarcane bagasse. The world's annual production of rice straw, wheat straw and corn straw that can be exploited for bioethanol production is 694.1, 354.3 and 203.6 million tons, respectively [20]. In the USA, 370 million and 350-450 million tons of forest biomass and agricultural wastes respectively are produced per year [17]. Although agrowastes are partly reutilized, e.g. as animal fodder, bedding, domestic fuel, used for cogeneration of electricity or reused in agriculture, a large fraction is still disposed as waste and is left in the fields; this can be utilized as a raw material for biofuels production. Sugarcane is nowadays one of the most important feedstocks for production of 1stgeneration bioethanol and also one of the plants with the highest photosynthetic efficiency, yielding around 55 tons of dry matter per hectare annually (approx. 176 kg/ha/day). Sugar cane bagasse, the fibrous lignocellulosic material remaining as waste is mostly used as a solid fuel in sugar mills or distilleries but due to its high cellulose content (Table 1) it can be reutilized as a feedstock for production of 2nd generation bioethanol. In the sugarcane season of 2010/11, the total sugar cane crop reached almost 1.627 billion tons (on 23 million hectares), which corresponds to 600 million tons of wet sugar cane bagasse [33]. Minor, but also important residues are the leaves, called sugarcane trash, amounting to 6-8 tons per hectare of sugarcane crop [34]. Another group of lignocellulosic biomass, herbaceous energy crops and grasses, which are represented predominantly by switch grass, alfalfa, sorrel or miscanthus [24], are interesting due to their low demands on soil quality, low-cost investments, fast growth, low moisture content, high yield per hectare (e.g. 20 t/ha for miscanthus) and high carbohydrate content (Table 1). Besides lignocellulosic plant materials, other low-cost large volume feedstocks such as municipal solid waste, municipal wastewater, food-processing waste or waste from the paper industry can be utilized for bioethanol production. Mixed municipal recovery solid waste (MSW) consists of approximately 55% mineral waste, 6% of metallic waste, 5% animal and vegetable waste (food residues, garden waste), 3% of paper and cardboard waste and 31% of others [35]. In the EU alone, the annual production of municipal wastes amounts 2.6 million tons, 65% of which is derived from renewable resources [35, 36]. The main challenge in its bioprocessing is its heterogeneous composition. To be used for ethanol production, degradable fractions of MSW should be separated after sterilization; cellulosic material (paper, wood or yard waste) represents approximately 60% of the dry weight of typical MSW as shown in Table 1 [25, 37]. Beside the solid wastes, lignocellulose extracted from municipal wastewater treatment processes can also be used as low-cost feedstock for biofuel production [26]. In Canada, 6.22 Mt of sugar could be annually produced using municipal sludge/biosolids and livestock manures [26]. Municipal wastewaters, which include faecal materials, scraps of toilet paper and food residues, should be pre-treated to separate solid and liquid fractions, the former of which is processed further to gain simple sugars. Primary sludge contains more cellulose compared to activated sludge (Table 1) because it is consumed in the activated sludge process and is further degraded by anaerobic digestion processes in the sewage disposal plant [26]. When talking about industrial wastes as 2nd generation raw materials for biofuels, wastes from cellulose/paper production cannot be neglected. Paper sludge is waste solid residue from wood pulping and papermaking processes and is represented by poor-quality paper fibres, which are too short to be used in paper machines. It is attractive as a raw material for bioprocessing mainly due to its low cost (it is currently disposed of in landfills or burned), its high carbohydrate content (Table 1) and its structure, which doesn´t require any pretreatment [8, 27, 28]. Another waste is represented by sulphite waste liquor (SWL), a solution of monomeric sugars formed during the sulfite pulping process by dissolution of lignin and most hemicelluloses. About 1 ton of solid waste is dissolved in SWL (11-14% solids) per ton of pulp and its annual production is around 90 billion litres [38]. SWL is usually burned after its concentration and evaporation, but since it's main components are sugars and lignosulfonates, its use as a raw material for bioethanol production has potential. Chemical composition of SWL (a spectrum of fermentable sugars, inhibitors, nutrients and minerals) differs significantly with the type of wood and technological procedures, e.g. concentration of the main sugars in SWL (% of dry matter) ranges for xylose from 3 to 5 % in soft wood (spruce, western hemlock) up to 21 % in eucalyptus, the highest concentration of galactose and glucose around 2.5 % is in soft wood SWL, content of mannose can reach values of almost 15 % in soft wood SWL [39-42]. SWL cannot be fermented without careful pretreatment - stripping off free sulfur dioxide and simultaneous concentration, steaming, removing inhibitors, adding nutrients, and adjusting the pH [43].

Table 1. Overview and composition of lignocellulosic biomass and other lignocellulosic sources

Biomass	Cellulose	Hemicellulose	Lignin	Reference
Hardwood	45-47	25-40	20-55	[17, 18]
Softwood	40-45	25-29	30-60	[17, 18]
Wheat straw	30-49	20-50	8-20	[19-22]
Rye straw	30.9	21.5	25.3	[21]
Corn fibre	15	35	8	[23]
Corn cobs	35-45	35-42	5-15	[22, 23]
Corn stover	39-42	19-25	15-18	[22, 23]
Corn straw	42.6	21.3	8.2	[20]
Rice straw	32-47	15-27	5-24	[20, 22, 23]
Rice hulls	24-36	12-19	11-19	[22]
Sugarcane bagasse	40	24-30	12-25	[20, 22, 23]
Switchgrass	30-50	10-40	5-20	[17, 23, 24]
Bermuda grass	25-48	13-35	6-19	[22, 23]
Cotton seed hairs, flax	80-95	5-20	0	[18, 22]
Municipal solid waste – separated fibre	49	16	10	[25]
Primary municipal sludge	29.3	not identified	not identified	[26]
Thickened waste activated sludge	13.8	not identified	not identified	[26]
Sawdust	45.0	15.1	25.3	[22]
Waste paper from chemical pulps	50-70	12-20	6-10	[17]
Newspaper	40-55	25-40	18-20	[1, 17]
Used office paper	55.7	13.9	5.8	[1]
Magazine	34.3	27.1	14.2	[1]
Magazine	34.3	27.1	14.2	[1]
Cardboard	49.6	15.9	14.9	[1]
Paper sludge	33-61	14.2	8.4-15.4	[27, 28]
Chemical pulps	60-80	20-30	2-10	[18]

Although lignocellulose biomass is cheap and predominantly comprises waste material, the logistics, handling, storage and transportation dramatically increases its cost and therefore its use directly on site is preferred over to its processing in a central plant [8]. Further price increases occur due to the character of material - most lignocelluloses mentioned above are not fermentable by common ethanol producers and must be decomposed and hydrolysed into simple sugars before fermentation is carried out.

3. BIOMASS DISRUPTION IN PRETREATMENT PROCESS

A prerequisite for ethanol production from lignocellulose is to break recalcitrant structure of material by removal of lignin, and to expose cellulose, making it more accessible to cellulolytic enzymes by modifying its structure; this happens in the pretreatment process. Basically, lignocellulose processing into fermentable sugars occurs in two steps: a) pretreatment yielding a liquid fraction that is mostly derived from hemicellulose and lignin and a solid fraction rich in cellulose, b) further enzymatic or chemical hydrolysis of the solid (wet) cellulose fraction to yield fermentable sugars.

Delignification (extraction of lignin by chemicals) is an essential prerequisite for enzymatic digestion of biomass; it disrupts the lignin polymeric structure, leading to biomass swelling and increase in its surface area and enables contact of cellulolytic enzymes with cellulose fibres. Although some pretreatment methods do not lead to a significant decrease in lignin content, all of them alter its chemical structure making biomass more digestible even though it may contain the same amount of lignin as non-pretreated biomass [29]. Hemicellulose is often dissolved during pretreatment because it is thermosensitive and easily acid-hydrolysed due to its amorphous branched structure; the liquid fraction obtained after pretreatment thus contains mainly pentose sugars (D-xylose, D-arabinose) originating from hemicelluloses, and strains fermenting pentose sugars must be used for its processing into ethanol as discussed later. The solid wet fraction obtained after pretreatment contains predominantly cellulose and needs further processing to yield fermentable sugars.

The conversion of lignocellulose into fermentable sugars is more difficult to achieve than conversion of starch; starchy material is converted from a crystalline to an amorphous structure at temperatures of 60-70°C, while lignocellulose is more resistant - a temperature of 320°C and a pressure of 25 MPa is needed to achieve its amorphous structure in water [17]. Therefore complete decomposition of cellulose is rarely attainable. Although lignocellulose pretreatment is an energy-intensive process, which contributes significantly to the price of the final product (18-20% of the total cost of lignocellulosic bioethanol is attributed to pretreatment) [8], it is a necessary expense because enzymatic hydrolysis of non-pretreated material provides less than 20% of the theoretical maximum yield of fermentable sugars for the majority of lignocellulose feedstocks [44]. The resistance of biomass to enzymatic attack is characterized by a number of physical variables such as lignin content, crystallinity index (ratio of crystalline to amorphous composition of cellulose), degree of polymerization, chain length, specific surface area, pore volume or particle size [31], which are material specific; e.g. pretreatment of woody biomass differs considerably from agriculture biomass, while paper sludge doesn´t need any processing.

Efficient pretreatment of biomass is characterized by an optimum combination of variables which leads to effective disruption of the complex lignocellulosic structure, removes most of the lignin, reduces cellulose crystallinity and increases the surface area of cellulose that is accessible to enzymatic attack. At the same time, it should minimize the loss of sugars, limit the formation of toxic compounds, enable the recovery of valuable components (e.g. lignin or furfural), use high solids loading, be effective for many lignocellulosic materials, reduce energy expenses, minimize operating costs and maximize the sugar yield in the subsequent enzymatic processing [45-47]. Pretreatment efficiency is usually assessed as: a) total amount of recoverable carbohydrates analysed as concentration of sugars released in the liquid and solid fraction after pretreatment, b) conversion of cellulose, expressed as the amount of sugars released by enzymatic hydrolysis of the solid phase, c) fermentability of released sugars, expressed as the amount of ethanol produced in the subsequent fermentation or d) its toxicity (concentration of inhibitory compounds released by sugar and lignin decomposition) analysed by HPLC or measured as the ability of test strains to grow.

Although it might seem that the problem of lignocellulose pretreatment has been solved by the chemical pulping process, which has been used commercially for a long time to produce various paper products, the opposite is true; despite most lignin is removed in these processes, they have been optimized to maintain the strength and integrity of cellulose fibres that are used for papermaking or as chemical feedstock and thus they are not easily hydrolysed by enzymes. The traditional sulfite pulping process was first reported in 1857 where treatment of wood with a mixture of sulfur dioxide in hot water considerably softened the wood; in 1900 the sulfurous acid process was patented [6]. Nowadays chemical pulp production based on the sulphite method [38] use sulfurous acid and its salts (Ca^{2+}, Mg^{2+}, Na^{+} and NH_4^{+}) in combination with SO_2 as a cooking liquor at temperatures of 120 - 150 °C. Sulfurous acid is an impregnation agent, improving the penetration of hydrolytic chemicals inside the wood structure [48], and importantly, promotes sulfonation of lignin leading to formation of lignosulfonic acid and its salts, that are soluble [49, 50]. Combinations of salts and cooking conditions produce different qualities of cellulose and different compositions of the sulfite waste liquors. Possibility to optimize old sulphite pulping process to obtain higher degree of saccharification of hard and softwoods had led to various modifications of process condition [48, 51-54]. So called SPORL technique is based on application of solution of bisulphate salts and sulfur dioxide (sulfurous acid) on biomass; sulfuric acid can also be added depending on lignin content (the higher amount of sulfuric acid is necessary for biomass with higher content of lignin, e.g. softwood, eucalyptus).

Many other processes have been investigated over the last decades in order to intensify lignocellulose pretreatment process by exploiting various physical, chemical and biological methods or their combination as reviewed elsewhere [29, 47] and summarized in Table 2.

Table 2. Overview and main characteristics of methods leading to biomass pretreatment

Pretreatment	Condition	Advantages	Disadvantages	Reference
	Physical pretreatment			
Mechanical (chipping, shredding, milling, grinding)	Normal temperature and pressure	Decreased cellulose crystallinity, increased surface area, decreased degree of polymerization	High energy demand, no lignin removal	[29, 47]
	Biological pretreatment			
Biological pretreatment – soft, brown or white rot fungi	Normal temperature and pressure	Low cost, low energy consumption, degradation of lignin and hemicellulose	Low efficiency, loss of carbohydrates (consumed by fungi), long residence times (10-14 days), need for carefully controlled growth condition, big space	[29, 45, 47]
	Chemical pretreatment			
Dilute acid pretreatment (H_2SO_4, HCl, H_3PO_4, HNO_3),	Concentration<4%, temperature 140-215 °C, pressure 0 5 MPa, reaction time seconds to minutes	High reaction rates, lignin disruption, increased accessibility of cellulose, improved digestibility, moderate temperatures	Little lignin removed, hemicellulose dissolved, sugar decomposition (inhibitors), need for acid recycling and pH adjustment	[29, 31, 45, 55]
Concentrated acid hydrolysis (H_2SO_4, H_3PO_4)	Concentration 70-77%, temperature 40-100 °C	Crystalline structure of cellulose completely destroyed, amorphous cellulose achieved, low temperature	Hemicellulose dissolved, equipment corrosion, sugar decomposition (inhibitors), need for acid regeneration, pH adjustment, environmental concerns	[45]
Alkali pretreatment (NaOH, KOH, $Ca(OH)_2$)	Temperature 25-130 °C	Decreased crystallinity of cellulose, decreased polymerization, lignin removal, few inhibitors	Hemicellulose dissolved, pH adjustment	[29, 55]
Ammonia pretreatment	Temperature 25-60 °C, reaction time several days	High delignification, cellulose swelling, high volatility of ammonia, low cost, ammonia	Cellulose crystallinity not reduced, environmental concerns	[45, 47]

		recycle, continuous process, short residence times		
Ozonolysis	Room temperature, normal pressure, reaction time - hours	Lignin degradation, no inhibitors, ambient temperature	Hemicellulose dissolved	[21]
Combined acid and alkali pretreatment (formic acid-aqueous ammonia, dilute sulphuric acid-sodium hydroxide)		Cellulose digestion, fractionation of lignocellulose, most of non-cellulosic components removed, high loading		[45]
Combined acid and organic solvent (concentrated H_3PO_4 + aceton),	Moderate temperatures	Cellulose crystalline structure disrupted, high yield of amorphous cellulose, lignin removed, reduced enzyme loading	Hemicellulose dissolved	[45]
Ionic liquid (IL)pretreatment	Temperature <100 °C, cellulose recovered by addition of water, ethanol or acetone	Lignin extraction, low temperature, high biomass loading, high lignin solubility, cellulose dissolution, solvents recovered and reused, environmentally friendly	Cellulose recovered by addition of acetone, deionized water or alcohol, IL denaturates enzymes, IL must be washed before reused	[29, 44, 45]
	Physicochemical pretreatment			
Steam explosion	Temperature 160-240 °C, pressure 0.7-4.8 MPa, reaction time 1-10 min followed by biomass explosion	Extensive redistribution of lignin, high cellulose digestibility, cellulose swelling, limited use of chemicals	Little lignin removed, incomplete destruction of biomass matrix, sugar decomposition (inhibitors), hemicellulose dissolved, high energy consumption	[29, 31, 45, 55]
Acid-catalyzed steam explosion	Steam explosion catalysed by addition of H_2SO_4 or SO_2	Decreased time and temperature compared to steam explosion	Inhibitors formation, hemicellulose dissolved, high temperature	[45]
Liquid hot water pretreatment	Temperature 180-230 °C, elevated pressure, pH 4-7, reaction time up to 15 min	Increased accessibility of cellulose, no inhibitors, no chemicals added, no need for pH adjustment and washing	Hemicellulose dissolved, lower loading	[45, 56]
Ammonia fiber explosion (AFEX)	Anhydrous liquid ammonia, temperature 60-120° C, pressure above 3 MPa, reaction time 30-60 min, followed decompression	Decreased crystallinity of cellulose, expanded fibre structure, increased accessible surface area, lignin depolymerisation and removal, low inhibitor concentrations, low temperature	Not suitable for softwood, hemicellulose dissolved, cost of ammonia, environmental concerns	[29, 31, 45, 55]

Ammonia recycle percolation	Aqueous ammonia (5-15%), temperature 150-180 °C, reaction time 10-90 min, flow 1-5 ml/min	Lignin removed, decreased crystallinity, low inhibitor concentrations, moderate temperatures	Hemicellulose dissolved, environmental concerns	[29]
Organosolv pretreatment	Organic (ethanol, methanol, ethylene glycol, glycerol, DMSO) or organic-aqueous mixtures, with catalyst at temperature >180 °C (HCl, H_2SO_4), temperature 100-250°C	Biomass fractionalization, pure cellulose, selectivity, effective for high-lignin biomass, organic solvents easily recovered (distillation) and reused, less energy	Hemicellulose dissolved, high cost of chemicals, inhibitors formation, need for containment vessels, explosion hazard, environmental concerns	[29]
Carbon dioxide explosion treatment	Supercritical CO_2, pressure 7-28 MPa, temperature 200 °C, time – several minutes	Increased surface area, low cost chemical, no inhibitors, high solid loading	Effectivity increased with moisture content, costly equipment	[29]
Wet oxidative pretreatment	Addition of oxidizing agent (oxygen, water, hydrogen peroxide)	Low concentration of inhibitors	High pressure and temperature, costly equipment and chemicals (oxygen)	[29]

Acid treatments lead mainly to hydrolysis of hemicelluloses (pentose and hexose fractions) while alkaline treatments bring about lignin removal. Concentrated acids such as sulphuric or hydrochloric have been used as powerful agents to treat lignocelluloses, but due to their toxicity, corrosivity and necessity of recovery after hydrolysis, attention has shifted to milder conditions e.g. 0.5 % (v/v) sulfuric acid [57]. To improve cellulose hydrolysis in dilute acid processes, higher temperatures are favoured [58] since at a moderate temperature, direct saccharification resulted in low yields. As demonstrated by Candido et al. [59] for bagasse, dilute acid hydrolysis is greatly influenced by reaction time; at 100°C in 10% v/v sulfuric acid, the loss of mass and hemicellulose content decreased with time while soluble lignin concentration increased. Several modifications of the dilute acid hydrolysis method have been reported, e.g. acid hydrolysis with 1 % H_2SO_4 to remove hemicellulose and lignin followed by an alkaline step to increase the yield of cellulose. Methods based on the use of organosolv, wet oxidation, steam explosion or steam enriched with various impregnating agents (SO_2, CO_2, NH_3) are also often used for lignocellulose pretreatment as summarized in Table 2. The principle of the organosolv is mild hydrolysis of lignocellulose catalysed by sulfuric acid or sodium hydroxide in the reactor followed by extraction into ethanol at temperatures around 175 °C. Taking sugar cane bagasse as an example, the solid to liquid ratio can vary from 1 to 5 kg/l or lower, and solubilized lignin and hemicellulose appear in the liquid phase [34]. Wet oxidation is widely used in research and development technologies. Martín et al. [60] compared wet oxidation of bagasse, which was mixed with water (ca. 6 % w/v dry bagasse) in a special autoclave under slightly alkaline conditions, with steam explosion. In the wet oxidation procedure, slightly lower solubilisation of lignin, higher solubilisation of hemicellulose and higher cellulose content in the solid phase (approx. 60 % w/w) was achieved in comparison with steam explosion (45 % w/w). The effect of steam enrichment with CO_2or SO_2 proved promising results as for enzymatic hydrolysis of cellulose and the low content of inhibitors, especially 2-furalaldehyde and 5-hydroxymethyl-2-furalaldehyde.

In summary, biomass pretreatment is a key bottleneck in the bioprocessing of lignocellulose biomass and even though all methods have distinct advantages, as summarized in Table 2, the main problems are high energy consumption and low substrate loading, leading to low sugar recovery. However, increasing the biomass concentration leads to high solid slurries which are very viscous, with a pasta-like behaviour, creating a challenge for mixing, pumping and handling; this increases energy demands reflected in a higher price for the ethanol as well as concentrates toxic compounds, thus counteracting any potential benefits [61].

Although the pretreatment process disrupts the complex structure of the material and causes partial hydrolysis of cellulose, the content of fermentable sugars is still very low; further enzymatic degradation of the cellulose polymeric chain must be carried out to increase the concentration of glucose, which is utilized (optimally together with hemicellulose-derived monomers) in fermentation as shown in Figure 1.

Most commercial enzyme preparations (the largest producers are Genencor, Novozymes or Spezyme) are produced by cultivation of *Trichoderma resei* as mixtures of enzymes with endo-1,4-ß-D-glucanase (EC 3.2.1.4, hydrolysis of (1→4) glucosidic linkages inside the chain), exo-1,4- ß-glucosidase (EC 3.2.1.74, hydrolysis of (1→4) linkage in (1→4)-β-D-glucans to remove successive glucose units), ß-glucosidase (EC 3.2.1.21, hydrolysis of terminal non-reducing β-D-glucosyl residues with release of β-D-glucose) and β-1,4-glucan cellobiohydrolase (EC 3.2.1.91, hydrolysis of (1→4)-β-D-glucosidic linkages in cellulose and cellotetraose releasing cellobiose from non-reducing ends of the chains) activities working in synergy.

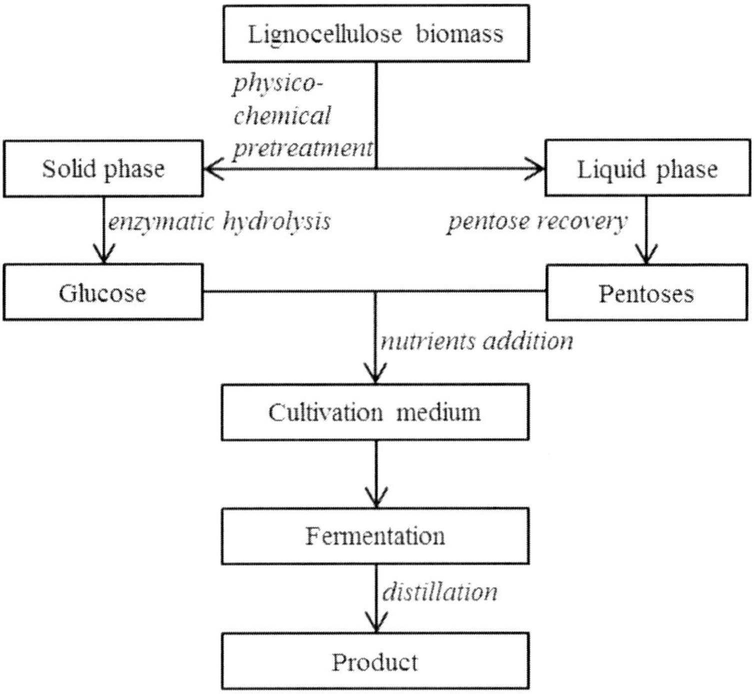

Figure 1. Simplified diagram of production of liquid biofuels from lignocellulose biomass

In recent years, the efficiency of commercial enzyme mixtures has rapidly increased and permits high conversions of cellulose to glucose; e.g. 85% and 91% yields of glucose were reported for ionic liquid pretreated poplar and switchgrass [62] and 85% and 83% yields were achieved for acid pretreated poplar and rice straws respectively [17, 63, 64]. Although the differential between the price of amylolytic and cellulolytic enzymes is currently reduced, the major difference is in dosing; about 40 -100 times more enzyme (based on protein weight) is required to breakdown cellulose compared to starch [29]. According to economic analyses, the conversion of biomass into fermentable sugars, which includes enzyme production and enzymatic hydrolysis together with indispensable pretreatment of biomass, comprises 33 % of the total cost [8, 17] and the estimated cost of cellulases is 50 cents per gallon (3.785 l) of ethanol, which is often comparable to the purchase cost of the feedstock [65]. For this reason attention has turned to further improvement of the composition and activity of enzyme cocktails, e.g. by constructing tailor-made multienzyme systems. It was shown that addition of xylanase and pectinase to alkali-pretreated biomass can reduce the negative effect of hemicellulose and pectin, which can restrict access of cellulases to the cellulose surface, while β-xylosidase can decompose xylobiose and polymerized xylooligomers to avoid inhibition of cellulolytic enzymes

[22,45]. Unfortunately, improved enzyme cocktails are not generally applicable, e.g. an enzyme complex enriched with β-mannanase and amyloglucosidase improved digestibility of dried distillers grains, but this was not required for corn stover [22]. Furthermore, the rate and efficiency of enzymatic hydrolysis can be affected by enzyme adsorption to non-cellulolytic substrates, e.g. lignin through phenolic groups and hydrophobic interactions, which limits the accessibility of cellulose to cellulases [45, 47]. To reduce this effect, "designer cellulosomes" have been recently constructed [45]. The cellulosome is a large complex of cellulolytic enzymes, originally produced by anaerobic bacteria [66], and has been engineered to comprise a recombinant chimeric scaffolding protein and many bound protein hybrids that have low lignin binding affinity. A different approach is represented by the addition of non-catalytic additives, e.g. surfactants (e.g. Tween, polyethylene glycol), polymers or proteins (bovine serum albumin, gelatine), which compete with cellulolytic enzymes for adsorption sites of lignin and thus prevent non-productive enzyme binding and can also facilitate enzyme recycling. Addition of expansins (plant proteins), expansin-like proteins or swollenin (fungal protein) promotes enhanced enzymatic hydrolysis by disrupting hydrogen bonding between cellulose and other cell-wall polysaccharides [45]. Recycling of enzymes, e.g. by ultrafiltration, re-adsorbtion onto fresh substrate, enzyme immobilization onto various materials e.g. chitosan-alginate composite, chitosan-clay composite, Eupergit C, mesoporous silicates, silicagel or kaolin are other approaches to reduce pretreatment costs [45].

The activity of cellulolytic enzymes can be reduced not only by ineffective binding, but also by feedback inhibition by glucose and cellobiose released by hydrolysis of cellulose as reviewed by Andric et al. [67] and by inhibitory effects of toxic products that may be released during pretreatment (type and concentration depends on biomass and process conditions) and can affect not only the rate and yield of saccharification but also substrate fermentability.

4. TOXIC COMPOUNDS RELEASED IN PRETREATMENT PROCESS

Toxic products can generally be divided into three main groups – aliphatic acids, furan derivatives and phenolic compounds [68-70] released by degradation of carbohydrates, and compounds arising from lignin. In acidic solutions, cellulose and hemicellulose are broken down into hexose and pentose sugars, which are further decomposed at high temperatures into furan derivatives represented mainly by 2-furaldehyde (furfural, FF) and 5-hydroxymethyl-2-furaldehyde (hydroxymethylfurfural, HMF). Free aliphatic acids, represented mainly by acetic, formic or levulinic acids, are created by substituents cleaved from lignin and hemicelluloses within the pretreatment, or are produced by cells during fermentation, while phenolic derivatives (4-hydroxybenzoic acid, 3,4-dihydroxybenzoic acid or vanilin) arise mainly from lignin decomposition in alkaline solution [71]. About 40 lignocellulose degradation products have been identified in various hydrolysates [71], the type and amount depending on type of biomass and pretreatment conditions [68]; e.g. furfural, hydroxymethylfurfural and levulinic acid occur in higher concentrations at low pH combined with high temperature and pressure [68, 71], while vanilin, vanilic, benzoic and 4-hydroxycoumaric acids are formed under alkaline conditions at elevated temperatures and acetic acid is produced in significant concentrations independent of the process and type of biomass [71]. Although many studies on the effect of inhibitors on cellulolytic enzymes have been published, a general conclusion is not easy to draw because it is influenced not only by the type and origin of the enzyme preparation, but also by its dosing and the concentration of inhibitors. However, in general, compounds exhibiting higher hydrophobicity tend to be more inhibitory to cellulolytic enzymes, the greatest inhibitory effect being caused by acetic and formic acids [72-74], while the activity of enzymes is not practically influenced by levulinic acid [73]. On the other hand, the presence of inhibitory compounds also affects ethanol productivity in the subsequent fermentation by influencing metabolic functions of ethanol producing strains. Inhibitory effects are described by type and concentration of toxic compounds (their effect is intensified when present in combination) and the strain used for ethanol production, but generally, fermentation is mainly influenced by the presence of furan derivatives together with phenolic compounds and weak acids (at low pH). As reviewed elsewhere [70, 75], low molecular weight compounds are able to penetrate the cell, while inhibitors with high molecular weights affect expression and activity of sugar and ion transporters. Growth and rate of ethanol production by *Saccharomyces cerevisiae*, the main ethanol producing strain, is significantly inhibited by furfural, while ethanol yield is almost not influenced [75] due to its ability to detoxify the broth by reduction of furfural to furfuryl alcohol, which is less toxic.

Surprisingly, in butanol production process, *C.beijerinckii* BA101, *C. acetobutylicum* P260, *C. acetobutylicum* ATCC 824, *Clostridium saccharobutylicum* 262 and *Clostridium butylicum* 592 were not sensitive towards sugar degradation products like furfural or hydroxymethylfurfural (up to concentrations of 2-3 g/l) but its growth and solvent production were inhibited by ρ-coumaric and ferulic acids present at a concentration of 0.3 g/l [76-78]. Solvent productivity and final solvent concentration in *C. beijerinckii* P260 were stimulated by addition of furfural or hydroxy methylfurfural (or both compounds) to the fermentation medium, at concentrations of up to 1 g/l [79].*C. acetobutylicum* ATCC 824 metabolized furfural and hydroxymethyl furfural into furfuryl alcohol and 2,5-bis-hydroxymethylfuran, respectively and these compounds positively influenced solvent production up to a concentration of 2 g/l. It was hypothesised that this biotransformation step, independent of initial furfural and HMF concentrations, might increase solventogenesis via an increased rate of regeneration of NAD^+ [80]. Another possible inhibitor of phenolic origin, syringaldehyde, caused inhibition of solvent production by *C. beijerinckii* NCIMB 8052 over the whole range tested (0.2-1 g/l). This inhibition was probably caused by decreased expression and activity of coenzyme A transferase, which participated in utilization of butyric and acetic acids, because these acids accumulated in the medium [81].

The inhibitory effects of toxic compounds released by sugars and lignin degradation can be reduced in several ways, e.g. optimization of pretreatment conditions to minimize the formation of inhibitors, use of specific detoxification methods, e.g. precipitation by calcium hydroxide (overliming) alone or in combination with sulphite addition, adsorption on charcoal, evaporation of the volatile fraction, extraction with ethyl acetate or diethyl ether, ion extraction, treatment with peroxidase (E.C. 1.11.7) and laccase (EC 1.10.3.2), or use of microbial strains with increased resistance to inhibitors (achieved by adaptation or prepared by genetic modification) [75, 82, 83]. Lignin degradation products, ρ-coumaric, ferulic and vanillic acids, together with vanillin, were effectively removed from a model solution of phenolic compounds by treatment with 0.01μM peroxidase (E.C. 1.11.7), resulting in improved growth and butanol production by *C. beijerinckii* NCIMB 8052 [84]. Sulphuric acid-hydrolysed corn fiber was treated with XAD-4 resin, resulting in an improvement of butanol yield achieved with *C. beijerinckii* BA101 [85]. Another popular approach for detoxification of acid hydrolysates for butanol production is “overliming” i.e. addition of $Ca(OH)_2$ in excess to hydrolysate [78, 85]. Although this detoxification method has been known for a long time, its mode of action, especially in the case of butanol production, is not completely clear. Addition of $Ca(OH)_2$ to an acid hydrolysate decreases furfural and HMF concentrations [86, 87] but does not affect acid concentrations; thus it is only possible to assume a beneficial neutralization effect. Furthermore it may be useful to treat hydrolysates with activated carbon [88].

5. FERMENTATION OF LIGNOCELLULOSIC SUBSTRATES

5.1. Ethanol Fermentation

Fermentation of lignocellulose hydrolysates is more complicated compared to fermentation of 1^st^generation feedstock (sugar cane juice, molasses, grains) for several reasons: a) pentose sugars (predominantly xylose) are present along with hexoses (mainly glucose, mannose, galactose) in the hydrolysate, b) toxic compounds released during pretreatment can influence metabolic activity of the fermentation strain, c) low concentrations of fermentable sugars hamper the attainment of a high ethanol concentration. Because lignocellulose hydrolysates are poor in some nutrients (phosphorus, trace elements, and vitamins) they are usually supplemented, e.g. by addition of corn steep or yeast extract before being used as a substrate for fermentation. For an efficient process it is necessary to identify a strain that utilizes both pentose and hexose sugars, produces ethanol with a high yield and productivity and is tolerant to both inhibitors and ethanol. One of the main challenges is to simultaneously co-ferment pentose and hexose sugars, but neither yeast *S. cerevisiae* nor the bacterium*Z. mobilis*, which are usually used for ethanol production, contain genes for expression of xylose reductase and xylitol dehydrogenase [89]. In order to enhance process effectiveness, co-fermentation or sequential fermentation of hexoses and pentoses has been examined by combining good ethanol producers with strains naturally utilizing pentoses e.g. *Pichia stipitis, Candida shehatae, Pachysolen tannophillus, Klebsiella oxytoca.* However, xylose utilization is the rate limiting step due to catabolite repression by hexoses and the low availability of oxygen, and inhibition of pentose-utilizing strains by ethanol [90, 91]. Moreover, the yield of ethanol by co-fermentation is usually lower than with separate processes, e.g. yields of 0.5 g ethanol per g glucose (98% of theoretical) and 0.15 g/g xylose (29% theoretical) were achieved by separate cultivation of *Z. mobilis* and *P. tannophillus* respectively, but in optimized co-fermentation, the yield was just 0.33 g ethanol/g sugar. The same yield was obtained in a 5-reactor process combining *P. stipitis* and *S. cerevisiae* [92], but it was enhanced to 0.49 g/g sugars (96% theoretical) by cultivation of an adapted co-culture of *S. cerevisiae*, *P. tannophilis* and recombinant *E. coli* in dilute-acid softwood hydrolysate [93]. In a subsequent process employing *P. stipilis* and *S. cerevisiae*, which was inactivated before *Pichia* inoculation to avoid oxygen competition, 75% of theoretical ethanol yield was achieved [94]. A different approach is represented by the use of a recombinant strain prepared either by cloning genes encoding xylose utilization into good ethanol producers or to construct synthetic pathways for ethanol production in pentose-utilizing hosts. Wild type yeasts can be genetically modified to utilize xylose by introducing fungal genes encoding xylose reductase and xylitol dehydrogenase or bacterial/fungal genes for xylose isomerase [95]. Yeast *S. cerevisiae* was transformed with the *xylA* gene from *Thermus thermophiles* and *Piromyces sp.* to produce xylose isomerase, but unfortunately, this enzyme was inhibited by xylitol, favouring instead, its formation. Recently a recombinant strain of *S. cerevisiae* expressing a heterologous *xylA* gene produced 0.42 g/g of ethanol from xylose [96]. A strategy using *xyl1* and *xyl2* genes from *P. stipitis*introduced into *S. cerevisiae* produced transformants that exclusively consumed xylose, but produced significant amounts of xylitol [97]. On the other hand, with recombinant *Z. mobilis*, which carried *E. coli* genes encoding for xylose isomerase, xylulokinase, transketolase and transaldolase, 86% ethanol yield from xylose was achieved. Another strain of *Z. mobilis*, expressing genes *araABD* from *E. coli,* encoding L-arabinose isomerase, L-ribulokinase, L-ribulose-5-P-4 epimerase together with genes for transketolase and transaldolase, was able to grow on arabinose with 98% ethanol yield. *E. coli*, which naturally utilizes a wide range of substrates including pentoses, was transformed by genes encoding pyruvate decarboxylase and alcohol dehydrodenase, resulting in enhanced ethanol production [96]. Adaptation of recombinant strains to inhibitors can further increase the yield of ethanol, e.g. the ethanol yield achieved with a genetically engineered strain of *S. cerevisiae* grown on bagasse hydrolysate was increased from 0.18 g/g to 0.38 g/g after adaptation [98]. Recombinant strains that not only consume pentoses but also hydrolyse hemicelluloses by co-expressing endoxylanase, β-xylosidase and β-glucosidase activities has recently been constructed [95] and yields of 0.41 g/g of ethanol were obtained from total sugars in a rice straw hydrolysate.

In addition to the wide range of sugars, their low concentration in hydrolysates is problematic. Since ethanol recovery by distillation is only economically viable on the industrial scale for yields greater than 4% (w/w), which for most hydrolysates requires a dry mass concentration greater than 20% [45], the use of high substrate loading is needed. Effect of substrate concentration (unbleached hardwood pulp and organosolve pretreated poplar) on glucose concentration resulting from the enzyme hydrolysis was studied in [52] and [99]. In laboratory scale after 48 h of enzymatic hydrolysis 158 g/l glucose in the hydrolyzate was reached, ethanol concentration after fermentation ranged between 50.4 and 63.1 g/l. The general problem for this kind of conversions is that high load of the pulp or pretreated lignocelulosic material gives rise to high viscosity and thus also to mixing and transport problems. These extremely high yields of glucose can be attributed to a very efficient peg mixer. Problems connected with use of such high viscosity slurries can be overcome by various strategies, e.g. maximizing dry matter by removing most hemicellulose and lignin, utilizing alternative bioreactors with novel mixing modes (e.g. peg mixer, shaking, gravitational tumbling, hand stirring) or gradual dosing of substrate into the bioreactor (fed-batch), which enables the use of more substrate and thus increases the yield of ethanol above values achievable in batch mode. Moreover, the actual concentration of toxic substrates is reduced and yield and/or productivity is enhanced by controlled dosing of substrate and prolonged cultivation time, thus shortening unprofitable periods between batches [45, 89]. Feed rates should reflect the type of hydrolysate and strain. Continuous cultures usually using immobilized cells (to prevent their wash out from the bioreactor at high dilution rates) is another strategy to increase process productivity [89].

Integration strategies, which replace classical separate hydrolysis and fermentation processes (SHF) by combining several process steps in one vessel represents another approach for lignocellulosic ethanol production. Simultaneous saccharification and fermentation (SSF), which combines enzymatic hydrolysis and fermentation in one step, permits an increased rate of cellulose hydrolysis by elimination of product inhibition (the released glucose is consumed by the microbial strain), an increased rate of sugar consumption, reduced contamination due to the presence of ethanol and a reduced number of reactors. However, SSF is constrained by different temperature optima for each process (the cellulase optimum is usually 40-50 °C, whereas the fermentation temperature usually cannot exceed 35 °C for most ethanol producers) and carbon source limitation in the early stages of the process. Several modifications of SSF to ease the problems and increase productivity have been published. These include the use of thermotolerant ethanol producers [100, 101], application of a pre-saccharification step [102] or the use of recombinant strains consuming both hexose and pentose sugars (a simultaneous saccharification and co-fermentation process (SSCF)) [103] in batch or fed-batch mode [104]. Consolidated bioprocessing (CBP), which combines cellulase production, cellulose hydrolysis and fermentation into a single step have been investigated as a way of reducing the cost of cellulolytic enzymes, increasing volumetric productivity and reducing capital investment [105]. Some biofuel companies (e.g. Mascoma and Qteros) have been founded based on this concept [105]. CBP microorganisms should combine high cellulase production and secretory capability, the ability to utilize a broad range of sugars, tolerance to high concentrations of salts, solvents and inhibitors, high ethanol productivity and yield, have a known genomic DNA sequence and developed recombinant technologies and ideally be usable as feed protein after fermentation [105]. There is a lack of native organisms that combine the ability to produce cellulolytic enzymes and be homoethanolic with high titres and yields. Although some thermophilic anaerobic bacteria e.g. *Clostridium thermocellum,* are high cellulase producers and utilize both pentose and hexose sugars, they have a low tolerance to ethanol ~30 g/l [106] and an insufficient yield ~0.2 g/g [107]. Therefore recombinant strains have been prepared by engineering cellulolytic microorganisms (e.g. *C. thermocellum, C. phytofermentans, C. cellulolyticum, T. reesei or F.oxysporum*) to produce ethanol. Knockout mutants of *Thermoanaerobacterium saccharolyticum* that lack lactic and acetic acid production exhibited an ethanol yield from xylose of 0.46 g/g [108], while recombinant *Geobacillus thermoglucosidasius* produced 0.42-0.47 g/g of ethanol from hexoses [65]. Another attempt, to create a recombinant cellulose-utilizing microorganism using non-cellulolytic strains with high ethanol production have not been very successful; although some recombinant ethanologenic strains secreting some active cellulases have been prepared [106, 109, 110], their requirement for a nutrient rich medium and often sensitivity to end-product inhibition hamper their use [105].

5.2. Abe (Acetone-Butanol-Ethanol) Fermentation

Different, so-called solventogenic species of the genus *Clostridium*, like *Clostridium acetobutylicum*,*Clostridium beijerinckii* or *Clostridium saccharoperbutylacetonicum,* can be used for 1-butanol production by ABE fermentation. The fermentation usually proceeds in two steps; at first butyric and acetic acids, along with hydrogen and carbon dioxide, are formed and then metabolic switching leads to the formation of solvents (mainly 1-butanol and acetone) and the cessation/slowdown of acid and gas production (for recent reviews see [111-114]). Industrial fermentative ABE (butanol) production, which has quite a long and impressive history connected with both World Wars, is nowadays carried out only in China and Brazil (estimated annual production of 100 000 t and 8 000 t from corn starch and sugar cane juice, respectively) [10]. However, many corporations such as BP, DuPont, Gevo, Green Biologics, Cobalt Technologies and others have declared their interest in this field. A unique example of the use of lignocellulosic hydrolysate on an industrial scale is the former Dukshukino plant (operated in the Soviet Union up to 1980s) producing acetone and butanol by fermentation. The plant was based on current "very modern" biorefinery concepts which assumed the conversion of complex feedstocks (hydrolysates of agricultural waste + molasses or corn) into many valuable products i.e. in addition to solvents (acetone, butanol and ethanol), it was possible to produce liquid CO_2, dry ice, H_2, fodder yeast, vitamin B_{12} and biogas [115].

The most interesting approach to fermentation of any lignocellulosic substrate is probably consolidated bioprocessing (CBP) i.e. a method in which a single microorganism is used for both substrate decomposition and fermentation to produce the required metabolites. Although some clostridial species such as *Clostridium thermocellum* can utilise cellulosic substrates and produce ethanol [116, 117], the ABE fermentation pattern unfortunately cannot be produced using clostridia. However,*C.acetobutylicum* ATCC 824 possesses genes for various cellullases and a complete cellulosome [118-120]. But even if production of some cellulases by *C. acetobutylicum* ATCC 824 was induced by xylose or lichenan [118], cellulose utilization was not achieved, possibly because of insufficient or deficient synthesis of an unknown specific chaperone that could be responsible for correct secretion of cellulases [119]. Nevertheless as solventogenic *Clostridium* species are soil bacteria that differ significantly in fermentative abilities and genome sizes, it is not excluded that in the future, some solventogenic species with cellulolytic activity will be isolated from an appropriate environment. Recently, a new strain of *Clostridium saccharobutylicum* with hemicellulolytic activity and ABE fermentation pattern was found amongst 50 soil-borne, anaerobic, sporulating isolates [121].

Table 3. Selection of batch ABE fermentations in laboratory scale using lignocellulosic hydrolysates as a substrate

Substrate	Pretreatment	Microbial strain	ABE concentration (g/l) /yield (%) /productivity (g/l/h)	Reference
Wheat straw	Diluted sulphuric acid+ enzyme	*C. beijerinckii* P260	13/25/0.14	[139]
Wheat bran	Diluted sulphuric acid	*C. beijerinckii* ATCC 55025	12/32/0.16	[140]
Corn fiber	Diluted sulphuric acid+ XAD-4 resin treatment + enzyme	*C. beijerinckii* BA101	8/32/0.11	[127]
Corn cobs	Steam explosion + enzyme	*C. acetobutylicum*	21/31/0.45	[131]
Rice straw	Alkali + $(NH_4)_2SO_4$ precipitation + activated carbon treatment + enzyme	*C. saccharoperbutylacetonicum* ATCC 27022	13/28/0.15	[88]
Sugar cane bagasse	Alkali + $(NH_4)_2SO_4$ precipitation + activated	*C. saccharoperbutylacetonicum* ATCC 27022	14/30/0.17	[88]
Sugar cane bagasse	Alkali + $(NH_4)_2SO_4$ precipitation + activated carbon treatment + enzyme	*C. saccharoperbutylacetonicum* ATCC 27022	14/30/0.17	[88]
Cassava bagasse	Heat + enzyme	*C. acetobutylicum* JB200	34/39/0.63	[130]
Domestic organic waste	Steam explosion, lyophilization + enzyme +4 fold concentration of released sugars	*C. acetobutylicum* DSM 792	9/26/0.08	[132]
Dried distiller`s grain and solubles	Diluted acid + overliming +enzyme hot water + overliming+ enzyme AFEX +overliming+ enzyme	*C. saccharobutylicum* 260 *C. butylicum* 592 *C. butylicum* 592	12/35/0.20 13/32/0.20 12/32/0.20	[78]
Sweet sorghum stem	Diluted acetic acid	*C. acetobutylicum* ABE 0801	19/32/0.10	[141]

All fermentations were run in SHF mode i.e. sugar release and fermentation were separate processes.

AFEX stands for ammonium fiber expansion process.

Until now, lignocellulosic substrates must be prehydrolysed for the ABE process. In the case of fermentation of lignocellulosic hydrolysate, usually containing low concentrations of fermentable sugars, one of the main bottlenecks in the ABE process, the low final titre of butanol (caused by severe butanol toxicity towards bacterial cells), is of minor importance. In fact hydrolysates are very good substrates for clostridia that express extensive fermentative abilities [122, 123]and can utilise not only cellulose-derived glucose but also hemicellulose monomers (xylose, arabinose, galactose, mannose). Co-fermentation of various sugar mixtures was described for *Clostridium beijerinckii* SA-1 (ATCC 35702) [124], *Clostridium acetobutylicum* DSM 792 [125], *C. acetobutylicum* ATCC 824 [126] and*C.beijerinckii* P260 [127] however, at the same time, catabolic repression of xylose utilization in the presence of glucose was demonstrated in *C. acetobutylicum* ATCC 824 [128, 129].

An overview of fermentation parameters achieved in batch ABE fermentations of different hydrolysates is presented in Table 3. The most promising results were obtained by Lu et al. [130] using cassava bagasse and a mutant strain, *C.acetobutylicum* JB200; the results of Marchal et al. [131] were unique at the scale used (48 m^3) as shown in Table 3. A frequent problem of lignocellulosic hydrolysates is a low final concentration of fermentable sugars caused by low density of the original substrate. This can be overcome by evaporation of the hydrolysate [132](see Table 3) or by addition of glucose and/or other carbohydrates present in the hydrolysate (this is only possible in laboratory scale experiments) [85,133-135]. In the case of glucose supplemented corn stover and switchgrass hydrolysates, final ABE concentrations of 26 and 15 g/l were achieved [135]. With *C.beijerinckii*P260, use of diluted and $Ca(OH)_2$ treated barley straw hydrolysate supplemented with glucose resulted in a solvent concentration of 27 g/l, a yield of 43% and productivity of 0.39 g/l/h [133]. In addition to materials presented in Table 3, other substrates like diluted sulfite spent liquor supplemented with glucose [134], palm empty fruit bunches [136, 137] or hardwood [138] were used in the ABE process but in these cases, additional optimizations were necessary.

In addition to a batch fermentation arrangement, semi-continuous fermentation of enzymatically hydrolyzed SO_2 pretreated pine wood using *C.acetobutylicum* P262 resulted in 18 g/l of solvents, a yield of 36% and solvent productivity of 0.73 g/l/h [142]. Further, fed-batch fermentation of wheat straw hydrolysate supplemented with varying concentrations of hydrolysate sugars (glucose, xylose, arabinose and mannose) using *C.beijerinckii* P260 yielded a solvent productivity of 0.36 g/l/h if gas stripping was used [127]. In the cases shown in Table 3, enzyme hydrolysis preceeded fermentation, however simultaneous saccharification and fermentation (SSF) was also tested. In SSF of acid pre-hydrolyzed wheat straw using *C.beijerinckii* P262 and solvent removal by gas stripping, 21 g/l of ABE was produced with a productivity of 0.31 g/l/h [127] Nevertheless, the solvent yield from hardwood using SSF was rather low, at 15% [138].

6. CONCLUSION

Intensive research over the last decades on lignocellulose-derived ethanol have focused mainly on intensification of biomass pretreatment, production of cellulolytic enzymes, and strain and process improvements, and have eliminated some of the main technological bottlenecks. Although a number of projects on 2nd generation bioethanol ended with the opening of pilot and demonstration plants around the world (production capacity in millions of gallons for the year 2012 given in brackets) e.g. the POET demonstration plant in Iowa (0.02 from corn stover and cobs), Abengoa in Kansas (0.01 from corn stover), Blue Sugarsin Wyoming (1.3 from stover and cobs), Chempolis in Finland (3.7 from paper waste), Fiberight in Iowa (6.0 MSW), Iogen in Canada (0.48 from stover), Praj MATRIX in India (0.01 from cellulose), UPM-Kymemene/Mesto in Finland (0.68 from mixed cellulose) and in spite of several proclamations, none of them is operating at the industrial scale [9]. To make this possible, further reductions in processing costs will be necessary to achieve a product that is competitive with 1st generation bioethanol. Further process integration is required, including decreased energy demand during pretreatment, increased sugar concentration, higher enzyme activity and strain recycling. By-products, e.g. lignin separated after pretreatment procedure can be used to generate energy for ethanol plant operations (lignin has higher caloric value (25.4 MJ/kg) then the biomass itself [8]) or used as a dispersant and binder in concrete admixtures, as an alternative to phenolic and epoxy resins, or as the principal component in thermoplastic blends, polyurethane foams or surfactants [143]. A combination of 1st and 2nd generation feedstocks (e.g. corn cobs together with stover) can eliminate bottlenecks and lead to product competitiveness. Higher bioethanol production costs can also be compensated for by political and economic instruments such as tax incentives (e.g. tax exemption on biofuels and higher excise taxes for fossil fuels) and legislation (mandatory blends) to enable ready access of 2nd generation biofuels to the market [30]. Butanol, as a second generation biofuel, might be produced via fermentation and used as an excellent fuel extender in addition to ethanol if the technological bottleneck of a low final concentration, yield and productivity could be overcome, and the assumption that suitable cheap waste pretreatments were possible.

ACKNOWLEDGEMENT

The. review was performed thanks to financial support of the projects Kontakt ME10146 of Ministry of Education, Youth and Sport of Czech Republic and BIORAF No.TE01020080 of the Technological Agency of the Czech Republic.

REFERENCES

1. Wang L., Sharifzadeh M., Templer R., Murphy R. J. Technology performance and economic feasibility of bioethanol production from various waste papers. Energy & Environmental Science 2012;5, 5717-5730.

2. Nigam P. S., Singh A. Production of liquid biofuels from renewable resources. Progress in Energy and Combustion Science 2011;37(1), 52-68.

3. Patáková P., Maxa D., Rychtera M., Linhová M., Fribert P., Mužíková Z., Lipovský J., Paulova L., Pospíšil M., Šebor G., Melzoch K. Perspectives of biobutanol production and use. In: Biofuel's Engineering Process Technology. Bernandes M., (ed.). Bernandes M. ed. Rijeka, Croatia: InTech; 2011. p. 243-266.

4. Braconnot H. Hydrolysis of cellulose into sugar. Gilbert´s Annalen der Physik 1819;63, 348.

5. U.S. Department of Energy - Energy Efficiency and Renewable Energy Biomass Program. Dilute acid hydrolysis.

6. Sherrard E. C., Kressman F. W. Review of processes in the United States prior to World War II. Industrial & Engineering Chemistry Research 1945;37(1), 5-8.

7. Lin Y., S. T. Ethanol fermentation from biomass resources: Current state and perspectives. Applied Microbiology and Biotechnology 2006;69, 627-642.

8. Banerjee S., Mudliar S., Sen R., Giri B. Commercializing lignocellulosic bioethanol: Technology bottlenecks and possible remedies. Biofuels Bioproducts & Biorefining 2010;4, 77-93.

9. Biofuel Digest. Free database download:207 advanced fiofuels&chems projets.http://www.biofuelsdigest.com/bdigest/2011/11/16/free-database-download-207-advanced-biofuels-chems-projects (accessed 13 August 2012).

10. Green E. M. Fermentative production of butanol - the industrial perspective. Current Opinion in Biotechnology 2011;22(3), 1-7.

11. The 2012 London Olympics, biofuels-style: BP to showcase its hhree most advanced biofuels.

12. Gray K. A., Zhao L., Emptage M. Bioethanol. Current Opinion in Chemical Biology 2006;10(2), 141-146.

13. Balat M., Ayar G. Biomass energy in the world, use of biomass and potential trends. Energy Sources 2005(29), 931-940.

14. Kim S., Dale B. E. Global potential bioethanol production from wasted crops and crop residues. Biomass and Bioenergy 2004(26), 361-375.

15. Ladanai S., Vinterback J. Global potential of sustainable biomass for energy. Uppsala: Swedish University of Agricultural Sciences; 2009. p. 29.

16. Moller R., Toonen M., Van Beilen J., Salentijn E., Clayton D. Crop platforms for cell wall biorefining: lignocellulose feedstocks, Outputs from the EPOBIO project: CPLpress Science publishers; 2007.

17. Limayem A., Ricke S. C. Lignocellulosic biomass for bioethanol production: Current perspectives, potential issues and future prospects. Progress in Energy and Combustion Science 2012(38), 449-467.

18. Balat M. Production of bioethanol from lignocellulosic materials via the biochemical pathway: a review. Energy Conversion and Management 2011(52), 858-875.

19. Volynets B., Dahman Y. Assesment of pretreatments and enzymatic hydrolysis of wheat straw as a sugar source for bioprocess industry. International journal of energy and environmnet 2011;2(3), 427-446.

20. Sarkar N., Ghosh S. K., Bannerjee S., Aikat K. Bioethanol production from agricultural wastes: an overview. Renewable Energy 2012(37), 19-27.

21. García-Cubero T., Gonzáles-Benito G., Indacoechea I., Coca M., Bolado S. Effect of ozonolysis pretreatment on enzymatic digestibility of wheat and rye straw. Bioresour Technol 2009(100), 1608-1613.

22. Van Dyk J. S., Pletschke B. I. A review of lignocellulose bioconversion using enzymatic hydrolysis and synergist cooperation between enzymes - factors afftecting enzymes, conversion and synergy. Bitechnology Advances 2012.

23. Saha B. C. Hemicellulose bioconversion. Journal Indian Microbiology and Biotechnology 2003;30, 279-291.

24. McKendry P. Energy production from biomass (part 1): Overview of biomass. Bioresour Technol 2002;83, 37-46.

25. Li S., Zhang X., Andresen J. M. Production of fermentable sugars from enzymatic hydrolysis of pretreated municipal solid waste after autoclave process. Fuel 2011.doi.10.1016/j.fuel.2011.07.012

26. Champagne P., Li C. Enzymatic hydrolysis of cellulosic municipal wastewater treatment process residuals as feedstocks for the recovery os simple sugars. Bioresour Technol 2009;100 (4), 5700-5706.

27. Peng L., Chen Y. Conversion of paper sludge to ethanol by separate hydrolysis and fermentation (SHF) using Saccharomyces cerevisiae. Biomass and Bioenergy 2011(35), 1600-1606,Yamashita Y., Kurosumi A., Sasaki C., Nakamura Y. Ethanol production from paper sludge by immobilized Zymomonas mobilis. Biochemical Engineering Journal 2008(42), 314-319.

28. Yamashita Y., Kurosumi A., Sasaki C., Nakamura Y. Ethanol production from paper sludge by immobilized Zymomonas mobilis. Biochemical Engineering Journal. 2008(42):314-319.

29. Agbor V. B., Cicek N., Sparling R., Berlin A., Levin D. B. Biomass pretreatment: Fundamentals toward application. Biotechnology Advances 2011(29), 675-685.

30. Bessou C., Ferchaud F., Gabrielle B., Mary B. Biofuels, greenhouse gases and climate change. A review. Agronomy for Sustainable Development 2011;31(1), 1-79.

31. Vidal B. C., Jr., Dien B. S., Ting K. C., Singh V. Influence of feedstock particle size on lignocellulose conversion--a review. Appl Biochem Biotechnol 2011;164(8), 1405-1421.

32. El Bassam N. Handbook of Bioenergy Crops: A complete reference to species, development and applications: Routledge; 2010.

33. OECD Agriculture outlook. http://www.keepeek.com/Digital-Asset-Management/oecd/agriculture-and-food/oecd-fao-agricultural-outlook-2011_agr_outlook-2011-en (accessed 24 July 2012).

34. Singh P., Suman A., Tiwari P., Arya N., Gaur A., Shivastava A. K. Biological pretreatment of sugarcane trash for its conversion to fermentable sugars. World Journal of Microbiology and Biotechnology 2008;24, 667-673.

35. Fodor Z., Klemeš J. J. Waste as alternative fuel - minimising emissions and effluents by advanced design. Process safety and environmental protection 2012;90, 263-284.

36. Jensen J. W., Felby C., Jorgensen H., Ronch G. O., Norholm N. D. Enzymatic processing of municipal solid waste. Waste Management 2010;30, 2497-2503.

37. Ballesteros M., Sáez F., Ballesteros I., Manzanares P., Negro M. J., Martínez J. M., Castaneda R., Dominguez J. M. O. Ethanol production

from the organic fraction obtained after thermal pretreatment of municipal solid waste. Appl Biochem Biotechnol 2009;161, 423-431.

38. Fernandes D. L. A., Pereira S. R., Serafim L. S., Evtugin D. V., Xavier A. M. R. B. Bioethanol. Rijeka: InTech; 2012.

39. Sjorostrom E. Wood Chemistry. Fundamentals and Applications. New York, USA: Academic Press; 1993.

40. Rodsrud G. Biorefining – expanding the sulfite pulping biorefinery concept. World Biofuels Markets; March 22 – 24, 2011; Rotterdam, NL2011.

41. Mulvany P. K., Agar H. D., Peniston Q. P., Mccarthy J. L. A chromatographic study of sulfite waste liquor sugars. 118th Meeting of the American Chemical Society; Sept. 3 – 8, 1950; Chicago, Illinois1950.

42. Lai L. X. Bioproducts from sulfite pulping: Bioconversion of sugar streams from pulp, sludge and spent sulfite liquor: University of Washington; 2010.

43. Harris E. E. Food-yeast production from wood-processing byproducts. Yeast in Feeding Symposium; Nov. 8 – 10, 1948; Milwaukee, Wisconsin, USA1948.

44. Vancov T., Alston A.-S., Brown T., Mcintosh S. Use of ionic liquids in converting lignocellulosic material to biofuels. Renewable Energy 2012;45, 1-6.

45. Huang R., Su R., Qi W., He Z. Bioconversion of lignocellulose into bioethanol: process intensification and mechanism research. Bioenergy Resources 2011;4, 225-245.

46. Modenbach A. A., Nokes S. E. The use of high-solids loadings in biomass pretreatment - a review. Biotechnology and Bioengineering 2012;109(6), 1430-1442.

47. Chandra R. P., Bura R., Mabee W. E., Berlin A., Pan X., Saddler J. N. Substrate pretreatment: the key to effective enzymatic hydrolysis of lignocellulosic? Advanced Biochemical Engineering/Biotechnology 2007;108, 67-93.

48. Wang G. S., Pan X. J., Zhu J. Y., Gleisner R., Rockwood D. Sulfite pretreatment to overcome recalcitrance of lignocellulose sport for robust enzymatic saccharification of hardwoods. Biotechnology Progress 2009;25(4), 1086-1093.

49. Nikitin N. I. The chemistry of cellulose and wood: Israel Program for Scientific Translations; 1966.

50. Sjorostrom E. Wood chemistry. Fundamental and applications. New York, USA: Academic Press; 1993.

51. Zhu J. Y., Pan X. J., Wang G. S., Gleisner R. Sulfite pretreatment (SPORL) for robust enzymatic saccharification of spruce and red pine. Bioresource Technology 2009;100, 2411-2418

52. Zhang X., Qin W., Paice M. G., Saddler J. N. High consistency enzymatic hydrolysis of hardwood substrates. Bioresource Technology 2009;100, 5890-5897.

53. Zhang X., Tu M., Paice M., Sacciadis G., Jiang Z., Jemaa N., Thibault A. Bioconversion of knot rejects from a sulphite pulp mill to ethanol. BioResources. 2010;5(1):23-42

54. Helle S. S., Petretta R. A., Duff S. J. B. Fortifying spent sulfite pulping liquor with hydrolyzed recect knots. Enzyme and Microbial Technology. 2007(41):44-50.

55. Modenbach A. A., Nokes S. E. The use of high-solids loadings in biomass pretreatment - a review. Biotechnology and Bioengineering 2012;102(6), 1430-1442.

56. Mosier N., Hendrickson R., Ho N., Sedlak M., Ladish M. Optimization of pH controlled liquid hot water pretreatment of corn stover. Bioresour Technol 2005(96), 1986-1993.

57. Hahn-Hagerdal B., Galbe M., Gorwa-Grauslund M. F., Lidén G., Zacchi G. Bio-ethanol - the fuel of tomorrow from the residues of today. Trends in Biotechnology 2006;24(12), 549-556.

58. Sun Y., Cheng J. Hydrolysis of lignocellulosic materials for ethanol production: a review. Bioresource Technology 2002;8, 1-11.

59. Candido R. G., Godoy G. G., Concalves A. R. Study of sugarcane bagasse pretreatment with sulfuric acid as a step of cellulose obtaining. World Academy of Science, Engineering and Technology. 2012; 61, 101-105.

60. Martín C., Marcet M., Thomsem A. B. Comparison between wet oxidation and steam explosion as pretreatment method for enzymatic hydrolysis of sugarcane bagasse. BioResources 2008;3(3), 670-683.

61. Modenbach A. A., Nokes S. E. The use of high-solids loadings in biomass pretreatment-a review. Biotechnololgy and Bioengineering 2012;109(6), 1430-1442.

62. Samayam I. P., Schall C. A. Saccharification of ionic liquid pretreated biomass with commercial enzyme mixtures. Bioresource Technology 2010;101, 561-3566.

63. Hsu T.-C., Guo G.-L., Chen W.-H., Hwang W.-S. Effect of dilute acid pretreatment of rice straw on structural properties and enzymatic hydrolysis. Bioresource Technology 2009;101, 4907-4913.

64. Kumar R., Wyman C. E. Effects of cellulase and xylanase enzymes on the deconstruction of solids from pretreatment of poplar by leading technologies. Biotechnology Progress. 2009;25(2):302-314.

65. Olson D. G., Mcbride J. E., Shaw A. J., Lynd L. R. Recent progress in consolidated bioprocessing. Current Opinion in Biotechnology 2012;23, 396-405.

66. Doi R. H., Kosugi A. Cellulosomes: plant-cell-wall degrading enzyme complexes. Nature reviews Microbiology 2004;2, 541-551.

67. Andric P., Meyer A. S., Jensen P. A., Dam-Johansen K. Reactor design for minimizing product inhibition during enzymatic lignocellulose hydrolysis: I. Significance and mechanism of cellobiose and glucose inhibition on cellulolytic enzymes. Biotechnology Advances 2010;28, 308-324.

68. Almeida J. R. M., Modig T., Petersson A., Hahn-Hagerdal B., Lidén G., Gorwa-Grauslund M. F. Increased tolerance and conversion of inhibitors in lignocellulosic hydrolysates by Saccharomyces cerevisiae. Journal of Chemical Technology and Biotechnology 2007;82, 340-349.

69. Palmqvist E., Hahn-Hagerdal B. Fermentation of lignocellulosic hydrolysates. I:Inhibition and detoxification. Bioresource Technology 2000;74, 17-24.

70. Palmquist E., Hahn-Hagerdal B. Fermentation of lignocellulosic hydrolysates. II: Inhibitors and mechanisms of inhibitions. Bioresource Technolology 2000;74, 25-33.

71. Du B., Sharma L. N., Becker C., Chen S.-F., Mowery R. A., Van Walsum G. P., Chambliss C. K. Effects of variying feedstock-pretreatment chemistry combinantions on the formation and accumulation of potentially inhibitory degradation products in biomass hydrolysates. Biotechnology and Bioengineering 2010;107(3), 430-440.

72. Panagiotou G., Olsoon L. Effect of compounds released during pretreatment of wheat straw on microbial growth and enzymatic hydrolysis rates. Biotechnology and Bioengineering 2007;96(2), 250-258.

73. Cantarella M., Cantarella L., Gallifuoco A., Spera A., Alfani F. Effect of inhibitors released during steam-explosion treatment of poplar wood on subsequent enzymatic hydrolysis and SSF. Biotechnology Progress 2004;20, 200-206.

74. Ximenes E., Kim Y., Mosier N., Dien B. S., Ladish M. Deactivation of cellulases by phenols. Enzyme and Microbial Technology. 2011(48):54-60.

75. Klinke H. B., Thomsen A. B., Ahring B. K. Inhibition of ethanol-producing yeast and bacteria by degradation products produced during pre-treatment of biomass. Applied Microbiology and Biotechnology 2004;66, 10-26.

76. Ezeji T., Qureshi N., Blaschek H. P. Butanol production from agricultural residues: impact of degradation products on *Clostridium beijerinckii*growth and butanol fermentation. Biotechnology and Bioengineering 2007;97(6), 1460-1469.

77. Ezeji T., Blaschek H. P. Butanol production from lignocellulosic biomass: Blackwell Publishing; 2010.p.19-37.

78. Ezeji T., Blaschek H. P. Fermentation of dried distillers' grains and solubles (DDGS) hydrolysates to sovents and value-added products by solventogenic *Clostridia*. Bioresour Technol 2008;99(12), 5232-5242.

79. Qureshi N., Bowman M. J., Saha B. C., Hector R., Berhow M. A., Cotta M. A. Effect of cellulosic sugar degradation products (furfural and hydroxymethyl furfural) on acetone-butanol-ethanol (ABE) fermentation using *Clostridium beijerinckii* P260. Food and Bioproducts Processing 2012;90(3), 533-540.

80. Zhang Y., Han B., Ezeji T. C. Biotransformation of furfural and 5-hydroxymethyl furfural (HMF) by *Clostridium acetobutylicum* ATCC 824 during butanol fermentation. New Biotechnology 2012;29(3), 345-351.

81. Richmond C., Ujor V., Ezeji T. C. Impact of syringaldehyde on the growth of *Clostridium beijerinckii* NCIMB 8052 and butanol production. Biotech 2012;2(2), 159-167.

82. Palmqvist E., Hahn-Hagerdal B. Fermentation of lignocellulosic hydrolysates. II: inhibitors and mechanism of inhibition. Bioresource Technology 2000(74), 25-33.

83. Palmqvist E., Hahn-Hagerdal B. Fermetation of lignocellulosic hydrolysates. I. Inhibition and detoxification. Bioresource Technology 2000;74:17-24.

84. Cho D. H., Lee Y. J., Um Y., Sang B.-I., Kim Y. H. Detoxification of model phenolic compounds in lignocellulosic hydrolysates with peroxidase for butanol production from *Clostridium beijerinckii*. Applied and Microbial Biotechnology 2009;83(6), 1035-1043.

85. Qureshi N., Ezeji T. C., Ebener J., Dien B. S., Cotta M. A., Blaschek H. P. Butanol production by *Clostridium beijerinckii*. Part I: Use of acid and enzyme hydrolyzed corn fiber. Bioresource Technology 2008;99(13), 5915-5922.

86. Larsson S., Reimann A., Nilvebrandt N.-O., Jönsson L. J. Comparison of different methods for the detoxification of lignocellulose hydrolyzates of spruce. Applied Biochemistry and Biotechnology 1999;77(1-3).

87. Martinez A., Rodriguez M. E., York S. W., Preston J. F., Ingram L. O. Effects of Ca(OH)2 treatments ("overliming") on the composition and toxicity of bagasse hemicellulose hydrolysates. Biotechnology and Bioengineering. 2000;69(5):526-536.

88. Soni B. K., Das K., Ghose T. K. Bioconversion of agro-wastes into acetone butanol. Biotechnology Letters 1982;4(1), 19-22.

89. Taherzadeh M. J., Karimi K. Acid-based hydrolysis processes for ethanol from lignocellulosic material: a review. BioResources 2007;2(3), 472-499.

90. Fu N., Peiris P. Co-fermentation of a mixture of glucose and xylose to ethanol by *Zymomonas mobilis* and *Pachysolen tannophilus*. World Journal of Microbiology and Biotechnology 2008(24), 1091-1097.

91. Lebeau T., Jouenne T., Junter G. A. Simultaneous fermentation of glucose and xylose by pure and mixed cultures of Saccharomyces cerevisiae and Candida shehatae immobilized in a two-chambered bioreactor. Enzyme and Microbial Technology. 1997(21):265-272.

92. Groothen D. R. J., Jansen M. L., Van Der Lans R. G. J. M., Luyben K. C. a. M. Reactors in series for the complete converstion of glucose/xylose mixtures by *Pichia stipitis* and *Saccharomyces cerevisiae*. Enzyme and Microbial Technology 1991;13(10), 828-833.

93. Qian M., Tian S., Li X., Zhang J., Pan Y., Yang X. Ethanol production from dilute-acid softwood hydrolysate by co-culture. Applied Microbiology and Biotechnology 2006;134(3), 27-283.

94. Li Y., Park J.-Y., Shiroma R., Tokuyasu K. Bioethanol production from rice straw by a sequential use of *Saccharomyces cerevisiae* and *Pichia stipitis* with heat inactivation of Saccharomyces cerevisiae cells prior to

xylose fermentation. Journal of Bioscience and Bioengineering 2011;111(6), 682-686.

95. Sakamoto T., Hasunuma T., Hori Y., Yamada R., Kondo A. Direct ethanol production from hemicellulosic materials of rice straw by use of an engineered yast strain codisplaying three types of hemicellulolytic enzymes on the surface of xylose-utilizing *Saccharomyces cerevisiae* cells. Journal of Biotechnology 2012(158), 203-210.

96. Dellomonaco C., Fava F., Gonzales R. The path to next generation biofuels: successes and challenges in the era of synthetic biology. Microbial Cell Factories 2010;9(3).

97. Scalcinati G., Otero J. M., Van Vleet J. R. H., Jeffries T. W., Olsson L., Nielsen J. Evolutionary engineering of *Saccharomyces cerevisiae* for efficient aerobic xylose consumption. FEMS Yeast Research 2012;12(5), 585-597.

98. Martín C., Marcet M., Almazán O., Jonsson L. J. Adaptation of a recombinant xylose-utilizing Saccharomyces cerevisiae strain to sugarcane bagasse hydrolysate with high concentration of fermentable inhibitors. Bioresource Technology 2007;98, 1767-1773.

99. Quin W. High consistency enzymatic hydrolysis of lignocelluloses. Vancouver: The University of British Columbia; 2010.

100. Suryawati L., Wilkins M. R., Bellmer D. D., Huhnke R. L., Maness N. O., Banat I. M. Simultaneous saccharification and fermentation of kanlow switchgrass pretreated by hydrothermolysis using *Kluyveromyces marxianus* IMB4. Biotechnology and Bioengineering 2008;101(5), 894-902.

101. García-Aparicio M. P., Oliva J. M., Manzanares P., Ballesteros M., Ballesteros I., González A., Negro M. J. Second-generation ethanol production from steam exploded barley straw by *Kluyveromyces marxianus* CECT 10875. Fuel. 2011;90,1624-1630.

102. de Souza C. J. A., Costa D. A., Rodrigues M. Q. R. B., Dos Santos A. F., Lopes M. R., Abrantes A. B. P., Dos Santos Costa P., Silveira W. B., Passos F. M. L., Fietto L. G. The influence of presaccharification, fermentation temperature and yeast strain on ethanol production from sugarcane bagasse. Bioresource Technology 2012;109, 63-69.

103. Chen M.-L., Wang F.-S. Optimization of a fed-batch simultaneous saccharification and cofermentation process from lignocellulose to ethanol. Industrial & Engineering Chemistry Research 2010;49, 5775-5789.

104. Olofsson K., Bertilsson M., Lidén G. A short review on SSF - an interesting process option for ethanol production from lignocellulosic feedstocks. Bitechnology for Biofuels 2008;1(7), 1-14.

105. Zhang X.-Z., Heng Y., Zhang P. One-step production of biocommodities from lignocellulosic biomass by recombinant cellulolytic *Bacillus subtilis*: Opportunities and chalenges. Engineering in Life Sciences 2010(5), 398-406.

106. Lee R. Lynd W. H. V. Z., John E. Mcbride, Mark Laser. Consolidated bioprocessing of cellulosic biomass: an update. Current Opinion in Biotechnology 2005(16), 577-583.

107. Chang T., Yao S. Thermophilic, lignocellulolytic bacteria for ethanol production: current state and perspectives. Applied Microbiology and Biotechnology 2011(92), 13-27.

108. Shaw A. J., Podkaminer K. K., Desai S. G., Bardsely J. S., Rogers S. R., Thorne P. G., Hogsett D. A., Lynd L. R. Metabolic engineering of a thermophilic bacterium to produce ethanol at high yield. Proceedings of the National Academy of Sciences 2008;105(37), 1769-13774.

109. Haan R. D., Rose S. H., Lynd L. R., Van Zyl W. H. Hydrolysis and fermentation of amorphous cellulose by recombinant *Saccharomyces cerevisiae*Metabolic Engineering 2007(9), 87-94.

110. Fujita Y., Ito J., Ueda M., Fukuda H., Kondo A. Synergistic saccharification, and direct fermentation to ethanol, of amorphous cellulose by use of an engineered yeast strain codisplaying three types of cellulolytic enzyme Applied and Environmental Microbiology. 2004; 20, 688-691.

111. Patáková P., Linhová M., Rychtera M., Paulová L., Melzoch K. Novel and neglected issues of acetone–butanol–ethanol (ABE) fermentation by*Clostridia*: *Clostridium* metabolic diversity, tools for process mapping and continuous fermentation systems. Biotechnology Advances. 2013; 31(1), 58-67.

112. Kumar M., Gayen K. Developments in biobutanol production: new insights. Applied Energy. 2011;88(6), 1999–2012

113. Tashiro Y., Sonomoto K. Advances in butanol production by *Clostridia*. In: Méndez-Vilas A . (ed.) Current Research, Technology and Education Topics in Applied Microbiology and Microbial Biotechnology. Bajadoz, Spain Formatex Research Center; 2010. p. 1383-1394.

114. Lee S. Y., Park J. H., Jang S. H., Nielsen L. K., Kim J., Jung K. S. Fermentative Butanol Production by *Clostridia*. Biotechnology and Bioengineering. 2008;101(2):209-228.

115. Zverlov V. V., Berezina O., Velikodvorskaya G. A., Schwarz W. H. Bacterial acetone and btanol poduction by idustrial frmentation in the Soviet Union: Use of hydrolyzed agricultural waste for biorefinery. Applied Microbiology and Biotechnology 2006;71(5), 587-597.

116. Raman B., Mckeown C. K., Rodriguez M., Brown S. D., Mielenz J. R. Transcriptomic analysis of *Clostridium thermocellum* ATCC 27405 cellulose fermentation. BMC Microbiology. 2011; 11, 134.

117. BMC Microbiology 2011 2011;11(134),Carere C. R., Sparling R., N. C., Levin D. B. Third generation biofuels via direct cellulose fermentation. Journal of Molecular Sciences 2008;9(7), 1342-1360.

118. López-Contreras A. M., Gabor K., Martens A. A., Renckens B. a. M., Claasen P. a. M., Van Der Oost J., De Vos W. M. Substrate-induced production and secretion of cellulases by *Clostridium acetobutylicum*. Applied and Environmental Microbiology 2004;70(9), 5238-5243.

119. Mingardon F., Chanal A., Tardif C., Fierobe H.-P. The issue of secretion in heterologous expression of *Clostridium cellulolyticum* cellulase-encoding genes in *Clostridium acetobutylicum* ATCC 824. Applied and Environmental Microbiology 2011;77(9), 2831-2838.

120. Cui Z., Li Y., Xiao Y., Feng Y., Cui Q. Resonance assignments of cohesin and dockerin domains from *Clostridium acetobutylicum* ATCC 824. Biomolecular NMR Assign 2012.

121. Berezina O. V., Brandt A., Yarotsky S., Schwartz W. H., Zverlov V. V. Isolation of a new butanol-producing *Clostridium* strain: High level of hemicellulosic activity and structure of solventogenesis genes of a new *Clostridium saccharobutylicum* isolate. Systematic and Applied Microbiology 2009;2(7), 449-459.

122. Tracy B. P., Jones S. W., Fast A. G., Indurthi D. C., Papoutsakis E. T. *Clostridia*: the importance of their exceptional substrate and metabolic diversity for biofuel and biorefinery applications. Current Opinion in Biotechnology. 2012;23(3), 364-381.

123. Shi Y., Li Y.-X., Li Y. Y. Large number of phosphotransferase genes in the *Clostridium beijerinckii* NCIMB 8052 genome and the study on their evolution. BMC Bioinformatics 2010(11 (Suppl 11)), S9,Tracy B. P., Jones S. W., Fast A. G., Indurthi D. C., Papoutsakis E. T. *Clostridia*: the importance of their exceptional substrate and

metabolic diversity for biofuel and biorefinery applications. Current Opinion in Biotechnology 2012;23(3), 364-381.

124. Heluane H., Evans M. R., Dagher S. F., Bruno-Bárcena J. M. Nutritional requirements of solventogenic *Clostridia* growing under butanol stress and co-utilization of D-glucose/D-xylose.A meta-analysis and a functional validation. Applied and Environmental Microbiology 2011;77(13), 4473-4485.

125. Survase S. A., Heiningen A., Granström T. Continuous bio-catalytic conversion of sugar mixture to aceton-butanol-ethanol by immobilized*Clostridium acetobutylicum* DSM 792. Applied Microbiology and Biotechnology 2012;93(6), 2309-2316.

126. Wang L., Chen H. Increased fermentability of enzymatically hydrolyzed steam-exploded cornstover for butanol production by removal of fermentation Inhibitors. Process Biochemistry 2011;46(2), 604-607.

127. Qureshi N., Saha B. C., Cotta M. A. Butanol production from wheat straw by simultaneous saccharification and fermentation using *Clostridium beijerinckii*: Part II-Fed-batch fermentation. Biomass and Bioenergy 2008;32(2), 176–183.

128. Ren C., Gu Y., Hu S., Wu Y., Wang P., Yang Y., Yang S., Jiang W. Identification and inactivation of pleiotropic regulator CcpA to eliminate glucose repression of xylose utilization in *Clostridium acetobutylicum*. Metabolic Engineering 2010;12(5), 446-454.

129. Grimmler C., Held C., Liebl W., Ehrenreich A. Transcriptional analysis of catabolite repression in *Clostridium acetobutylicum* growing on mixtures of D-glucose and D-xylose. Journal of Bacteriology. 2010;150(3):315-323.

130. Lu C., Zhao J., Yang S.-T., Wei D. Fed-batch fermentation for n-butanol production from cassava bagasse hydrolysate in a fibrous bed bioreator with continuous gas stripping. Bioresource Technology 2012;104, 80-387.

131. Marchal R., Ropars M., Pourquié J., Fayolle F., Vandecasteele J. P. Large-scale enzymatic hydrolysis of agricultural lignocellulosis biomass. Part 2: Conversion into acetone-butanol. Bioresource Technology 1992;42(3), 205-217.

132. Claassen P. a. M., Budde M. a. W., López-Contreras A. M. Acetone, butanol and dthanol production from domestic organic waste by solventogenic*Clostridia*. Journal of Molecular and Microbiological Biotechnology 2000;2(1), 39-44.

133. Qureshi N., Saha B. C., Dien B., Hector R. E., Cotta M. A. Production of butanol (a biofuel) from agricultural Residues: Part I- Use of barley straw hydrolysate. Biomass and Bioenergy 2010;34(4), 559-565.

134. Survase S. A., Sklavounos E., Jurgens G., Heiningen A., Granström T. Continuous acetone-butanol-ethanol fermentation using SO2-ethanol-water spent liquor from spruce. Bioresource Technology 2011;102(23), 10996-11002.

135. Qureshi N., Saha B. C., Hector R. E., Dien B., Hughes S., Liu S., Iten L., Bowman M. J., Sarath G., Cotta M. A. Production of butanol (a biofuel) from agricultural residues: Part II- use of corn stover and switchgrass hydrolysates. Biomass and Bioenergy 2010;34(4), 566-571.

136. Noomtim P., Cheirsilp B. Production of Butanol from Palm Empty Fruit Bunches Hydrolyzate by *Clostridium acetobutylicum*. Energy Procedia 2011(9), 140-146.

137. Ibrahim M. F., Abd-Aziz S., Razak M. N., Phang L. Y., Hassan M. A. Oil palm empty fruit bunch as alternative substrate for acetone-butanol-ethanol production by *Clostridium butyricum* EB6. Appl Biochem Biotechnol. 2012;166(7):1615-1625.

138. Shah M. M., Song S. K., Lee Y. Y., Torget R. Effect of pretreatment on simultaneous saccharification and fermentation of hardwood into acetone/butanol. Appl Biochem Biotechnol 1991;28-29(1), 99-109.

139. Qureshi N., Saha B. C., Hector R. E., Hughes S. R., Cotta M. A. Butanol production from wheat straw by simultaneous saccharification and fermentation using *Clostridium beijerinckii*: Part I- Batch Fermentation Biomass and Bioenergy 2008;32(2), 176-183.

140. Liu Z., Ying Y., Li F., Ma C., Xu P. Butanol production by *Clostridium beijerinckii* ATCC 55025 from wheat bran. Journal of Industrial Microbiology and Biotechnology 2010;37(5), 495-501.

141. Yu J., Zhang T., Zhong J., Zhang X., Tan T. Biorefinery of sweet sorghum stem. Biotechnology Advances 2012;30(4), 811-816.

142. Parekh S. R., Wayman M. Ethanol and butanol production by fermentation of enzymatically saccharified SO2-prehydrolysed lignocellulosics. Enzyme and Microbial Technology 1988;10(11), 660-668.

143. Lora J. H., Gasser W. G. Recent industrial applications of lignin: a sustainable alternative to nonrenewable materials. Journal of Polymers and the Environment 2002;10(1/2), 39-48.

CHAPTER 5

Lignocellulosic agriculture wastes as biomass feedstocks for second-generation bioethanol production: concepts and recent developments

Jitendra Kumar Saini , Reetu Saini, Lakshmi Tewari

Department of Microbiology, College of Basic Sciences and Humanities, GB Pant University of Agriculture and TechnologyDBT-IOC Centre for Advanced Bio-Energy Research, Research and Development Centre, Indian Oil Corporation Ltd.

Department of Microbiology, M.S. Garg P.G. CollegeDBT-IOC Centre for Advanced Bio-Energy Research, Research and Development Centre, Indian Oil Corporation Ltd.

Department of Microbiology, College of Basic Sciences and Humanities, GB Pant University of Agriculture and Technology

ABSTRACT

Production of liquid biofuels, such as bioethanol, has been advocated as a sustainable option to tackle the problems associated with rising crude oil prices, global warming and diminishing petroleum reserves. Second-generation bioethanol is produced from lignocellulosic feedstock by its saccharification, followed by microbial fermentation and product recovery. Agricultural residues generated as wastes during or after processing of agricultural crops are one of such renewable and lignocellulose-rich biomass resources available in huge amounts for bioethanol production. These agricultural residues are converted to bioethanol in several steps which are described here. This review enlightens various steps involved in production of the second-generation bioethanol. Mechanisms and recent advances in pretreatment, cellulases production and second-generation ethanol production processes are described here.

KEYWORDS

Lignocellulose, Bioethanol, Cellulase, Agricultural wastes, Residues

1. INTRODUCTION

One of the greatest challenges of twenty-first century is to meet the growing demand of energy for transportation, heating and industrial processes, and to provide raw materials for chemical industries in sustainable ways. Biofuels have emerged as an ideal option to meet these requirements in a sustainable manner. Several primary drivers underlie the increasing interests in biofuels, such as increasing uncertainty of petroleum supplies due to rising demand, decline in known reserves, and concerns over global warming and green house gas emissions associated with fossil fuels usage and this has led to various government programs promoting biofuels. Moreover, biofuels are unique among available alternative energy sources in their general compatibility with existing liquid transport fuel. The global production and use of biofuels have increased dramatically in recent years, from 18.2 billion liters in 2000 to 60.6 billion liters in 2007, with about 85 % of this being bioethanol. Bioethanol is the most common and one of the practically important liquid biofuel and can be produced from a variety of cheap substrates. According to an estimate, it can reduce greenhouse gas emissions by approximately 30–85 % compared to gasoline, depending on the feedstock used (Fulton et al. 2004). The USA and Brazil are currently the primary producers of fuel ethanol, producing 49.6 and 38.3 % of 2007 global production, respectively (Coyle 2007). Worldwide increasing interest in the production of bioethanol is exemplified by production of 85 billion liters of bioethanol in 2011 (Singh and Bishnoi 2012; Avci et al. 2013).

The present review is a concise overview of the basic concepts and some recent advances in ethanol production with special emphasis on lignocellulosic agricultural residues/wastes and their sources, pretreatment methods, enzymatic hydrolysis and fermentation to generate bioethanol in ecologically sustainable and cost-effective manner. Some challenges still existing in economic production of second-generation bioethanol and their potential solutions are discussed in brief towards the end.

2. BIOETHANOL: AN ECO-FRIENDLY BIOFUEL

Bioethanol is made biologically by fermentation of sugars derived from a variety of sources. The use of ethanol as a motor fuel began with its use in the internal combustion engine invented by Nikolas Otto in 1897 (Ahindra 2008). Alcohols have been used as fuels since the inception of the automobile. The term alcohol often has been used to denote either ethanol or methanol as a fuel. With the oil crises of 1970s, ethanol became established as an alternative fuel. Many countries started programs to study and develop fuels in an economic way from available raw materials. Countries including Brazil and the USA have long promoted domestic bioethanol production. "First generation bioethanol" is made from sugar feedstock such as cane juice (in Brazil) and molasses (in India) or from starch-rich materials such as corn (in US). Though bioethanol production from 'first generation technologies' is estimated to increase to more than 100 billion liters by 2022 (Goldemberg and Guardabasi 2009), these raw materials compete with food, are insufficient to meet the increasing demand for fuels, have negative impact on biodiversity and may even lead to deforestation to gain more farmland (Hahn-Hägerdal et al. 2006). The cumulative impact of these concerns have increased the interests in developing "second generation ethanol" from non-food lignocellulosic materials such as agricultural residues, wood, paper and municipal solid waste, and dedicated energy crops (viz. miscanthus, switchgrass, sweet sorghum, etc.), which constitute the most abundant renewable organic component in the biosphere (Claassen et al. 1999).

Bioethanol is widely recognized as a unique transportation fuel with powerful economic, environmental and strategic attributes. As bioethanol can be produced from biomass of crop plants, it offers opportunities to improve the income levels of smallholder farmers. At a community level, farmers can cultivate energy crops that fetch an income while also meeting their food needs. Ethanol derived from biomass is the only liquid transportation fuel that does not contribute to the green house gas effect. Ethanol represents closed carbon dioxide cycle because after burning of ethanol, the released carbon dioxide is recycled back into plant material as plants use it to synthesize cellulose during photosynthesis. Ethanol production process only uses energy from renewable energy sources; no net carbon dioxide is added to the atmosphere, making ethanol an environmentally beneficial energy source. Ethanol contains 35 % oxygen that helps complete combustion of fuel and thus reduces particulate emission that poses health hazard to living beings. The toxicity of the exhaust emissions from ethanol is lower than that of petroleum sources (Wyman and Hinman1990). Thus, the use of even 10 % ethanol blends reduces GHG emissions by 12–19 % compared with conventional fossil fuels. Burning E 85 (85 % ethanol) reduces the nitrogen oxide, particulate and sulfate emissions by 10, 20 and 80 %, respectively, compared to conventional gasoline.

Bioethanol market is expected to reach 10×10^{10} l in 2015 (Licht 2008). The largest bioethanol producers in the world are the US, Brazil, and China. In 2009, US produced 39.5×10^{9} l of ethanol using corn as a feedstock while the second largest producer, Brazil, created about 30×10^{9} l of ethanol using sugarcane. China is nowadays investing heavily in ethanol production and is one of its largest producers (Ivanova et al.2011). In India, the interest in biofuels is growing so as to substitute oil for achieving energy security and promote agricultural growth. Indian government has planned to achieve a target of 20 % blending of fossil fuels with biodiesel and bioethanol by 2017. In addition, a national policy for biofuel has been framed including promotion of biofuel production, particularly on wastelands (Ravindranath et al. 2011).

2.1 Feedstocks for bioethanol: Agricultural residues

The varied raw materials used in the manufacture of bioethanol are conveniently classified into three main types: sugars, starches, and cellulose materials. Sugars (such as cane or sweet sorghum juice, molasses) can be used directly for ethanol production via fermentation. Starches (from corn, cassava, potatoes, and root crops) must first be hydrolyzed to fermentable sugars by the action of enzymes from malt or molds. Cellulose (from wood, agricultural residues, waste sulfite liquor from pulp, and paper mills) must likewise be converted into sugars, generally by the action of acids or cellulolytic enzymes (Franks et al. 2006).

There are various forms of biomass resources in the world, which can be grouped into four categories, viz. wood product industry wastes, municipal solid waste, agriculture residues and dedicated energy crops. These biomass resources seem to be the largest and most promising future resources for biofuels production. This is because of the ability to obtain numerous harvests from a single planting, which significantly reduces average annual costs for establishing and managing energy crops, particularly in comparison to conventional crops (Franks et al. 2006). The global production of plant biomass, of which over 90 % is lignocellulose, amounts to about 200×10^{9} tons/year, where about $8–20 \times 10^{9}$ tons of the primary biomass remains potentially accessible (Kuhad and Singh 1993). Lignocellulosic material represents a promising option as feedstock for ethanol production considering their output/input energy ratio, availability, low cost and higher ethanol yields. For second-generation biofuel production, utilization of renewable biomass resources has received major focus in the world. Renewable 'plant biomass' refers particularly to cheap and abundant non-food lignocellulose-rich materials available from the plants. Biomass to bioethanol process could help in mitigation of global climate change by reducing emissions (mainly CO_2) as well as decreasing dependence upon fossil fuels. Thus, deployment of biomass resources has been projected to play an important role in sustainable development. The second-generation biofuels include hydrogen, natural gas, bio-oils, producer gas, biogas, alcohols and biodiesel. In countries like India, agricultural production of various crops like cotton, mustard, chilli, sugarcane, sorghum, sweet sorghum, pulses, oilseeds, etc. results in generation of huge amounts of wastes that do not find any alternative use and are either left in the fields or are burned. Hence, these could be used as good alternative resources to generate biofuels such as bioethanol, in an environmentally friendly manner. Use of agricultural residues helps in reduction of deforestation by decreasing our reliance on forest woody biomass. Moreover, crop residues have short harvest period that renders them more consistently available to bioethanol production (Knauf and Moniruzzaman 2004; Kim and Dale 2004; Limayema and Ricke 2012).

Maize, wheat, rice, and sugarcane are the four agricultural crops with maximum production as well as area under cultivation. These four crops are responsible for generating majority of lignocellulosic biomass in agriculture sector and rest of the agrowastes constitute only a minor proportion of the total agrowaste production in the world. Corn stover is the left over residue after harvesting corn kernel and comprises stalks, leaves, cobs, and husks. Its annual production is approximately 1 kg/kg corn grain or 4 tons/acre (Kim and Dale 2004; Heaton et al. 2008; Cheng and Timilsina 2011). Straw is generated during wheat grain harvest at a rate of 1–3 tons/acre annually under rigorous farming conditions. Rice straw is the leftover of rice production and includes stems, leaf blades, leaf sheaths, and the remains of the panicle after threshing. It is one of the most abundant lignocellulosic waste materials in the world. Out of the annual global production of 731 million tons of rice straw Asia alone produces 667.6 million tons. Bagasse is produced in huge amounts during sugarcane processing. It is also a cheap renewable agricultural resource for ethanol production (Bhatia and Paliwal 2011). Most of the agricultural residues have similar contents of cellulose, hemicelluloses, and lignin, but rice straw has more silica content while wheat straw contains significant amount of pectin and proteins (Sarkar et al. 2012).

Lignocellulosic agricultural wastes have cellulose as a major component, but their chemical composition varies considerably (Table 1). Global production of major agrowastes and their bioethanol production potential are shown in Table 2. Maximum rice straw and wheat straw are generated in Asia and corn straw and sugarcane bagasse are mainly produced in America.

According to an estimate, lignocellulosic biomass can be used to generate approximately 442 billion liters of bioethanol per year and if total crop residues and wasted crops are also considered, this figure can rise to 491 billion liters, about 16 times higher production than the actual global production (Kim and Dale 2004; Sarkar et al. 2012). In US alone, a total of 1,368 MT biomass are available for bioethanol production, out of which agrowastes with 428 MT constitute major proportion, followed by forestry wastes, energy crops, grains and corn, municipal and industrial wastes and other wastes contributing 370, 377, 87, 58 and 48 MT, respectively (Perlack et al. 2005; USDOE Biomass Program 2009; RFA 2010).

Table 1. Composition of various agricultural and other lignocellulosic residues

Material	Cellulose[a]	Hemicellulose	Lignin	Ash	Extractives
Algae (green)	20–40	20–50	–	–	–
Cotton, flax, etc.	80–95	5–20	–	–	–
Grasses	25–40	25–50	10–30	–	–
Hardwoods	45 ± 2	30 ± 5	20 ± 4	0.6 ± 0.2	5 ± 3
Hardwood barks	22–40	20–38	30–55	0.8 ± 0.2	6 ± 2
Softwoods	42 ± 2	27 ± 2	28 ± 3	0.5 ± 0.1	3 ± 2
Softwood barks	18–38	15–33	30–60	0.8 ± 0.2	–
Cornstalk	39–47	26–31	3–5	12–16	–
Wheat straw	37–41	27–32	13–15	11–14	–
Newspaper	40–55	25–40	18–30	–	–
Chemical pulp	60–80	20–30	2–10	–	–
Sorghum stalks	27	25	11	–	–
Corn stover	38–40	28	7–21	3.6–7.0	–
Coir	36–43	0.15–0.25	41–45	2.7–10.2	–
Bagasse	32–48	19–24	23–32	1.5–5	–
Rice straw	28–36	23–28	12–14	14–20	–
Wheat straw	33–38	26–32	17–19	6–8	–
Barley straw	31–45	27–38	14–19	2–7	–
Sorghum straw	32	24	13	12	–
Sweet sorghum Bagasse	34–45	18–28	14–22	–	–

Ref Kuhad et al. (1997), Reddy and Yang (2005), Li et al. (2010)

[a] Composition represented in %wt on dry matter basis

Table 2. Worldwide availability of major agricultural wastes and their bioethanol production potential

Agricultural wastes	Availability[a] (million tons)	Estimated bioethanol potential[a] (Gl)
Wheat straw	354.34	104
Rice straw	731.3	205
Corn straw	128.02	58.6
Sugarcane bagasse	180.73	51.3

[a] Calculated from Sarkar et al. (2012)

2.2 Structural organization of lignocellulosic feedstocks

Agricultural residues such as wheat straw, rice straw, bagasse, cotton stalk and wheat bran are rich in lignocellulose and primarily contain cellulose, lignin, hemicellulose, and extractives. Cellulose forms a skeleton that is surrounded by hemicellulose and lignin functioning as matrix and encrusting materials, respectively (Ingram and Doran 1995). Table 1 presents the biochemical composition of major lignocellulosic feedstocks that are being used worldwide for bioethanol production.

Cell wall polysaccharides

Classically, cell wall polysaccharides have been grouped into three fractions: cellulose, hemicellulose and pectic polysaccharides, proteins and other miscellaneous compounds (Chesson and Forsberg 1988) as discussed below.

Cellulose

Cellulose, the major structural component in the plant cell wall, is a linear homo-polysaccharide consisting of anhydrous glucose units (500–15,000) that are linked by β-1,4-glycosidic bonds, with cellobiose as the smallest repetitive unit. The β-1,4 orientation of the glucosidic bonds results in the potential formation of intramolecular and intermolecular hydrogen bonds, which make native cellulose highly crystalline, insoluble, and resistant to enzyme attack. The highly crystalline regions of cellulose in the plant cell wall are separated by less ordered amorphous regions (Chesson and Forsberg 1988).

Hemicellulose

Hemicellulose is a short, highly branched polymer of pentoses (e.g. D-xylose and L-arabinose) and hexoses (e.g. D-manose, D-galactose, and D-glucose) with 50–200 units. Its acetate groups were randomly attached with ester linkages to the hydroxyl groups of the sugar rings. The role of hemicellulose is to provide a linkage between lignin and cellulose (Holtzapple 1993).

Pectic compounds and proteins

Pectic polysaccharides make up approximately 35 % of the primary cell walls, the main components being galactosyluronic residues. Its other major components are rhamnose, arabinose, and galactose. Pectic substances are hydrophillic and therefore have certain adhesive properties. Proteins are a minor component of the plant cell wall which may be covalently cross-linked with lignin and polysaccharides (Cassab and Varner 1988).

Phenolic compounds

Three types of phenolic compounds viz. lignin, tannins and phenolic acids are found in plant cell walls. Lignin is a heterogeneous, amorphous, and cross-linked aromatic polymer where the main aromatic components are trans-coniferyl, trans-sinapyl and trans-*p*-coumaryl alcohols. Lignin is covalently bound to side groups on different hemicelluloses, forming a complex matrix that surrounds the cellulose micro-fibrils. In plant cell wall it varies from 2 to 40 %. The existence of strong carbon–carbon (C–C) and ether (C–O–C) linkages in the lignin gives the plant cell wall strength and protection from attack by cellulolytic microorganisms (Mooney et al. 1998). Tannins are high molecular weight (500–3,000) polyphenolic compounds, composed of either hydroxyflavans, leucoanthocyanidin (flavan-3,4-diol) and catechin (flavan-3-ol) or glucose. Phenolic acids are structural components of the lignin core in plant cell wall. The presence of carboxyl and phenolic groups in phenolic acids enable such compounds to link to lignin and carbohydrates by ether or ester bonds.

Bioconversion of lignocellulosic biomass to bioethanol

Biomass to ethanol bioconversion process consists of several steps, including pretreatment of biomass, enzymatic hydrolysis, fermentation and product recovery. Proper combination of each step is important for achieving higher bioethanol yield in a cost-effective and sustainable manner.

2.3 Processing and pretreatments

The main processing challenge in the ethanol production from lignocellulosic biomass is the feedstock pretreatment. During pretreatment, the matrix of cellulose and lignin bound by hemicellulose should be broken to reduce the crystallinity of cellulose and increase the fraction of amorphous cellulose, the most suitable form for enzymatic attack. The yield of cellulose hydrolysis after pretreatment often exceeds 90 % of theoretical as compared to 20 % when pretreatment is not carried out (Lynd 1996). For pretreatment of lignocellulosics, several physical, physico-chemical and biological processes have been developed that improve lignocellulose digestibility in very different ways (Aden et al. 2002; Sun and Cheng 2002; Wyman et al. 2005a, b). These processes are summarized in Table 3.

2.3.1 Physical pretreatment methods

Lignocellulosic biomass can be pulverized by chipping, grinding, shearing, or milling, which reduces the particle size and increases surface area, facilitating the access of cellulases to the biomass surface and increasing the conversion of cellulose. Primary size reduction employs hammer mills or Wiley mills to produce particles that can pass through 3- to 5-mm diameter sieve. Other useful physical treatment methods include pyrolysis, irradiation with gamma radiation, microwave, infrared, or sonication (Brown 2003; Mosier et al. 2005).

Table 3. Pretreatment methods of lignocellulosic biomass for fuel ethanol production

Methods	Procedure/agents	Remarks	Examples of pretreated materials	References
I. *Physical methods*				
Mechanical size reduction	Chipping, grinding, milling	Milling: vibratory ball mill Wiley mill (final size: 0.2–2 mm), knife or hammer mill (final size 3–6 mm)	Hardwood, straw, corn stover, timothy, alfalfa, cane and sweet sorghum bagasse	Sun and Cheng (2002)
Pyrolysis	T > 300 °C, then cooling and condensing	Formation of volatile products and char Residues: produce 80–85 % reducing sugars (>50 % glucose); can be carried out under vacuum	Wood, Waste cotton, corn stover	Khiyami et al. (2005)
II. *Physico-chemical methods*				
Steam explosion	Saturated steam at 160–290 °C, p = 0.69–4.85 MPa for several sec or min, then decompression until atm. Pressure	It can handle high solid loads: size reduction with lower energy input compared to comminution. 80–100 % hemicellulose hydrolysis, destruction of a portion of xylan fraction, 45–65 % xylose recovery; Inhibitors formation; addition of H_2SO_4, SO_2, or CO_2 improves efficiency of further enzymatic hydrolysis; cellulose depolymerization	Poplar, aspen, eucalyptus softwood (Douglas fir) bagasse, corn stalk, wheat straw, rice straw, barley straw, sweet sorghum bagasse, *Brassica carinata* residue, olive stones, Timothy grass, alfalta, reed canary grass	Ballesteros et al. (2001, 2002, 2004), Hamelinck et al. (2005), Lynd et al. (2002), Soderstrom et al. (2003), Sun and Cheng (2002)
Liquid hot water (LHW)	Pressurized hot water, p > 5 MPa, T = 170–230 °C, 1–46 min; solids load <20 %	Lignin is not solubilized, but redistributed. 80–100 % hemicellulose hydrolysis, 88–98 % xylose recovery; low or no formation of inhibitors, cellulose conversion >90 %; partial solubilization of lignin (20–50 %)	Bagasse, corn stover, olive pulp, Alfalfa fiber	Ballesteros et al. (2002), Koegel et al. (1999), Lynd et al. (2002)
Ammonia fiber explosion (AFEX)	1–2 kg ammonia/kg dry biomass, 90 °C, 30 min, p = 1.12–1.36 MPa	Ammonia recovery is required 0–60 % hemicellulose hydrolysis, no inhibitor formation; further cellulose conversion can be >90 %, for high-lignin biomass (<50 %); 10–20 % lignin solubilization	Aspen wood chips bagasse, wheat straw, barley straw, rice hulls, corn stover switchgrass, coastal bermudagrass, alfalfa newsprint	Lynd et al. (2002), Sun and Cheng (2002)
CO_2 explosion	4 kg CO_2/kg fiber, p = 5.62 MPa	No inhibitors formation Further cellulose conversion can be >75 %	MSW Bagasse Alfalfa recycled paper	Sun and Cheng (2002)
Ozonolysis	Ozone, room temperature and pressure	No inhibitors formation further cellulose conversion can be >57 % lignin degradation	Poplar, sawdust, pine, bagasse, wheat straw, cotton straw, green hay, peanut	Sun and Cheng (2002)
Dilute-acid hydrolysis	0.75–5 % H_2SO_4, HCL or HNO3, p = 1 MPa, continuous process for low solids loads (5–10 wt% substrate/mixture), T = 160–200 °C, batch process for high solids loads (10–40 % substrate/mixture), T = 120–160 °C	pH neutralization is required that generates gypsum as a residue. 80–100 % hemicellulose hydrolysis, 75–90 % xylose recovery; high temperature favors further cellulose hydrolysis lignin is not solubilized, but it is redistributed	Poplar wood bagasse, corn stover, wheat straw, rye straw, rice hulls, switchgrass, Bermudagrass	Hamelinck et al. (2005), Lynd et al. (2002), Sun and Cheng (2002), Wooley et al. (1999)
Concentrated-acid hydrolysis	10–30 % H_2SO_4, 170–190 °C, 1:1.6 solid-liquid ratio 21–60 % peracetic acid, silo-type system	Acid recovery is required; residence time greater compared to dilute-acid hydrolysis; peracetic acid provokes lignin oxidation	Poplar sawdust, bagasse	Cuzens and Miller (1997)
Alkaline hydrolysis	Dilute NaOH, 24 h, 60 °C, $Ca(OH)_2$, 4 h, 120 °C, it can be complemented by adding H_2O_2 (0.5–2.15 vol %) at lower temperature (35 °C)	Reactor costs are lower compared to acid pretreatment >50 % hemicellulose hydrolysis, 60–75 % xylose; recovery low inhibitors formation; cellulose swelling, further cellulose conversion can be >65 %; 24–55 % lignin removal for hardwood, lower for softwood	Hardwood, bagasse, corn stover, straws with low lignin content (10–18 %), cane leaves	Hamelinck et al. (2005), Kaar and Holtzapple (2000), Lynd et al. (2002), Saha and Cotta (2008), Sun and Cheng (2002)
Organosolv process	Organic solvents (methanol, ethanol, acetone, ethylene glycol, triethylene glycol) or their mixture with 1 % of H_2SO_4 or HCl 185–98 °C, 30–60 min, pH = 2.0–3.4	Solvent recovery required, almost total hydrolysis of hemicellulose; high yield of xylose almost total lignin solubilization and breakdown of internal lignin and hemicellulose bonds	Poplar wood mixed softwood (spruce, pine, Douglas fir)	Lynd et al. (2002), Pan et al. (2005), Sun and Cheng (2002)
III. *Biological methods*				
Fungal pretreatment	Brown-, white- and soft-rot fungi; Cellulase and hemicellulase production by solid-state fermentation of biomass	Fungi produces cellulases, hemicellulases, and lignin-degrading enzymes: ligninases, lignin peroxidases, polyphenoloxidases, laccase and quinone-reducing enzymes, very slow process	Corn stover, wheat straw	Sun and Cheng (2002)
Bioorganosolv pretreatment	*Ceriporiopsis subvermispora* for 2–8 weeks followed by ethanolysis at 140–200 °C for 2 h	Fungi decompose the lignin network ethanol action allows hemicellulose hydrolysis biological pretreatment can save 15 % of the electricity needed for ethanolysis ethanol can be reused, environmentally friendly process	Beech wood	Itoh et al. (2003)

2.3.2 Physico-chemical methods

Physico-chemical methods are considerably more effective than physical methods of pretreatment. Different chemical agents employed during these processes are ozone, acids, alkali, peroxide and organic solvents. Several physico-chemical methods (Table 2) are employed for pretreament of biomass before its saccharification, such as ammonia fiber explosion (AFEX) (Sun and Cheng 2002; Mosier et al. 2005), autohydrolysis (steam explosion) (Grous et al. 1986; Ramos and Fontana 1996), SO_2 steam explosion (Sipos et al. 2009), acid treatment (Sun and Cheng 2002; Gamez et al. 2006) and alkali treatment (Chang and Holtzapple 2000; Kaar and Holtzapple 2000).

2.3.3 Biological treatment

The brown rot, white rot and soft-rot fungi such as *Phanerochaete chrysosporium*, *Trametes versicolor*, *Ceriporiopsis subvermispora*, *and Pleurotus ostreatus* are employed for biological pretreatment of lignocellulosic biomass. Besides lignin peroxidases and manganese-dependent peroxidases, polyphenol oxidases, laccases and quinosine-reducing enzymes also degrade lignin by producing aromatic radicals. Biological treatment requires low energy and normal environmental conditions but the hydrolysis yield is low and requires long treatment times (Brown 2003).

2.4 Enzymatic saccharification of pretreated biomass

Cellulose hydrolysis, also known as saccharification, is the process in which the cellulose is converted into glucose. Enzymatic hydrolysis is the key to cost-effective ethanol production from lignocellulosic substrates in the long run, as it is very mild process, gives potentially high yields, and the maintenance costs are low compared to acid or alkaline hydrolysis (Kuhad et al. 1997). The process is compatible with many pretreatment methods, but materials poisonous to the enzymes need to be removed or detoxified when chemical pretreatment precedes enzymatic hydrolysis. Factors affecting enzymatic saccharification process involve substrate concentration, enzyme loading, temperature and time of saccharification (Tucker et al.2003).

Cellulase enzyme complex

The cellulose-degrading enzymes were discovered by Reese (1976). The term cellulase complex normally refers to a set of enzymes involved in complete cellulose hydrolysis. Cellulose and the modified cellulose-degrading enzymes are divided into three major groups of enzymes: endo-glucanases (EG), exoglucanases (cellobiohydrolases, CBH) and β-glucosidase (BGL) (Fig. 1) which belong to the EC 3.2.1.X class.

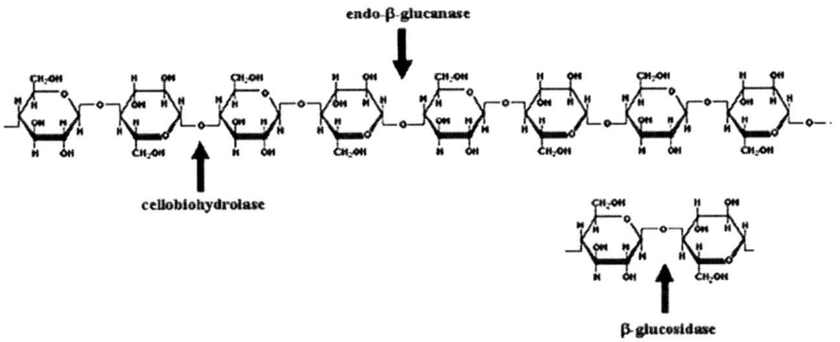

Figure. 1. Sites of action of cellulases on cellulose polymer

Endoglucanases

Endo-β-(1,4)-glucanases (or 1,4-β-D-glucan-4-glucanohydrolases, EC 3.2.1.4), commonly referred to as endoglucanases, are characterized by their random hydrolysis of β-(1,4)-glucosidic linkages (Wood and McCrae 1979). Acting on soluble cellulose derivatives, their random cleavage causes rapid decrease in chain length and hence changes in viscosity relative to the release of reducing end groups. When acting on cellodextrins, the rate of hydrolysis increases with the degree of polymerization within the limits of substrate solubility, with cellobiose and cellotriose being the major final product.

Exoglucanases

Exo-β-(1-4)-glucanase (or 1,4-β-D-glucan cellobiohydrolases, EC 3.2.1.91) cleave cellobiose units from the non-reducing ends of cellulose molecules. Exo-β-(1,4)-glucosidase (or 1,4-β-D-Glucan glucohydrolases, EC 3.2.1.74) cleaves glucose units successively from the non-reducing end of the glucan. They are distinguished from β-glucosidase by their preference for substrates of longer chain length and by the inversion of their products.

β-glucosidases

β-glucosidase (or β-D-glucoside glucohydrolase, EC 3.2.1.21) hydrolyzes cellobiose and other very short chain β-1,4-oligoglucosides up to cellohexaose to form glucose. Most β-glucosidases are active on a range of β-dimers of glucose. Unlike exoglucosidases, the rate of hydrolysis of cellobiose decreases markedly as the degree of polymerization of the substrate increases.

Hemicellulolytic enzymes

In xylan degradation, endo-1,4-β-xylanase, β-xylosidase, α-glucuronidase, α-L-arabinofuranosidase and acetylxylan esterase act on different heteropolymers, while during glucomannan degradation, β-mannanase and β-mannosidase cleave the polymer backbone (Niehaus et al. 1999).

Synergism between cellulases

When the combination of two enzymes is more efficient than the sum of the enzymes acting alone, the two enzymes have synergy. The complete hydrolysis of cellulose to glucose requires a combination of enzymes (endo-, exo-glucanse and β-glucosidase) which work in a synergistic manner for hydrolysis of both native and modified cellulose (Irwin et al. 1993). Presumably the EG makes internal cuts in the cellulose chain, and thereby provides new accessible chain ends for the cellobiohydrolase/exoglucanase and β-glucosidase to work on to gain increased hydrolytic activity (Fig. 2) and due to synergistic effect, each enzyme speeds up the action of the other, with a resulting increase of hydrolysis yield. This model for the synergy between endoglucanases and exoglucanases is called the endo-exo model (Beguin and Aubert 1994).

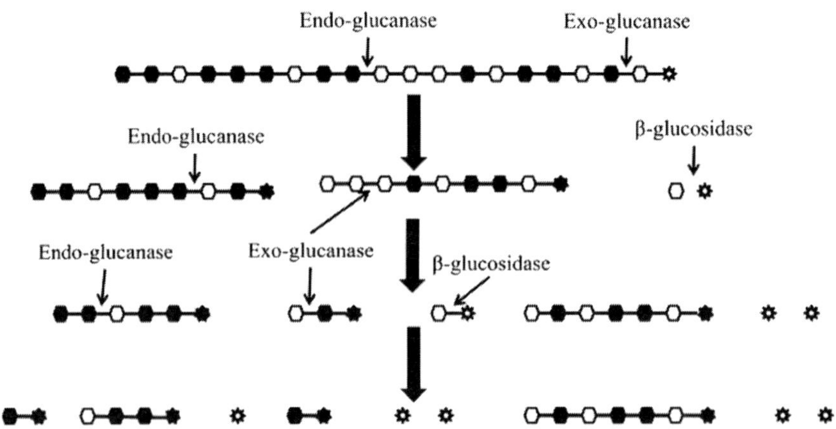

Figure. 2. The endo-exo model for synergy between endoglucanase, exoglucanase and β-glucosidase in a cellulolytic system during cellulose hydrolysis. = reducing end; = modified reducing end; - = β(1,4) linkage; = modified glucose; = unmodified glucose

Microbial cellulases

In nature there are many microorganisms that produce cellulolytic enzymes (cellulases). The cellulolytic organisms can be sorted into two different subcategories depending on their enzyme organization in the cell: (a) the microorganism with their cellulases organized into multi-enzyme complexes called cellulosomes, e.g.*Clostridium thermocellum* and *Cellulomonas.* (b) The cellulolytic organisms producing non-complexed cellulase that are not attached to one another, and act individually and cooperatively on cellulose, and by doing this gain strong synergy effects. Examples of fungi from this class are *T. reesei* and *Humicola grisea*and of bacteria, *Streptomyces lividans* and *Cellulomonas fimi* (Bayer et al. 1998). *Trichoderma* spp. (e.g. *T. reesei*, *T. viride*, *T. longibrachiatum*, *T. pseudokoningii* and *T. harzianum*) are ideal cellulolytic model organisms for studying cellulose degradation since these secrete large amounts of cellulases. To date, two CBHs (Cel6A and Cel7A), and at least five EGs (Cel5A, Cel7B, Cel12A, Cel45A, and Cel61A), have been found in the cellulolytic system of *T. reesei.* These enzymes belong to six different GH families, 5, 6, 7, 12, 45, and 61 (Table 4). Today several species of cellulase producing *Penicillium* spp. are known (e.g. *P. citrinum P. occiantalis*, *P. italicum*, etc.). Moreover, many species of *Aspergillus* such as *A. nidulans*, *A. niger* and *A. oryzae* are also known as potential cellulase producers (Pocas-Fonseca and Maranhac 2005).

Table 4. Properties of *T. reesei* Cellulases

Enzyme	New name	Molecular mass (kDa)	pI	Conc (%)[b]	Stereo-selectivity	No. of residues	Position of CBM
CBH I	Cel 7A	59–68	3.5–4.2	50–60	Retaining	497	C
CBH II	Cel 6A	50–58	5.1–6.3	15–18	Inverting	447	N
EG I	Cel 7B	50–55	4.6	12–15	Retaining	436	C
EG II	Cel 5A	48	5.5	9–11	Retaining	397	N
EG III	Cel 12A	25	7.4	0–3	Retaining	218	Na[b]
EG IV	Cel 45A	37[a]	Na	na	Na	344	C
EG V	Cel 61A	23[a]	2.8–3.0	0–3	Inverting	270	C
BGL I	Cel 3A	71	8.7	Na	Na	Na	Na
BGL II	Cel 1A	114	4.8	Na	Na	Na	Na

Ref Tolan (2002)

CBH cellobiohydrolase, *EG* endoglucanase, *BGL* beta-glucosidase

[a] calculated according to the amino acid sequence deducted from gene sequence

[b] Cel 12A does not have a Cellulose binding module

Optimization of culture conditions for cellulase production

Cellulase production using fungal cultures is a complex system. Many factors affect cellulase production including nutrient availability, pH, temperature, dissolved oxygen concentration, agitation speed, etc.

Medium composition

No general medium composition can be given for growth and optimum cellulase production by all microbes, since the medium must be adapted to the organism in use. Basal medium after Mandels and Weber (1969) has been most frequently used for cellulase production by *T. viridae*, either directly or with slight modifications. Among the cellulosic materials, sulfite pulp, printed papers, mixed waste paper, wheat straw, paddy straw, sugarcane bagasse, jute stick, carboxymethylcellulose corncobs, groundnut shells, cotton, ball milled barley straw, delignified ball milled oat spelt xylan, larch wood xylan, etc. have been used as the substrates for cellulase production. Carboxymethycellulose or cereal straw (1 %, w/w) has been reported as the best carbon source for CMCase and β-glucosidase production using *Chaetomium globosum* as the cellulolytic agent. Further, 3 % malt extract or water hyacinth has also been optimum for CMCase, FPase and β-glucosidase as observed with lactose as the additional carbon sources. However, the saccharification of alkali-treated bagasse at higher substrate levels (up to 4 % w/v) has also been reported. Addition of Ammonium sulfate (0.5 g/l) leads to maximum production of cellulases. However, an increase in the level of β-glucosidase but decrease in endoglucanase and exoglucanase levels was reported when corn steep liquor (0.8 % v/v) was added to synthetic cellulose, wheat straw and wheat bran as the substrates. Phosphorus is an essential requirement for fungal growth, metabolism and several intracellular processes. Different phosphate sources such as potassium dihydrogen phosphate, tetra-sodium pyrophosphate, sodium β-glycerophosphate and dipotassium hydrogen phosphate have been evaluated for their effect on cellulase production (Garg and Neelakantan 1982).

Temperature

Temperature has a profound effect on lignocellulosic bioconversion. The temperature range for cellulase production is generally within 25–35 °C for a variety of microbial strains, e.g. *T. reesei*, *Thielavia terrestris*, *Mycelieopthora fergussi*, *Aspergillus wentii*, *Penicillum rubrum*, *Aspergillus niger*, *Aspergillus ornatus* and*Neurospora crassa.* Some of the thermophilic fungi, having maximum growth at or above 45–50 °C, had produced cellulase with maximum activity at 50–78 °C (Li et al. 2011).

pH of medium

pH has been known to affect enzyme and cells metabolism tremendously. When the environmental pH is over the operational pH (pH 2.0–pH 7.0), the intracellular pH and enzyme activity are greatly influenced. For cellulase production by *T. reesei* using sugarcane bagasse, an optimum pH range of 5.0–6.0 has been suggested. pH cycling method has been used successfully in the past to obtain high cellulase productivity from 25.0 to 38.75 IU/l/h using 3 % cellulose and *T. reesei* QM 9414. In addition, the fungal cellulases have been found to have the highest catalytic capability between pH 3.5 and 6.0 at 50 °C (Bracey 1998; Kansoh1998).

Oxygen concentration

Oxygen is required for cell growth of most eukaryotes; therefore, cell growth is affected by agitation tremendously. For cellulase production, the percentage of dissolved oxygen is typically maintained above 30 %. Cells die when oxygen is not enough and they stop growing afterwards.

Cellulolytic fungal consortium

Although significant advances have been made, a considerable amount of work is still required to enhance the production efficiency of cellulase enzymes. One possible way of improving this situation is to have a mixture containing enzymes of different origins (fungal and/or bacterial). Improved enzyme production by coculturing of two or more microbial strain is being used increasingly for enhanced enzyme production (Garcia-Kirchner et al. 2005). Enhanced cellulose hydrolytic activities have also been observed by the co-cultivation of *A. ellipticus* and *A. fumigatus* (Gupte and Madamwar 1997), co-culture of *A. flavus* and *A. niger*(Saini et al. 2013), co-culture of *A. niger* and *T. reesei* and co-culture of *T. reesei* and *A. phoenicis*(Gutierrez-Correa and Tengerdy 1997).

Purification and characterization of fungal cellulases

Culture filtrates from fungal growth often contain a mixture of several extracellular enzymes besides cellulases and hemicellulases and present considerable purification problems. Therefore, multiple purification steps, including different chromatographic runs, are needed to purify cellulase components (Stahlberg et al.1988). Various chromatographic techniques have been described in the literature for purification of cellulase enzyme such as molecular exclusion, affinity chromatography, ion-exchange chromatography, chromatofocusing, fast protein liquid chromatography (FPLC) and hydrophobic interaction chromatography (HIC) (Tomaz and Queiroz 1999). The biochemical characteristics of β-glucosidase and cellulase are summarized in Table 5.

Commercial cellulases

The cost of cellulase enzymes is widely considered an important factor in the commercialization of lignocellulosic biomass-to-ethanol processes (Wright 1988). A large number of industries are manufacturing cellulase enzymes globally (Table 6). The enzyme producers Genencor International and Novozymes A/S have achieved increased enzyme activity and reduced 30-fold cost of production. Genencor developed a blend of genetically enhanced enzymes that act in synergy to convert cellulose to sugars (http://www.nrel.gov/awards/2004hrvtd.html). Cellic CTec3 is a state-of-the-art cellulase and hemicellulase complex that allows for the most cost-efficient conversion of pretreated lignocellulosic materials to fermentable sugars compared to any other cellulase or enzyme complex available in the market for cellulosic ethanol production (http://www.bioenergy.novozymes.com/en/cellulosic-ethanol/CellicCTec3/Pages/default.aspx).

Table 5. Biochemical properties of fungal β-glucosidase and cellulases

Source	Mr (kDa)	Quaternary structure	Opt. pH	Opt. Temp. (°C)
β-glucosidases				
Aspergillus niger	105	Dimer	5	55
Aspergillus niger	330	Tetramer	4.6–5.3	70
Candida peltate	43	Monomer	5	50
Ceriporiopsis subvermispora	110	NR	3.5	60
Fomitipsis pinicola	105	Monomer	4.5	50
Melanocarpus sp.	92	Monomer	6	60
Phanerochaete chrysosporium	114	NR	4–5.2	NR
Penicillium purpurogenum	110	Monomer	5	65
Cellulases				
T. reesei	48	Monomeric	4.0–5.0	50
A. niger	31	Monomeric	4.0	30
A. nidulans	83	NR	5.0	50
T. harzianum	78	Monomeric	7.7–8.0	40
Geotrichum sp.	80	Monomeric	5.5	55
Trichoderma sp.	51	Monomeric	5	50
F. oxysporum	42.7	Monomeric	5.0	75

Ref Rashid and Siddiqui (1998)

NR not reported

Visualizing structural changes in hydrolyzed bagasse

To detect morphological and structural changes in polymers, some physico-chemical (thermal analysis, X-ray diffraction, gel permeation chromatography), spectroscopic (Infrared and Raman spectroscopy, nuclear magnetic resonance and mass spectroscopy) and microscopic [scanning electron microscopy (SEM), atomic force microscopy (AFM), transmission electron microscopy (TEM), and chemical force microscopy (CFM)] (Samir et al. 2005) methods are consistently being used in the literature (Volke-Spulveda 1998). Standard methods generally employed to examine biodegradation of biopolymers are: visual observations, weight loss measurements through determination of residual polymer, changes in mechanical properties and molar mass and radio-labeling. A number of other techniques have also been used to assess the biodegradability of polymeric material. These include differential scanning colorimetry (DSC), nuclear magnetic resonance spectroscopy (NMR), X-ray photoelectron spectroscopy (XPS), X-ray Diffraction (XRD), contact angle measurements and water uptake Carmen et al. (2009). More recently, Fourier transform infrared spectroscopy (FTIR) and simultaneous TG-DTG-DTA have been used to study biodegradation of polymers (Hadad et al. 2005).

Table 6 Commercially available cellulases

Product name	Company	Source	pH	Temp (°C)	Form
Biocellulase TRI	Quest Intl. (USA)	*T. reesei*	4.0–5.0	50	Liquid
Biocellulase A	Quest Intl. (USA)	*A. niger*	5.0	55	Powder
Celluclast 1.5L	Novo Nordisk, (Danbury, CT)	*T. reesei*	5.0	50	Liquid
Cellulase TAP10[6]	Amano Enzyme (Troy, VA)	*T. viride*	5.0	50	Powder
Cellulase AP30 K	Amano Enzyme (Troy, VA)	*A. niger*	4.5	60	Powder
Cellulase TRL	Solvay Enzymes (Elkhart, IN)	*T. reesei*	4.5	50	Liquid
Econase CE	Alko-EDC (USA)	*T. reesei*	5.0	50	Liquid
Multifect CL	Genencor Intl. (USA)	*T. reesei*	4.5	50	Liquid
Multifect GC	Genencor Intl. (USA)	*T. reesei*	4.0	50	Liquid
Spezyme	Genencor Intl. (USA)	*T. reesei*	4.0	50	Liquid
Ultra-Low Microbial (ULM)	Iogen, (Ottawa, Canada)	*T. reesei*	NA	NA	Liquid
Cellic CTec 2	Novozymes (Bagsvaerd, Denmark)	Enzyme cocktail	NA	NA	Liquid
Cellic CTec 3	Novozymes (Bagsvaerd, Denmark)	Enzyme cocktail	NA	NA	Liquid

Ref Nieves et al. (1998)

2.5 Fermentation and product recovery

The biomass is hydrolyzed by cellulolytic enzymes into fermentable sugars (pentoses or hexoses), which are fermented to ethanol by several microorganisms. For making ethanol production commercially viable, an ideal microorganism should utilize broad range of substrates, with high ethanol yield, titre and productivity, and should have high tolerance to ethanol, temperature and inhibitors present in hydrolysate. Some of the major characters of a viable ethanol process are listed in Table 7.

Table 7. Important traits for bioethanol fermentation process

Trait	Requirement
Bioethanol yield	>90 % of theoritical maximum
Bioethanol tolerance	>40 g/l
$Y_{P/S}$	Close to 0.5 g/g
Q_P	>1 g/l/h
Robust growth and simple growth requirement	Inexpensive medium formulation
Culture growth conditions retard contaminants	Acidic pH or higher temperatures

Ref Dien et al. (2003), Balat et al. (2008)

2.6 Integrated bioprocesses for saccharification and fermentation

As shown in Fig. 3, various saccharification and fermentation bioprocess integrations have been reported. First is separate (or sequential) hydrolysis and fermentation (SHF), a two-stage process involving saccharification of the substrate, followed by the fermentation of saccharified fluid, separately. Main features of SHF include optimal operating conditions for each step and minimal interactions between hydrolysis and fermentation steps. However, SHF process is limited by end-product inhibition and chances of contaminations, which may decreases ethanol yield (Balat et al. 2008; Sanchez and Cardona 2008; Neves et al. 2007; Sarkar et al. 2012). Second process configuration is simultaneous saccharification and fermentation (SSF), in which hydrolyzes of cellulose is consolidated with the direct fermentation of the produced glucose, avoiding the problem of product inhibition associated with enzymes. Main advantages of simultaneous saccharification and fermentation process are comparatively lower costs, higher ethanol yields due to removal of feedback inhibition on enzymatic saccharification and reduction in the required number of vessels or reactors. Some of the disadvantages of SSF are different optimum conditions for enzyme hydrolysis and fermentation processes (Bjerre et al. 1996; Hamelinck et al. 2005; Neves et al. 2007; Balat et al. 2008; Sarkar et al. 2012). The most common and robust fermenting microorganisms employed in ethanol production are *S. cerevisiae* and *Z. mobilis*. Ethanol production from sugars derived from starch and sucrose has been commercially dominated by this yeast. However, *S. cereviseae* is capable of converting only hexose sugars to ethanol. The most promising yeasts that have the ability to use both pentoses and hexoses are *Pichia stipitis*, *Candida shehatae* and *Pachysolan tannophilus*. Thermotolerant yeasts, such as*Kluyveromyces marixianus*, could be more suitable for ethanol production at industrial level, because of their ability to ferment at higher temperatures. In high-temperature process energy savings can be achieved through a reduction in cooling costs. Hence, thermotolerant yeasts are highly desirable in SSF process. Another strategy for ethanol fermentation is simultaneous saccharification and co-fermentation (SSCF), in which co-fermentation of hexoses and pentoses is carried out. In SSCF the co-fermenting microorganisms need to be compatible in terms of operating pH and temperature (Neves et al. 2007). However, the ability to ferment pentoses along with hexoses is not widespread among microorganisms and lack of ideal co-fermenting microorganism is one of the greatest obstacles in industrial production of second-generation ethanol (Talebnia et al. 2010). Sometimes co-culture technique proves to be a useful technology whereby a combination of hexose and pentose fermenting microorganisms is utilized for complete utilization of biomass sugars. For example, co-culture of *Candida shehatae* and *Saccharomyces cerevisiae* was reported as suitable for the SSCF process (Neves et al. 2007). One more configuration for ethanol fermentation is consolidated bioprocessing (CBP). In this process, ethanol and all required enzymes are produced by a single microorganism or microbial community, in the same reactor. The process is also known as direct microbial conversion (DMC). The main advantage of CBP is that its application avoids the cost involved in purchase or production of enzymes (Hamelinck et al. 2005; Lynd et al. 2005). Approached pathways in the development of CBP organisms are described by Lynd et al. (2002). Bacteria such as *Clostridium thermocellum* and some fungi including *Neurospora crassa*, *Fusarium oxysporum* and *Paecilomyces* sp. have shown this type of activity. However, CBP is a less-efficient process with poor ethanol yields and longer fermentation time of more than 3–4 days. Significant cost reductions are expected when progressing from improved SSF via SSCF to CBP. Table 8 summarizes the bioethanol production from some major agroresidues.

2.7 Optimization of saccharification and fermentation bioprocess

Experimental design and statistical analysis for optimization of process conditions are some of the most critical stages in the development of an efficient and economic bioprocess. Classical (such as one factor at a time) and statistical methodologies are available for optimizing process conditions (such as response surface methodology, RSM). RSM is an efficient statistical technique for optimization of multiple variables to predict the best performance conditions with a minimum number of experiments. These designs are used to find improved or optimal process settings, troubleshoot process problems and weak points and make a product or process more robust against external and noncontrollable influences. Full factorial, partial factorial, Box-Behnken and central composite rotatable designs (CCRD) are the most common techniques used for process analysis and monitoring (Sasikumar and Viruthagiri 2008). This method has been in use for hydrolysis of a wide variety of materials to find the optimum conditions for different lignocellulosic biomasses (Talebnia et al. 2010; Saini et al. 2013) and for standardizing SSF for production of ethanol from pretreated sugarcane bagasse by cellulase and yeast *Kluyveromyces fragilis* (Sasikumar and Viruthagiri2008).

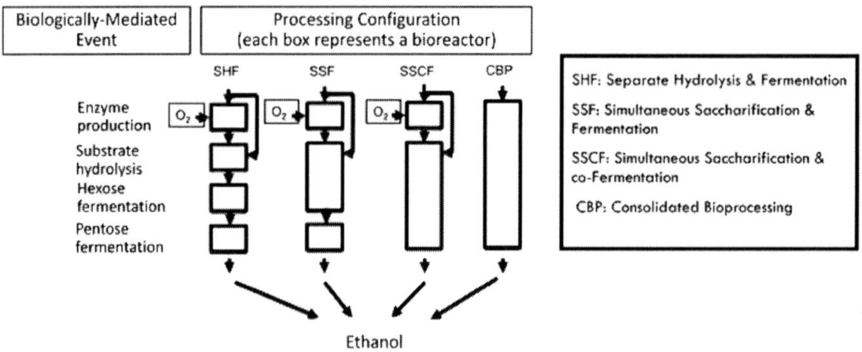

Figure. 3 Process configurations for conversion of lignocellulosic biomass to bioethanol

Table 8. Bioethanol production from major agroresidues

Biomass	Fermenting microorganism	Ethanol yield or titre
Wheat straw	*Pichia stipitis* NRRL Y-7124 (strain adapted to acid hydrolysate inhibitors)	0.35 g/g yield
	Pichia stipitis A	0.41 g/g yield
Rice straw	*Candida shehatae* NCL-3501 (co-ferment glucose and xylose)	0.45 g/g and 0.37 g/g from autohydrolysate and acid hydrolysate, respectively
	S. cerevisiae ATCC 26603 (only ferment glucose)	
SCB	*Pichia stipitis* BCC15191 (glucose–xylose co-fermenting strain)	0.29 g/g yield
	Genetically modified *E. coli* KO11 (glucose and xylose co-fermenting strain)	91.50 % yield and 3.15 % (w/v) ethanol titre

Ref Sarkar et al. (2012)

2.8 Economic considerations for cellulosic ethanol production

To be competitive, and economically acceptable, the cost for bioconversion of biomass to liquid fuel must be lower than the current gasoline prices (Subramanian et al. 2005). It seems, however, much more attainable because of increasing efforts of researchers working towards improvisation in the efficiency of biomass conversion technologies. However, there is still huge scope to bring down the cost of biomass-to-ethanol conversion. The cost of feedstock and cellulolytic enzymes are the two important parameters for low-cost ethanol production. Biomass feedstock cost represents around 40 % of the ethanol production cost. An analysis of the potential of bioethanol in short and long term (2030) in terms of performance, key technologies and economic aspects such as cost per kilometer driven has been conducted recently by Hamelinck et al. (2005).

The choice of feedstock for ethanol production depends upon its availability and the ongoing uses. Some dedicated energy sources like damaged rice, sorghum grains and sweet sorghum bagasse, sunflower stalks and hulls, *Eicchornia crassipies*, *P. brava,* alfalfa fibers, residual starch and crushed wheat grains, agro waste, and *Saccharum spontaneum* are more feasible sources for bioethanol production. The use of integrated approach (Process engineering, fermentation and enzyme and metabolic engineering) could improve the ethanol production economics. Aristidou and Penttila (2000) reported that the total cost of cellulosic ethanol will be dropped from more than $1.0 to ~$0.3–0.5/l, with a projected cost of less than $0.25/l in the near future. Wooley et al. (1999) have explained the further economic analysis of bioethanol ($ 0.78/gallon) and suggested a projected cost of as low as $ 0.20/l by 2015 if enzymatic processing and biomass improvement targets are met. The projected cost of ethanol production from cellulosic biomass as per the earlier estimates ($4.63/gallon in 1980) has been reduced by almost a factor of four ($1.22/gallon) over the last 20 years.

The distillation cost of per unit amount of ethanol produced is substantially higher at low ethanol concentrations; the researchers have dealt with the idea of concentrating sugar solutions prior to fermentation. Ethanol distillation cost can be further improved using membrane distillation process. It has the lowest operational cost, simple to use and is easy to maintain and is the most efficient and cost-effective option among the available distillation processes (Camacho et al. 2013).

3. CHALLENGES AND FUTURE OUTLOOK

Lignocellulosic biomass has long been advocated as a key feedstock for cost-effective bioethanol production in an environment-friendly and sustainable manner. Lignocellulose-rich agricultural wastes/residues are abundant and renewable resources for second-generation bioethanol production. Till now research on utilization of agricultural residues for second-generation bioethanol production has shown very promising results worldwide. Several lab and pilot scale as well as demonstration studies for cellulosic ethanol production from agrowastes have been reported successful but still there exists a huge gap between the projected and actual bioethanol production at industrial level. Therefore, to make full use of these cheap, abundant and renewable resources for economically feasible bioethanol production, several difficulties have yet to be overcome. These challenges include (1) collection, harvesting, supply and handling of agrowastes; (2) cost-effective pretreatment technology; (3) reduction in cost of cellulolytic enzymes; (4) achieving efficient depolymerization of cellulose and hemicellulose into fermentable monomeric sugars by development of more efficient enzyme blends/cocktails; (5) use of higher biomass loadings for achieving higher yields of fermentable sugars and thus high titers of ethanol; (6) use of efficient thermotolerant yeast strains capable of fermentation at temperatures more close to optimum for cellulolytic enzymes; and finally, (7) xylose and glucose co-fermentation, and the use of recombinant/metabolically engineered microbial strains. Considering the huge availability of feedstocks from agriculture and other sources and tremendous efforts being carried out to make second-generation biofuel production more cost-effective, there seems huge scope for the large-scale production of second-generation biofuels in near future. This will certainly involve elimination of the current technology hurdles of lignocellulose to bioethanol conversion process by making microbial processes more efficient.

4. CONCLUSIONS

Lignocellulosic biomass-derived second-generation biofuels are promising alternatives to petroleum-based fossil fuels. The utilization of agricultural residues and wastes for bioethanol production is a cost-effective and environmental-friendly approach for sustainable development. Considering the recent research progress in the fields of enzyme production, pretreatment, as well as metabolic engineering of yeasts, production of bioethanol from lignocellulosic agricultural wastes will certainly prove to be a feasible technology to achieve energy security in very near future.

ACKNOWLEDGMENTS

Authors JKS and LT thank Director, Experiment Station, GBPUA&T, Pantnagar, India, for providing necessary research facilities.

REFERENCES

1. Aden A, Ruth M, Ibsen K, Jechura J, Neeves K, Sheehan J, Wallace B, Montague L, Slayton A, Lukas J (2002) Lignocellulosic biomass to ethanol process design and economics utilizing co-current dilute acid prehydrolysis and enzymatic hydrolysis for corn stover. NREL, Golden, Colorado. In: Proceedings of NREL technical report, pp 510–532

2. Ahindra N (2008) Biofuels refining and performance. McGraw Hill, USA

3. Aristidou A, Penttila M (2000) Metabolic engineering applications to renewable resource utilization. Curr Opin Biotechnol 11:187–198

4. Avci A, Saha BC, Dien BS, Kennedy GJ, Cotta MA (2013) Response surface optimization of corn stover pretreatment using dilute phosphoric acid for enzymatic hydrolysis and ethanol production. Bioresour Technol 130:603–612

5. Balat M, Balat H, Oz C (2008) Progress in bioethanol processing. Prog Energy Combust Sci 34:551–573

6. Ballesteros I, Oliva JM, Sanchez F, Ballesteros M (2001) Ethanol production from lignocellulosic byproducts of olive oil extraction. Appl Biochem Biotechnol 91:237–252

7. Ballesteros I, Oliva JM, Negro MJ, Manzanares P, Ballesteros M (2002) Ethanol production from olive oil extraction residue pretreated with hot water. Appl Biochem Biotechnol 98:717–732

8. Ballesteros M, Oliva JM, Negro MJ, Manzanares P, Ballesteros I (2004) Ethanol from lignocellulosic materials by a simultaneous saccharification and fermentation process (SFS) with *Kluyveromyces marxianus* CECT 10875. Process Biochem 39:1843–1848

9. Bayer EA, Chanzy H, Lamed R, Shoham Y (1998) Cellulose, cellulases and cellulosomes. Curr Opin Struct Biol 8:548–557

10. Beguin P, Aubert JP (1994) The biological degradation of cellulose. FEMS Microbiol Rev 13:25–58

11. Bhatia L, Paliwal S (2011) Ethanol producing potential of *Pachysolen tannophilus* from sugarcane bagasse. Int J Biotechnol Bioeng Res 2(2):271–276

12. Bjerre AB, Olesen AB, Fernqvist T (1996) Pretreatment of wheat straw using combined wet oxidation and alkaline hydrolysis resulting in convertible cellulose and hemicellulose. Biotechnol Bioeng 49:568–577

13. Bracey D (1998) Determination of the intracellular pH (pHi) of growing cells of *Saccharomyces cerevisiae* the effect of reduced-expression of the membrane H^+-ATPase. J Microbiol Methods 31:113–124

14. Brown RC (2003) Proceedings of Biorenewable resources, 1st edn. Iowa State Press, Ames, Iowa

15. Camacho LM, Dumee L, Zhang J, de Li J, Duke M, Gomez J, Gray S (2013) Advances in membrane distillation for water desalination and purification applications. Water 5:94–196

16. Carmen MO, Muller Laurindo JB, Yamashita F (2009) Effect of cellulose fibers on the crystallinity and mechanical properties of starch-based films at different relative humidity values. Carbohydr Polym 77(2):293–299

17. Cassab CI, Varner JE (1988) Cell wall proteins. Ann Rev Plant Physiol 39:321–353

18. Chang VS, Holtzapple MT (2000) Fundamental factors affecting biomass enzymatic reactivity. Appl Biochem Biotechnol 84–86:5–37

19. Cheng JJ, Timilsina GR (2011) Status and barriers of advanced biofuel technologies: a review. Renew Energy 36:3541–3549

20. Chesson A, Forsberg CW (1988) Polysaccharide degradation by rumen microorganisms. In: Hobson (ed) The rumen microbial ecosystem, pp 251–284

21. Claassen PAM, Contreras AML, Sijtsma L, Weusthuis RA, Van Lier JB, Van Niel EWJ, Stams AJM, de Vries SS, Weusthuis RA (1999) Utilisation of biomass for the supply of energy carriers. Appl Microbiol Biotechnol 52:741–755CrossRef

22. Coyle W (2007) The future of biofuels: a global perspective. Amber Waves 5(5). Available online at www.ers.usda.gov/AmberWaves/November07/Features/Biofuels.htm

23. Cuzens JC, Miller JR (1997) Acid hydrolysis of bagasse for ethanol production. Renew Energy 10(2):285–290

24. Dien BS, Cotta MA, Jeffries TW (2003) Bacteria engineered for fuel ethanol production current status. Appl Microbiol Biotechnol 63:258–266

25. Franks CD, Burow GB, Burke JJ (2006) A comparison of US and Chinese sorghum germplasm for early season cold tolerance. Crop Sci 46:1371–1376

26. Fulton L, Howes T, Hardy J (2004) Biofuels for transport: an international perspective. International Energy Agency, Paris, France. Available online at http://www.iea.org/textbase/nppd/free/2004/biofuels2004.pdf

27. Gamez S, Cabriales JJG, Ramirez JA, Garrote G, Vazquez M (2006) Study of hydrolysis of sugar cane bagasse using phosphoric acid. J Food Eng 74:78–88

28. Garcia-Kirchner O, Segura-Granados M, Rodriguez-Pascual P (2005) Effect of media composition and growth conditions on production of beta -glucosidase by *Aspergillus niger* C-6. Appl Biochem Biotechnol 121(1):347–360

29. Garg SK, Neelakantan S (1982) Effect of nutritional factors on cellulase enzyme and microbial protein production by *Aspergillus terreus* and its evaluation. Biotechnol Bioeng 24(1):109–125

30. Goldemberg J, Guardabasi P (2009) Are biofuels a feasible option? Energy Policy 37:10–14

31. Grous WR, Converse AO, Grethlein HE (1986) Effect of steam explosion pretreatment on pore size and enzymatic hydrolysis of poplar. Enz Microb Technol 8:274–280

32. Gupte A, Madamwar D (1997) Solid state fermentation of lignocellulosic waste for cellulose and β-glucosidase production by co-cultivation by *Aspergillus ellipticus* and *Aspergillus fumigatus*. Biotechnol Prog 13:166–169

33. Gutierrez-Correa M, Tengerdy RP (1997) Production of cellulase on sugar cane bagasse by fungal mixed culture solid substrate fermentation. Biotechnol Lett 19:665–667

34. Hadad D, Geresh S, Sivan A (2005) Biodegradation of polyethylene by the thermophilic bacterium *Brevibacillus borstelensis*. J Appl Microbiol 98(5):1365–2672

35. Hahn-Hägerdal B, Galbe M, Gorwa-Grauslund MF, Liden G, Zacchi G (2006) Bio-ethanol: the fuel of tomorrow from the residues of today. Trend. Biotechnol 24(12):549–556

36. Hamelinck CN, Hooijdonk GV, Faaij APC (2005) Ethanol from lignocellulosic biomass: techno-economic performance in short-, middle- and long-term. Biomass Bioenerg 28:384–410

37. Heaton EA, Flavell RB, Mascia PN, Thomas SR, Dohleman FG, Long SP (2008) Herbaceous energy crop development: recent progress and future prospects. Curr Opin Biotechnol 19:202–209

38. Holtzapple MT (1993) Hemicellulose. In: Macrae R, Robinson RK, Sadler MJ (eds) Encyclopedia of food science, food technology and nutrition, vol 2. Academic Press, London, pp 2324–2334

39. Ingram LO, Doran JB (1995) Conversion of cellulosic materials to ethanol. FEMS Microbiol Rev 16:235–241

40. Irwin DC, Spezio M, Walker LP, Wilson DB (1993) Studies of eight purified cellulases specificity synergism and binding domain effects. Biotechnol Bioeng 42:1002

41. Itoh H, Wada M, Honda Y, Kuwahara M, Watanabe T (2003) Bioorganosolve pretreatments for simultaneous saccharification and fermentation of beech wood by ethanolysis and white-rot fungi. J Biotechnol 103(3):273–280

42. Ivanova V, Petrova P, Hristov J (2011) Application in the ethanol fermentation of immobilized yeast cells in matrix of alginate/magnetic nanoparticles, on chitosan-magnetite microparticles and cellulose-coated magnetic nanoparticles. Int Rev Chem Eng 3:289–299

43. Kaar WE, Holtzapple MT (2000) Using lime pretreatment to facilitate the enzymic hydrolysis of corn stover. Biomass Bioenerg 18:189–199

44. Kansoh AL (1998) Biodegradation and utilization of bagasse with *Trichoderma reesei*. Polym Degrad Stab 63:273–278

45. Khiyami MA, Pometto AL, Brown RC (2005) Detoxification of corn stover and corn starch pyrolysis liquors by*Pseudomonas putida* and *Streptomyces setonii* suspended cells and plastic compost support biofilms. J Agric Food Chem 53:2978–2987

46. Kim S, Dale BE (2004) Global potential bioethanol production from wasted crops and crop residues. Biomass Bioenerg 26:361–375

47. Knauf M, Moniruzzaman M (2004) Lignocellulosic biomass processing: a perspective. Int Sugar J 106:147–150

48. Koegel RG, Sreenath HK, Straub RJ (1999) Alfalfa fiber as a feedstock for ethanol and organic acids. Appl Biochem Biotechnol 92(6):105–115

49. Kuhad RC, Singh A (1993) Lignocellulose biotechnology: current and future prospects. Crit Rev Biotechnol 13:151–172

50. Kuhad RC, Singh A, Ericksson KE (1997) Microorganisms and enzymes involved in the degradation of plant fiber cell walls. Adv Biochem Eng Biotechnol 57:45–125

51. Li BZ, Balan V, Yuan YJ, Dale BE (2010) Process optimization to convert forage and sweet sorghum bagasse to ethanol based on ammonia fiber expansion (AFEX) pre-treatment. Bioresour Technol 101:1285–1292

52. Li DC, Li AN, Papageorgiou AC (2011) Cellulases from Thermophilic fungi: recent insights and biotechnological potential. Enzyme Res.

53. Licht FO (2008) World fuel ethanol production. Renewable Fuels Association. http://www.ethanolrfa.org/resource/facts/trade/

54. Limayema A, Ricke SC (2012) Lignocellulosic biomass for bioethanol production: current perspectives, potential issues and future prospects. Prog Energy Combust Sci 38:449–467

55. Lynd LR (1996) Overview and evaluation of fuel ethanol from cellulosic biomass: technology, economics, the environment, and policy. Ann Rev Energy Environ 21:403–465

56. Lynd LR, Weimer PJ, van Zyl WH, Pretorius IS (2002) Microbial cellulose utilization: fundamentals and biotechnology. Microbiol Mol Biol Rev 66(3):506–57

57. Lynd LR, van Zyl WH, Mcbride JE, Laser M (2005) Consolidated bioprocessing of cellulosic biomass: an update. Curr Opin Biotechnol 16(5):577–583

58. Mandels M, Weber J (1969) Production of cellulases. Adv Chem Series 95:391–414

59. Mooney CA, Mansfield SD, Touhy MG, Saddler JN (1998) The effect of initial pore volume and lignin content on the enzymatic hydrolysis of softwoods. Bioresour Technol 64:113–119

60. Mosier N, Wyman C, Dale B, Elander R, Lee YY, Holtzapple M, Ladisch M (2005) Features of promising technologies for pretreatment of lignocellulosic biomass. Bioresour Technol 96(6):673–686

61. Neves MA, Kimura T, Shimizu N, Nakajima M (2007) State of the art and future trends of bioethanol production, dynamic biochemistry, process biotechnology and molecular biology. In: Proceedings of global science books, pp 1–13

62. Niehaus F, Bertoldo C, Kahler M, Antranikian G (1999) Extremophiles as a source of novel enzymes for industrial application. Appl Microbiol Biotechnol 51:711–729

63. Nieves RA, Ehrman CI, Adney WS, Elander RT, Himmel ME (1998) Technical communication: survey and analysis of commercial cellulase preparations suitable for biomass conversion to ethanol. World J Microbiol Biotechnol 14:301–304

64. Pan XJ, Arato C, Gilkes N, Gregg D, Mabee W, Pye K, Xiao ZZ, Zhang X, Saddler J (2005) Biorefining of softwoods using ethanol organosolv pulping: preliminary evaluation of process streams for manufacture of fuel-grade ethanol and co-products. Biotechnol Bioeng 90(4):473–481

65. Perlack RD, Wright L, Turhollow LA, Graham RL, Stokes B, Erbach DC (2005) Biomass as feedstock for a bioenergy and bioproducts industry: the technical feasibility of a billion-ton annual supply. In: Proceedings of Oak Ridge National Laboratory Report ORNL/TM-2005/66. US Department of Energy, Oak Ridge, TN

66. Pocas-Fonseca MJ, Maranhac AQ (2005) Diversity among Cellulolytic Fungi. In: Satyanarayan T, Johri BN (eds) Microbial diversity: current perspectives and potential applications. I. K. International Pvt. Ltd., New Delhi (India), pp 165–180

67. Ramos DF, Fontana J (1996) Pretreatment of sugarcane bagasse for enhanced ruminal digestion. Appl Biochem Biotech 57:171–182

68. Rashid MH, Siddiqui KS (1998) Thermodynamic and kinetic study of stability of the native and chemically modified β-glucosidase from *Aspergillus niger*. Process Biochem 33(2):109–115

69. Ravindranath NH, Lakshmi CS, Manuvie R, Balachandra P (2011) Biofuel production and implications for land use, food production and environment in India. Energy Policy 39:5737–5745

70. Reddy N, Yang Y (2005) Biofibers from agricultural byproducts for industrial applications. Trend Biotechnol 23:22–27

71. Reese ET (1976) History of the cellulase program at the US Army Natick Development Center. Biotechnol Bioeng Symp 9:451–467

72. RFA (2010) Ethanol industry outlook: climate of opportunity. http://www.ethanolrfa.org/page/-/objects/pdf/outlook/RFAoutlook2010_fin.pdf?nocdn=1

73. Saha BC, Cotta MA (2008) Lime pretreatment, enzymatic saccharification and fermentation of rice hulls to ethanol. Biomass Bioenerg 32:971–977

74. Saini JK, Anurag RK, Arya A, Kumbhar BK, Tewari L (2013) Optimization of saccharification of sweet sorghum bagasse using response surface methodology. Ind Crop Prod 44:211–219

75. Samir MASA, Alloin F, Dufresne A (2005) Review of recent research into cellulosic whiskers, their properties and their application in nanocomposite field. Biomacromolecules 6:612–626

76. Sanchez OJ, Cardona CA (2008) Trends in biotechnological production of fuel ethanol from different feedstocks. Bioresour Technol 99:5270–5295

77. Sarkar N, Ghosh SK, Bannerjee S, Aikat K (2012) Bioethanol production from agricultural wastes: an overview. Renew Energy 37:19–27C

78. Sasikumar E, Viruthagiri T (2008) Optimization of process conditions using response surface methodology (RSM) for ethanol production from pretreated sugarcane bagasse: kinetics and modeling. BioEnergy Res 1(3–4):239–247

79. Singh A, Bishnoi NR (2012) Enzymatic hydrolysis optimization of microwave alkali pretreated wheat straw and ethanol production by yeast. Bioresour Technol 108:94–101

80. Sipos B, Reczey J, Somorai Z, Kadar Z, Dienes D, Reczey K (2009) Sweet sorghum as feedstock for ethanol production: enzymatic hydrolysis of steam-pretreated bagasse. Appl Biochem Biotechnol 153:151–162

81. Soderstrom J, Galbe M, Zacchi G (2005) Separate versus simultaneous saccharification and fermentation of two-step steam pretreated softwood for ethanol production. J Wood Chem Technol 25(3):187–202

82. Stahlberg J, Johansson G, Pettersson G (1988) A binding-site-deficient, catalytically active, core protein of endoglucanase III from the culture filtrate of *Trichoderma reesei*. Eur J Biochem 173:179–183

83. Subramanian KA, Singal SK, Saxena M, Singhal S (2005) Utilization of liquid biofuels in automotive diesel engines: an Indian perspective. Biomass Bioenerg 29:65–72

84. Sun Y, Cheng J (2002) Hydrolysis of lignocellulosic materials for ethanol production: a review. Bioresour Technol 83(1):1–11

85. Talebnia F, Karakashev D, Angelidaki I (2010) Production of bioethanol from wheat straw: an overview on pretreatment, hydrolysis and fermentation. Bioresour Technol 101(13):4744–4753

86. Tolan J (2002) Iogen's process for producing ethanol from cellulosic biomass. Clean Technol Enviro Policy 3:339

87. Tomaz CT, Queiroz JA (1999) Studies on the chromatographic fractionation of *Trichoderma reesei* cellulases by hydrophobic interaction. J Chromatogr A 865(1–2):123–128

88. Tucker MP, Kim KH, Newman MM, Nguyen QA (2003) Effects of temperature and moisture on dilute-acid steam explosion pretreatment of corn stover and cellulase enzyme digestibility. Appl Biochem Biotechnol 10:105–108

89. US Department of Energy Biomass Program (2009).

90. Volke-Spulveda T (1998) Microbial degradation of thermo- oxidized low-density polythene. J Appl Polym Sci 79:1435–1440

91. Wood TM, McCrae SI (1979) Synergism between enzymes involved in the solubilization of native cellulose. Adv Chem Ser 181:181

92. Wooley R, Ruth M, Sheehan J, Ibsen K, Majdeski H, Galvez A (1999) Lignocellulosic biomass to ethanol process design and eco-nomics utilizing co-current dilute acid prehydrolysis and enzymatic hydrolysis. In: Proceedings of current and futuristic scenarios. Technical report NREL/TP-580-26157. Golden, CO (USA), National Renewable Energy Laboratory, p 123

93. Wright JD (1988) Ethanol from lignocellulosics: an overview. Energy Prog 84:71–80

94. Wyman CE, Hinman ND (1990) Ethanol: fundamentals of production from renewable feedstocks and use as transportation fuel. Appl Biochem Biotechnol 24:735–775

95. Wyman C, Dale B, Elander R, Holtzapple M, Ladish M, Lee Y (2005a) Comparative sugar recovery data from laboratory scale application of leading pretreatment technologies to corn stover. Bioresour Technol 96(18):2026–2032

96. Wyman C, Dale B, Elander R, Holtzapple M, Ladish M, Lee YY (2005b) Coordinated development of leading biomass pretreatment technologies. Bioresour Technol 96(18):1959–1966

CHAPTER 6

Metabolic Engineering of Hydrocarbon Biosynthesis for Biofuel Production

Anne M. Ruffing[1]

[1] Sandia National Laboratories, Department of Bioenergy and Defense Technologies, Albuquerque, NM, USA

1. INTRODUCTION

The world's supply of petroleum hydrocarbons, which serve as feedstock for the fuel and chemical industries, is rapidly diminishing to satisfy the global demand for energy and consumer goods. In response to this increasing demand and limited supply, the cost of crude oil has risen to over $100 per barrel in 2012, a 10-fold increase compared to prices in the late 1990s [1]. As fossil fuels are nonrenewable resources, the price of oil is only expected to increase in the future. This unavoidable reality necessitates the development of renewable energy sources in order to maintain the current standard of living. Among the alternative energy options under development, biofuels are anticipated to supplement and eventually replace the petroleum-based fuels that supply the transportation and chemical industries. Currently, first generation biofuels like corn-based ethanol are blended into conventional petroleum fuels, with biofuels supplying 2.7% of the world's transportation fuel in 2010 [2]. It appears that biofuels are on their way to becoming a viable renewable energy source, yet technological and biological advancements are necessary for sustainable and economical biofuel production at the scales necessary to support the world's energy needs.

The current practice of using food crops, like corn or soybean, as feedstocks for biofuel production is not a viable, long-term solution to the energy crisis. In fact, to replace our current petroleum usage with crop-based ethanol production, the entire surface area of land on Earth would be needed for corn production [3]. In addition to this shortcoming, first generation biofuels compete with food production for arable land, require significant nutrient resources (fertilizer and fresh water), and typically have low net energy yields due to the low energy density of the product fuel (i.e. ethanol) and the energy input required to harvest the feedstock and convert it into fuel [4]. Second and third generation biofuels address these limitations. Second generation biofuels use lignocellulosic biomass as the feedstock for fuel production. Lignocellulose, the main component of plant biomass, is the most abundant form of renewable carbon on the Earth, making it an ideal feedstock for renewable hydrocarbon production. The cellulose and hemicellulose components of lignocellulose can be degraded into fermentable sugars to serve as the carbon source for microbial-based fuel production. The carbon feedstocks for both first and second generation biofuels are ultimately derived from carbon dioxide (CO_2) fixation through the process of photosynthesis. Third generation biofuels use photosynthetic microorganisms (i.e. microalgae) to directly convert CO_2 into fuel molecules or fuel precursors, eliminating the biomass intermediate (Figure 1). While both second and third generation biofuels require land, nutrients, and energy investment for harvesting and fuel production, the fuel production yields from these processes are predicted to be capable of meeting energy needs. However, these technologies have yet to be demonstrated at scale and still require further improvement before they can be economically competitive with fossil fuels.

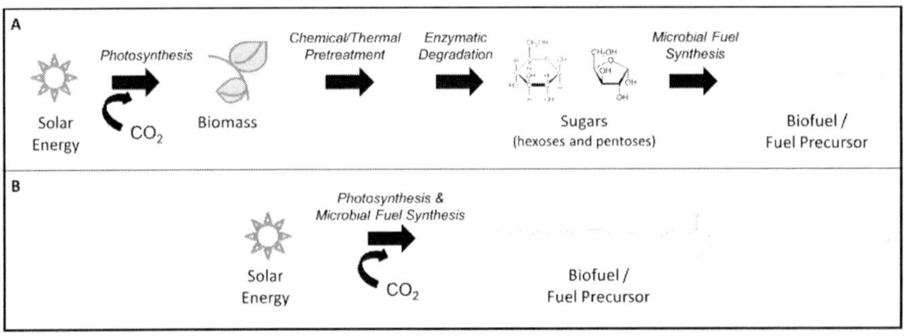

Figure 1. Process steps for (A) second (i.e. lignocellulosic feedstock) and (B) third (i.e. inorganic carbon feedstock) generation biofuels.

Both second and third generation biofuels rely on microbes to convert the carbon feedstock into the desired hydrocarbon fuels. Microorganisms have been identified that are capable of producing a range of fuel molecules and fuel precursors, yet the natural rates of microbial fuel synthesis are typically too low to support industrial-scale production. Metabolic engineering is a powerful tool to improve microbial fuel production, either through engineering the metabolic pathways within the native microorganism to encourage high fuel synthesis or though transferring the fuel production pathway into a model organism for optimization. This chapter will focus on the application of metabolic engineering to increase hydrocarbon fuel production. Within this chapter, hydrocarbon-based fuels are defined to include oxygen-containing fuel molecules with long hydrocarbon chains, such as fatty alcohols and fatty acid ethyl esters (FAEE), in addition to pure hydrocarbons like alkanes, alkenes, and isoprenoid-based molecules: hemiterpene (C5), monoterpenes (C10), and sesquiterpenes (C15). Hydrocarbon-based fuel precursors will also be considered, including free fatty acids (FFAs) and triacylglycerol (TAG). The structures of these hydrocarbon-based fuels and precursors are illustrated inFigure 2. Hydrocarbon-based fuels and precursors can be produced by both second and third generation biofuel processes. Therefore, the first section in this chapter will discuss the metabolic pathways for hydrocarbon fuel production and common metabolic engineering strategies for improving fuel synthesis. Because second and third generation biofuel processes rely on different carbon sources, sugars and CO_2 respectively, the remaining sections will focus on the use of organic carbon (heterotrophy) and inorganic carbon (autotrophy) as feedstocks for biofuel production. This division, based on carbon source, is important from both the biofuel production and metabolic engineering perspectives. The chapter will conclude with a discussion of the future outlook for microbial-based, hydrocarbon fuel synthesis.

2. ENGINEERING HYDROCARBON BIOSYNTHESIS PATHWAYS

The hydrocarbon-based biofuels considered in this chapter (Figure 2) are all derived from two metabolites: fatty acids and isoprenoids. Thus, the two metabolic pathways commonly targeted by metabolic engineering strategies are the fatty acid biosynthesis pathway and the two pathways for isoprenoid production (Figure 3).

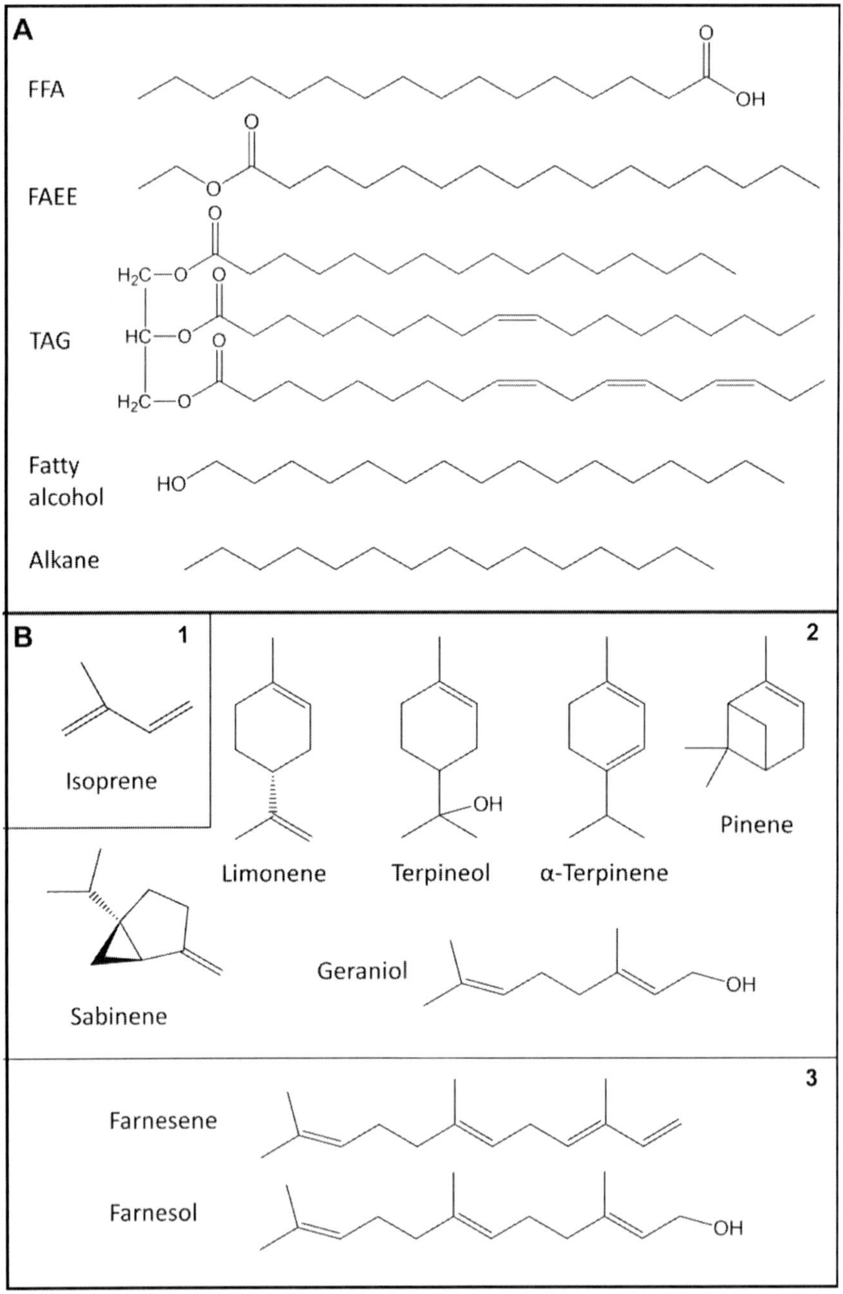

Figure 2. Chemical structures of hydrocarbon-based biofuels and fuel precursors. (A) Fuels derived from fatty acid biosynthesis and (B) fuels derived from isoprenoid biosynthesis, including (1) hemiterpene, (2) monoterpenes, and (3) sesquiterpenes.

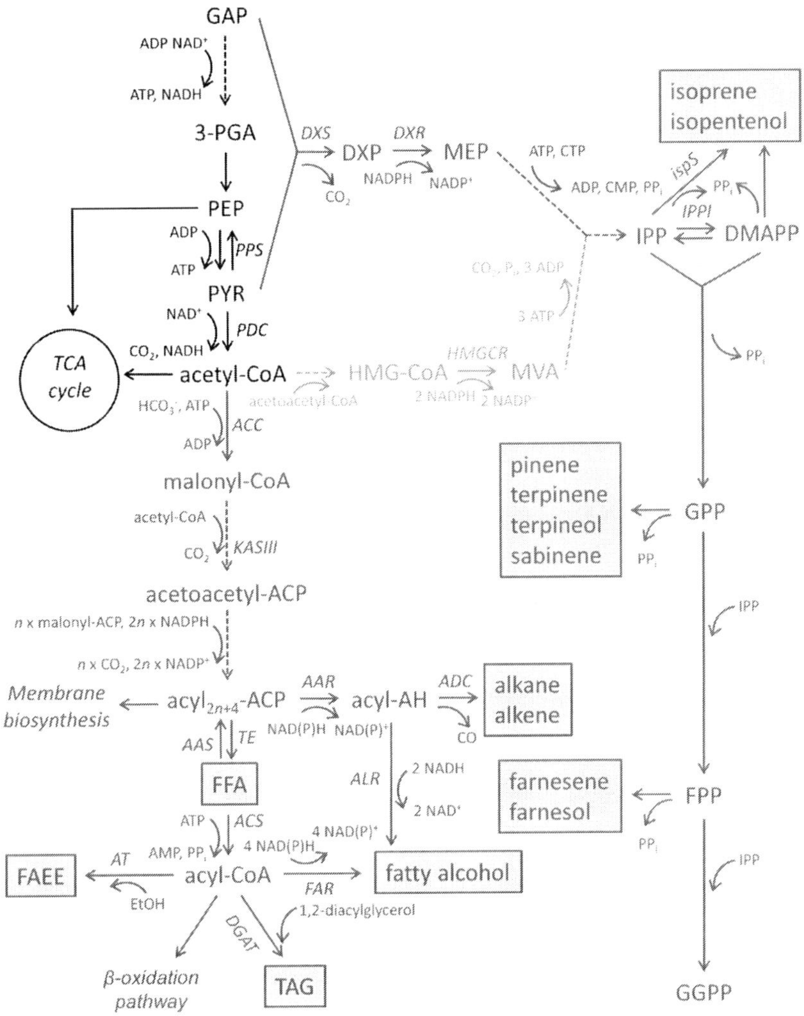

Figure 3. Hydrocarbon biosynthesis pathways for the production of biofuels, with the fatty acid biosynthesis pathway in blue, isoprenoid pathway in red, mevalonate pathway in green, and methylerythritol phosphate pathway in purple. Biofuels and biofuel precursors are highlighted in the colored boxes. Enyzmes are in italics. Solid arrows represent a single enzymatic step, while dashed arrows represent multiple enzymatic steps. Abbreviations for metabolites and enzymes are listed at the end of the chapter.

2.1. Fatty Acid Derived Biofuels

As shown in Figure 3, fatty acid biosynthesis interfaces with the primary metabolism at the acetyl-CoA node. Fatty acid biosynthesis is initiated by the formation of acetoacetyl-ACP, the substrate for fatty acid chain elongation. The conversion of acetyl-CoA to acetoacetyl-ACP includes two key enzymatic steps: (1) the conversion of acetyl-CoA to malonyl-CoA, catalyzed by acetyl-CoA carboxylase (ACC) and (2) the conversion of malonyl-ACP to acetoacetyl-ACP via β-ketoacyl-ACP synthase III (KASIII). These two enzymes are common metabolic engineering targets for improving fatty acid biosynthesis. In fact, ACC has been shown to be a rate-limiting step of fatty acid synthesis in *Escherichia coli*, and overexpression of ACC has been shown to yield more than a 5-fold increase in FFA production [5]. Overexpression of KASIII in *E. coli* also improved FFA synthesis, increasing lipid production by 20-60% [6]. After acetoacetyl-ACP formation, fatty acid chain elongation proceeds by an iterative process, whereby the hydrocarbon chain is elongated in increments of 2 carbons. Once the elongation process terminates, the final acyl-ACP is divided among three possible paths: one leading to membrane biosynthesis, an essential pathway for cell growth, and the other two yielding hydrocarbon fuels or fuel precursors (Figure 3).

To produce biofuels with an even-numbered carbon chain, the acyl-ACP is cleaved by a thioesterase (TE), releasing the FFA. The TE is yet another key target for metabolic engineering. The final fuel properties, including viscosity, cloud point, flash point, oxidative stability, ignition delay, and combustion quality, are largely determined by the hydrocarbon chain length and degree of saturation [7]. Accordingly, numerous TEs have been cloned and characterized, predominantly from plant sources, to control the carbon chain length of the FFAs. Engineering strategies often exploit this collection of TEs to tailor the biofuel product. Favored TEs include a truncated TE (‘*tesA*) from *E. coli* and acyl-ACP TEs from *Umbellularia californica* and *Cuphea hookeriana*, producing FFAs with carbon lengths of 16:0, 12:0, and 10:0 and 8:0, respectively [8-10]. The FFAs themselves can be extracted as fuel precursors and converted into biodiesel (FAMEs or FAEEs) using acid-catalyzed chemical processes [11]. To allow for FFA accumulation, the β-oxidation pathway and free fatty acid recycling are often eliminated by gene knockout of acyl-CoA synthetase (*acs*) and acyl-ACP synthetase (*aas*) [12]. An alternative strategy was recently demonstrated, whereby FFAs were synthesized through an engineered reversal of the β-oxidation cycle [13]. In this strategy, acetyl-CoA is used directly for fatty acid chain elongation, allowing for improved carbon and energy efficiency compared to the fatty acid biosynthesis pathway which requires activation of acetyl-CoA to malonyl-CoA. Engineering a reversed β-oxidation cycle required modification of multiple regulatory mechanisms, knockout of other fermentative pathways, expression of a TE or other fuel producing enzyme, and overexpression of key enzymes in the β-oxidation pathway [13]. While this strategy yielded the highest reported concentration of FFAs in *E. coli* (7 g/L), its application to other host organisms may be restricted by inadequate knowledge of the native regulatory mechanisms.

With an intact *acs*, FFAs can be converted into acyl-CoA, a precursor for other fuel products including the biodiesel precursor, TAG, and fuels such as FAEEs and fatty alcohols (Figure 3). The conversion of acyl-CoA to TAG requires the provision of 1,2-diacylglycerol and a diacylglycerol acyltransferase (DGAT) to catalyze transfer of the acyl chain. While DGAT has been overexpressed to improve TAG production in plants [14], the utility of this strategy still remains to be tested in microorganisms. Most metabolic engineering strategies for microbial TAG synthesis focus on improving the supply of the precursors: FFA and glycerol-3-phosphate (G3P) [15, 16]. Microbial production of FAEEs typically involves heterologous expression of both the pathway for ethanol production and an acyltransferase (AT) [17-19]. Selection of the two genes required for ethanol synthesis, pyruvate decarboxylase (*pdc*) and alcohol dehydrogenase (*adh*), will largely depend on the host organism, but generally, efforts involving prokaryotic hosts such as *E. coli* and cyanobacteria will use *pdc* and *adh* from *Zymomonas mobilis* due to their capacity for high ethanol production [20]. To date, only one AT has been heterologously expressed for FAEE production: the wax synthase gene (*aftA*) from *Acinetobacter baylyi* ADP1 [17-19]. A third biofuel product derived from acyl-CoA is fatty alcohols. The enzymatic conversion of acyl-CoA to a fatty alcohol is dependent upon whether the fatty acyl-CoA reductase (*far*) is of prokaryotic or eukaryotic origin. Most prokaryotic FARs reduce acyl-CoA to a fatty aldehyde, requiring another enzyme, fatty aldehyde reductase (ALR), for conversion to the fatty alcohol product. On the other hand, eukaryotic FARs catalyze the direct conversion of acyl-CoA to fatty alcohol without release of an aldehyde intermediate [21]. Metabolic engineering strategies for fatty alcohol production include: expression of a prokaryotic FAR, *acr1* from *Acinetobacter calcoaceticus* BD413, with reliance on native fatty aldehyde reductases for fatty alcohol synthesis [19]; expression of 5 different eukaryotic FAR homologs from the model plant organism *Arabidopsis thaliana* [22]; and expression of a eukaryotic FAR, *far1* from mouse [23]. The recent discovery of a prokaryotic FAR from *Marinobacter aquaeolei* VT8, capable of catalyzing the direct conversion of acyl-CoA to fatty alcohol, may be a beneficial alternative to the use of eukaryotic FARs for fatty alcohol production in prokaryotic hosts such as *E. coli* and cyanobacteria [24]. An alternative strategy used by Dellomonaco and colleagues identifies surrogates for *far* and *adh* in the native *E. coli* genome based on sequence homology [13]. With the numerous biofuel products derived from acyl-CoA and the natural enzymatic diversity for these conversions, we have only just begun to explore and develop the metabolic engineering tools essential to enable large-scale synthesis.

In addition to oxygen-containing biofuels, acyl-ACP can also be converted into pure hydrocarbon fuels in the form of alkanes and alkenes (Figure 3). In 2010, the discovery of an alkane synthesis pathway in cyanobacteria provided the genetic knowledge necessary for engineering microbial alkane production [25]. The pathway consists of two enzymatic steps: (1) reduction of acyl-ACP to a fatty aldehyde by means of an acyl-ACP reductase (AAR) and (2) decarbonylation of the aldehyde to an alkane or alkene, catalyzed by an aldehyde decarbonylase (ADC). Due to the recent discovery of this pathway, few metabolic engineering strategies have been applied for alkane production. Some strategies focus on improving supply of the acyl-ACP precursor, relying on the native cyanobacterial pathway for alkane synthesis [23], while others have simply transferred the alkane pathway (AAR and ADC) into another host organism [25-27]. With the rapidly growing database of genome sequence information, numerous homologs of AAR and ADC have been identified [26, 27], representing a diverse range of targets for metabolic engineering. Future optimization of the alkane biosynthesis pathway may result in the high alkane yields needed for biofuel production.

2.2. Isoprenoid-Based Biofuels

The chemical composition of petroleum-based fuels: gasoline, diesel, and jet fuel, includes linear, branched, and cyclic alkanes, aromatics, and chemical additives [28]. Isoprenoid-based biofuels have the structural diversity to mimic these petroleum compounds, with up to 50,000 known isoprenoid structures including branched and cyclic hydrocarbons with varying degrees of unsaturation [29, 30]. Isoprenoids reported to be potential fuel candidates include: the hemiterpene (C5) isoprene; monoterpenes (C10): terpinene, pinene, limonene, and sabinene; the sesquiterpene (C15) farnesene, and their associated alcohols: isopentenol, terpineol, geraniol, and farnesol [12, 31]. Two metabolic pathways are capable of producing the isoprenoid building blocks isopentenyl pyrophosphate (IPP) and dimethylallyl diphosphate (DMAPP): the mevalonate (MVA) pathway [32] and the methylerythritol phosphate (MEP) pathway, also known as the 1-deoxy-D-xylulose-5-phosphate (DXP) pathway and the non-mevalonate pathway (Figure 3) [33]. In general, the MVA pathway is found in eukaryotes and archaea while the MEP pathway is utilized by prokaryotes. In agreement with the proposed evolutionary origin of plants, they contain both isoprenoid pathways with the MEP pathway localized in the plastid and the MVA pathway in the cytosol [34]. The MVA and MEP pathways differ with respect to their requirement for carbon, energy, and reducing equivalents; this is illustrated by the net balances for IPP biosynthesis from glyceraldehyde-3-phosphate (GAP):

$$\text{MVA:} 3\,\text{GAP} + 3\,\text{ADP} + 4\,\text{NAD(P)}^{+} + 2\,\text{P}_{i} \rightarrow \text{IPP} + 4\,\text{CO}_2 + 3\,\text{ATP} + 4\,\text{NAD(P)H} \quad (1)$$

$$\text{MEP: } 2\ GAP + ADP + CTP + P_i \rightarrow IPP + CO_2 + ATP + CMP + PP_i \quad (2)$$

Based on these balances, IPP production via the MEP pathway is more efficient at carbon utilization, as only 2 GAPs are required and 1 CO_2 is emitted, compared to 3 GAPs and 4 CO_2 for the MVA pathway. On the other hand, IPP production via the MVA pathway is more energy efficient overall, resulting in ATP generation and yielding a net gain in reducing equivalents (NAD(P)H). These carbon, energy, and reducing equivalent requirements should be considered when designing a metabolic engineering strategy for isoprenoid biosynthesis.

The MVA pathway interfaces with the primary metabolism at the acetyl-CoA node (Figure 3), and it can be divided into two parts: the top, which involves 3 enzymatic steps to convert acetyl-CoA to MVA, and the 3 enzymatic conversions of the bottom portion to produce IPP from MVA. One novel metabolic engineering strategy compared the efficiencies of the top and bottom portions of the MVA pathway in *E. coli* using heterologously expressed pathways from 5 different eukaryotic sources. The most efficient top and bottom portions were combined to maximize the yield of isoprenoid building blocks [35]. Accumulation of an intermediate metabolite, 3-hydroxy-3-methyl-glutaryl-CoA (HMG-CoA), is a known bottleneck in the top MVA pathway, and HMG-CoA was also shown to inhibit cell growth in *E. coli* [36]. Thus, overexpression of the HMG-CoA reductase (HMGCR) increased MEV production and synthesis of subsequent FPP-derivatives in both *E. coli* and *S. cerevisiae* [36-38]. Whole pathway expression and elimination of the HMGCR bottleneck have proven to be successful techniques for enhancing the metabolic throughput of the MVA pathway.

The MEP pathway requires two primary metabolites as precursors: GAP and pyruvate (PYR) (Figure 3). Compared to the 6 enzymatic steps of the MVA pathway, the MEP pathway is comprised of 7 steps. Metabolic engineering strategies for the MEP pathway have primarily focused on the first two enzymatic steps. Overexpression of 1-deoxy-D-xylulose-5-phosphate synthase (*dxs*), catalyzing the conversion of GAP and PYR to 1-deoxy-D-xylulose-5-phosphate (DXP), resulted in 6-10-fold increases in the final isoprenoid product [39, 40]. Targeting the next enzymatic step through overexpression of DXP reductoisomerase (*dxr*) was shown to have little effect on isoprenoid production using the native gene; however, expression of *dxs* and *dxr* from *Bacillus subtilis* improved isoprenoid production 2.3-fold in *E. coli* [41]. The final step of the MEP pathway was also shown to be rate-limiting, as heterologous expression of IPP isomerases (IPPI) enhanced isoprenoid production in *E. coli* [42]. Based on its rate-limiting steps, the MEP pathway is a prime candidate for a push-pull metabolic engineering strategy, whereby overexpression of the first step 'pushes' carbon flux into the MEP pathway and overexpression of the final step 'pulls' the metabolic flux towards the end product. This strategy yielded nearly 2-fold improvements in isoprenoid production in *E. coli* [43, 44]. Lastly, overexpression of the entire MEP pathway can increase isoprenoid biosynthesis. In fact, Leonard and colleagues demonstrated that 5 additional copies of the MEP pathway genes yielded the highest production, while further increasing the gene copy number to 10 produced lower titers [45].

While targeted gene overexpression may alleviate pathway bottlenecks, the pathway is still subject to native regulatory mechanisms which may limit isoprenoid biosynthesis from either the MVA or MEP pathways. A highly successful strategy for overcoming regulatory limitations is overexpression of the non-native isoprenoid pathway. Expression of the MVA pathway from *Saccharomyces cerevisiae* in *E. coli* has enabled higher levels of isoprenoid synthesis compared to engineering the native MEP pathway as the sole isoprenoid pathway [46-50]. The success of this strategy has made it a favorite among metabolic engineers seeking to improve isoprenoid biosynthesis. Farmer and Liao presented a clever approach for regulating the carbon flux into an engineered MEP pathway in *E. coli* [51]. In this work, a native regulatory circuit was used to control the carbon flux into and through the MEP pathway by regulating expression of two key enzymes: phosphoenolpyruvate synthase (PPS) and isopentenyl diphosphate isomerase (IPPI). Under excess carbon flux, expression of *pps* and *idi* was activated using the regulatory circuit, redirecting carbon flux into and through the MEP pathway, yet when the carbon flux was growth limiting, expression of these genes was reduced. This strategy allows for high isoprenoid production without negatively impacting cell growth. As evidence, the regulated pathway improved isoprenoid titers by 50%, while simply placing *pps* and *ippi* under control of strong*tac* promoters resulted in growth inhibition [51]. Native regulatory mechanisms are often obstacles limiting isoprenoid biosynthesis, yet they can also be exploited to optimize the flux balance to support both cell growth and isoprenoid production.

Additional targets for improving isoprenoid-based fuel production include precursor supply, cofactor supply, and optimization of the downstream fuel synthesis pathway. Acetyl-CoA is the precursor for isoprenoid production via the MVA pathway. Overexpression of acetaldehyde dehydrogenase (ALDH) and acetyl-CoA synthetase (ACS), both of which produce acetyl-CoA, increased the acetyl-CoA supply and subsequently isoprenoid biosynthesis in *S. cerevisiae* [52]. On the other hand, the MEP pathway requires two precursors from the glycolysis pathway: PYR and GAP. The supply of these metabolites is complicated by the fact that PYR is derived from GAP, and consequently, the PYR/GAP balance is an important metabolic engineering target. The supply of GAP was shown to be limiting in *E. coli*, as modifying the conversion between PEP and PYR to redistribute the flux toward GAP synthesis increased isoprenoid production [53]. In addition to the carbon precursors, co-factors in the form of energy (ATP, CTP) and reducing equivalents (NADPH) are also required for isoprenoid synthesis. Co-factor supply is often overlooked in strategies for isoprenoid production, yet by improving the availability of NADPH in *S. cerevisiae*, isoprenoid synthesis through the MVA pathway increased by 85% [54]. This result emphasizes the importance of co-factor availability. Despite optimizing production of the isoprenoid building blocks, the downstream efficiency of assembling the final fuel product may still limit the overall yield. Successful strategies for improving downstream efficiency include overexpression of GPP and FPP synthases [47], overexpression and codon optimization of hemiterpene, monoterpene, and sesquiterpene synthases [41, 47, 48], fusion proteins to localize FPP synthesis and its conversion to sesquiterpene [47], and downregulation of competing products like squalene [37, 48]. The optimized production of isoprenoid-based fuels requires strategies to address limitations throughout the metabolic pathway, from precursor and co-factor supply to end product synthesis.

3. INFLUENCE OF FEEDSTOCK ON HYDROCARBON-BASED BIOFUEL PRODUCTION

While hydrocarbon-based biofuel production relies on the biosynthetic pathways discussed in the previous section, the source of feedstock plays an important role in the overall production process. As discussed in the Introduction to this chapter, there are two main feedstocks for biofuel production: lignocellulosic biomass and gaseous CO_2, supporting the production of second and third generation biofuels, respectively (Figure 1). Both processes ultimately rely on CO_2 and sunlight as the carbon and energy source, but the microbial conversion processes are distinctly different between the two feedstocks. Lignocellulosic biomass deconstruction produces organic carbon, mostly in the form of hexoses and pentoses (C5 and C6 sugars); this feedstock requires heterotrophic microorganisms to convert the organic carbon into biofuel. Alternatively, the fixation of inorganic carbon feedstock (CO_2/HCO_3^-) into biofuel is reliant upon autotrophic microbes. The heterotroph vs. autotroph requirement of the respective feedstocks is an important distinction from both the metabolic engineering and biofuel production perspectives. Only a few model microorganisms are capable of both heterotrophy and autotrophy, resulting in different host candidates for second and third generation biofuel production. The feedstock will also influence the metabolic engineering targets, as heterotrophs utilize glycolysis and oxidative phosphorylation pathways for carbon consumption and energy production while oxygen-generating autotrophs utilize the Calvin-Benson-Bassham cycle and photosynthesis under light conditions (Figure 4). This section will discuss the host organisms, engineering strategies, and biofuel production processes specific to each carbon feedstock.

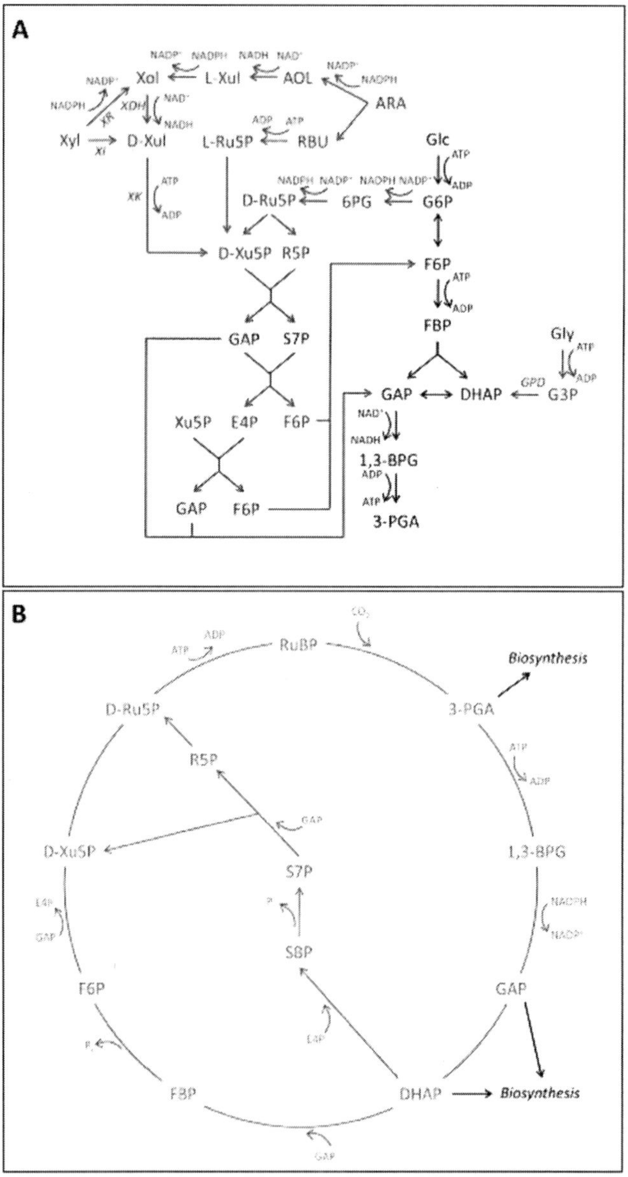

Figure 4. Heterotrophic (A) and autotrophic (B) pathways for carbon utilization, with the Embden-Meyerhof-Parnas (EMP) pathway (glycolysis) in black, the pentose phosphate pathway (PPP) in blue, pentose utilization pathways in red, glycerol metabolism in purple, and the Calvin-Benson-Bassham cycle in green. Abbreviations for metabolites and enzymes are listed at the end of the chapter.

3.1. Hydrocarbon Biofuel Production From Organic Carbon Feedstocks

The release of C5 and C6 sugars from lignocellulosic biomass deconstruction supports the growth of heterotrophic microorganisms and the metabolic conversion of sugars into biofuel. Representative hydrocarbon-based fuel titers produced by engineered, heterotrophic hosts are listed in Table 1. The most common heterotrophic hosts for biofuel production are the model organisms *Escherichia coli* and*Saccharomyces cerevisiae*. These hosts are attractive candidates for fuel production due to their fast growth rates, well-known genetics and regulation, advanced molecular tools for genetic engineering, and established use in the industrial setting. Neither *E. coli* nor *S. cerevisiae* naturally produce significant amounts of hydrocarbon-based fuels, necessitating the application of metabolic engineering techniques. Heterotrophic organisms that naturally produce hydrocarbon-based fuels are also potential hosts for large-scale biofuel production. For example, *Bacillus subtilis* naturally produces higher concentrations of isoprene than other commonly known bacteria like *E. coli* [55]. *B. subtilis* is also a model organism for Gram-positive bacteria with established tools for genetic modification, advancing its appeal as a host for isoprene production. Similarly, heterotrophic algae can produce significant quantities of TAG. This has motivated some preliminary investigation into engineering the model green alga, *Chlamydomonas reinhardtii*, for TAG production [56-58]. While most metabolic engineering efforts have focused on these model heterotrophic hosts, genetic tools can be developed for other organisms with desirable fuel production traits.

Most heterotrophic hosts for biofuel production utilize the Embden-Meyerhof-Parnas (EMP) pathway for sugar catabolism (Figure 4). The EMP pathway has evolved for efficient carbon utilization and is typically not rate-limiting for fuel production. As such, EMP pathway enzymes are not often targeted for genetic manipulation. However, the organic feedstock from lignocellulose deconstruction is comprised of a range of sugars, including hexoses: glucose, mannose, and galactose, and pentoses: xylose and arabinose [79]. A major concern in converting these sugars into fuel is the efficient utilization of all available hexoses and pentoses. While some organisms like *E. coli* can naturally metabolize these different forms of sugar, others, like *S. cerevisiae*, can only utilize specific forms [80]. *S. cerevisiae* does not naturally express pathways for catabolizing pentoses. There are two known pathways for xylose catabolism, both of which have been expressed in *S. cerevisiae* [81-83]. Xylose can be converted into xylulose-5-phosphate (Xu5P), an intermediate in the pentose phosphate pathway (PPP), through expression of a xylose isomerase (XI) and xylulose kinase (XK) [82]. Alternatively, the XI can be replaced by a xylose reductase (XR) and xylitol dehydrogenase (XDH) [81, 82]. Complications in these two xylose utilization pathways include the inhibition of XI by xylitol

(Xol) and the reducing equivalents required by XR and XDH [80]. Successful strategies for engineering xylose utilization in *S. cerevisiae* include expression of a fungal XI from *Piromyces* sp. E2 along with overexpression of the non-oxidative PPP pathway [84] and expression of XR and XDH from the xylose-fermenting yeast *Pichia stipitis* [85]. Two pathways have also been expressed in *S. cerevisiae*for arabinose utilization [86, 87]. The bacterial pathway for arabinose catabolism consists of 3 enzymatic steps, while the fungal pathway involves 5 enzymatic steps, 4 of which require cofactors of NADPH or NAD^+ (Figure 4). Efficient arabinose utilization in *S. cerevisiae* has been achieved through heterologous expression of a bacterial arabinose catabolism pathway along with overexpression of the non-oxidative PPP and evolutionary engineering [88]. While most of these metabolic engineering examples focus on utilizing sugars for fermentation to ethanol, the strategies for engineering carbon utilization can also be applied for hydrocarbon-based fuel production.

Unlike *S. cerevisiae*, *E. coli* can utilize the hexoses and pentoses derived from lignocellulose; however, the carbon catabolite repression (CCR) system in *E. coli* leads to inefficient, diauxic growth [89]. Through CCR, *E. coli* sequentially consumes different sources of organic carbon based on substrate preference, leading to delayed and often incomplete utilization of unpreferred sugars like xylose and arabinose. This translates into lower productivities and yields along with downstream complications due to the presence of unmetabolized sugars [80]. As a result, CCR is often targeted by metabolic engineering to alleviate these undesired effects. A common engineering strategy is to use mutants of the transcriptional activator CRP (cyclic AMP receptor protein) which have been modified to eliminate the allosteric requirement for cAMP, thereby leading to expression of the pentose catabolizing pathways in the presence of the preferred substrate, glucose [90]. The phosphotransferase system (PTS), responsible for the preferential uptake of glucose, has also been deleted to encourage simultaneous utilization of mixed sugars [91]. Lastly, deletion of methylglyoxal synthase was shown to improve the co-metabolism of sugars, ostensibly due to elimination of methylglyoxal, an inhibitor of sugar metabolism [92]. Through modifying the components of CCR, *E. coli* can be engineered to efficiently utilize the organic carbon mixture resulting from lignocellulose degradation.

In addition to the hexoses and pentoses derived from lignocellulosic biomass, glycerol may soon become an inexpensive organic carbon source for fuel production. Glycerol is a byproduct of the conversion of TAG into biodiesel during algal biofuel processing, and thus, large quantities of glycerol may be available for use as an organic carbon source. The main pathway for aerobic glycerol utilization involves a two-step conversion to produce the glycolytic metabolite DHAP [93]. The glycerol utilization pathway is not a common target for metabolic engineering, yet glycerol has been reported as a supplementary carbon source for the production of isoprenoid-based fuels, farnesol and α-farnesene [47, 48]. Future metabolic engineering efforts may focus more on glycerol utilization as the availability of glycerol increases.

Table 1. Hydrocarbon fuels and fuel precursors produced by genetically engineered microorganisms.

Hydrocarbon Fuel/ Fuel Precursor	Concentration Range	Microbial Hosts	References
		Heterotrophic Production	
FFA	0.5 – 7 g/L	*Escherichia coli*	[5, 12, 13, 19, 59, 60]
	0.024 – 0.2 g/L	*Saccharomyces cerevisiae*	[61, 62]
TAG	20 - 32.6% dcw, 0.12 g/L	*Chlamydomonas reinhardtii*	[56-58]
	0.4 – 0.7 g/L	*Saccharomyces cerevisiae*	[63, 64]
FAEE	0.07 – 1.5 g/L	*Escherichia coli*	[18, 19, 65-67]
	N/A	*Saccharomyces cerevisiae*	[17]
Fatty alcohols	0.001 – 1.67 g/L	*Escherichia coli*	[13, 19, 22, 27, 59, 66, 68]
Alkanes/Alkenes	0.042 – 0.32 g/L	*Escherichia coli*	[25, 27]
Other Isoprenoids (lycopene, β-carotene, amorphadiene, levopimaradiene, cubebol)	0.002 – 1 g/L	*Escherichia coli*	[35, 39, 42, 45, 50, 69]
	0.01 g/L	*Saccharomyces cerevisiae*	[37, 52]
Isoprene	0.31 – 0.53 g/L	*Escherichia coli*	[41, 49]
	0.002 g/L	*Bacillus subtilis*	[55]
Farnesol	N/A	*Escherichia coli*	[48]
	0.009 – 0.15 g/L	*Saccharomyces cerevisiae*	[37, 38, 70, 71]
Farnesene	0.38 – 1.1 g/L	*Escherichia coli*	[47, 72]
		Autotrophic Production	
FFA	0.11 - 0.20 g/L	*Synechocystis* sp. PCC 6803	[73-75]
	0.015 - 0.06 g/L	*Synechococcus elongatus* PCC 7942	[73, 75, 76]
	0.051 g/L	*Synechococcus* sp. PCC 7002	[77]
TAG	28.5% dcw	*Chlamydomonas reinhardtii*	[57]
FAEE	0.077 – 0.086 g/L	*Synechococcus* sp. PCC 7002	[77]
Fatty alcohols	200 μg/L	*Synechocystis* sp. PCC 6803	[23]
Alkanes/Alkenes	150 μg/L/OD730	*Synechocystis* sp. PCC 6803	[23]
	0.05 g/L	*Synechococcus* sp. PCC 7002	[26]
	N/A	*Thermosynechococcus elongatus* BP-1	[26]
Isoprene	0.5 mg/L	*Synechocystis* sp. PCC 6803	[78]

Second generation biofuel production still remains to be demonstrated at large scales, yet the overall process is easily integrated with current technologies. Equipment and practices used for agricultural harvesting can be directly applied to harvesting lignocellulosic biomass. In fact, some agricultural processes already produce biomass waste streams that can be utilized for feedstock, such as corn stover. Moreover, commercial fermenters can be employed as bioreactors for the microbial fuel conversion. The main technical difficulties in large-scale lignocellulosic fuel production center on provision of the carbon source. The quantities of biomass needed to support industrial-scale fuel production will require a significant investment of land and nutrient resources, and the supply will be subject to varying climate conditions. A supply chain infrastructure must also be constructed to harvest the biomass and transport it to the production facilities. A primary technical focus of current research on lignocellulosic-derived fuels is the deconstruction of biomass into useable sugars. The thermal, chemical, and enzymatic processes for biomass deconstruction have been a limiting factor for economical second generation biofuel production [94, 95]. As the cost of biomass deconstruction is reduced with new technology, the large-scale production of second generation biofuels will begin to contribute to the world's supply of renewable energy.

3.2. Hydrocarbon Biofuel Production From Inorganic Carbon Feedstocks

The direct conversion of CO_2 into hydrocarbon-based fuels could greatly simplify the overall production process and reduce the cost of biofuel production (Figure 1). The search for autotrophic microorganisms capable of performing this CO_2-to-fuel conversion started in the late 1970's with the U.S. Department of Energy's Aquatic Species Program (ASP) [96]. The ASP isolated and screened over 3,000 species of microalgae from a diverse range of environmental habitats. The program focused mainly on eukaryotic algae, as they naturally produce significant amounts of TAG. During the course of the program, the recombinant DNA technology used in metabolic engineering was developed, yet due to the infancy of this technology, it was not applied to microalgae for fuel applications until near the end of the ASP [15]. With the development of recombinant DNA technology, prokaryotic microalgae (i.e. cyanobacteria, previously known as blue-green algae) were recognized as potential hosts for fuel production, and the successful engineering of cyanobacteria for ethanol production confirmed their potential [97]. Unfortunately, research funding for microalgal fuel production waned as crude oil prices fell in the 1990's. However, in the late 2000's, the cost of crude oil soared, spurring a resurgence of interest in microalgae for fuel production and in the application of metabolic engineering to enhance fuel yields. In general, both eukaryotic microalgae (referred to as algae in the subsequent text) and prokaryotic microalgae (referred to as cyanobacteria in the subsequent text) utilize photosynthesis for energy generation and the Calvin-Benson-Bassham cycle for CO_2 fixation (Figure 4). However, due to the cellular differences between algae and cyanobacteria, the strategies for engineering autotrophic fuel production will be discussed based on this host division.

3.2.1. Engineering Algae For Biofuel Production

Algae are predicted to have first appeared approximately 1.5 billion years ago from an endosymbiotic event in which a eukaryotic cell engulfed a cyanobacterium [98]. The cyanobacterium evolved into the modern day chloroplast, the algal organelle responsible for photosynthesis and carbon fixation. Today, algae can be found in a wide-range of environmental habitats from freshwater lakes and oceans to deserts and even the snow of the Antarctic [99]. Along with this diversity of habitat, algae have evolved diverse cellular physiologies and genetics, resulting in a wealth of potential hosts and genetic sources for engineering fuel production. Many types of algae are currently under consideration for fuel production due to their natural TAG synthesis, including diatoms, green algae, eustigmatophytes, prymnesiophytes, and red algae [100]. While many types of algae produce the fuel precursor TAG, few algal species have well-developed genetic tools available for engineering improved lipid production [101, 102]. Consequently, there are only a few reported examples of engineering algae for biofuel production.

To date, the only genetic mutation shown to improve lipid production in algae is the elimination of starch biosynthesis, a competing carbon sink. The generation of mutants with impaired starch synthesis using random mutagenesis techniques resulted in up to a 10-fold increase in cellular lipid production in*C. reinhardtii* [56-58, 103]. Other targeted metabolic engineering attempts, such as overexpression of ACC in the diatoms *Cyclotella cryptic* and *Navicula saprophila*, failed to improve TAG biosynthesis [15, 96]. In addition to targeting overall TAG production, metabolic engineering strategies have been applied to influence the chemical composition of the fatty acid side chains. By expressing two heterologous TEs, the diatom *Phaeodactylum tricornutum* produced TAG with increased levels of lauric acid (C12:0) and myristic acid (C14:0) [104]. These shorter chain length fatty acids are more desirable for fuel production, and this demonstrates the potential to control the chemical composition of the fuel product and its associated properties with metabolic engineering. While examples of engineering algal TAG production are sparse, many engineering strategies have proven successful at improving the fatty acid content in plants. These strategies include expression of ACC and KASIII involved in fatty acid biosynthesis, expression of G3P dehydrogenase (GPD) for production of the glycerol backbone of TAG, expression of ATs such as DGAT, expression of TEs to release FFAs, and deletion of desaturases to alter the fatty acid composition [105]. Similar strategies may also be successful at improving TAG production in algae.

The metabolic engineering of algae is complicated by several factors. Most algae have a rigid cell wall structure that makes transformation difficult. A common transformation technique uses glass beads (or silicon carbide whiskers) along with a cell wall-deficient algal strain [106]. The cell wall can be removed using enzymatic techniques or through genetic mutation. Alternatively, a microparticle bombardment technique has been applied successfully to transform many different algal species [107]. In this technique, the recombinant DNA is coated onto a metal microparticle and 'shot' into the algal cell using a helium-powered 'gun'. Other transformation methods include electroporation and the traditional plant transformation technique of *Agrobacterium tumefaciens* T-DNA-mediated transfer [107]. Once the recombinant DNA enters the cell, it must integrate into one of 3 algal genomes: nuclear, chloroplast, or mitochondrial (assuming the transformed DNA is not a stably maintained plasmid). DNA has been successfully integrated into the chloroplast genome via homologous recombination, whereby the recombinant gene and marker are flanked by homologous (i.e. matching) regions of the targeted chloroplast DNA, and the recombinant DNA replaces the matching region in the chloroplast. Unfortunately, homologous recombination does not occur in the nuclear genomes of many algae [108], and instead, the recombinant DNA is randomly integrated into the nuclear genome. This complicates metabolic engineering strategies due to the possibility of detrimental genetic effects resulting from the random integration and the lack of a technique for targeted gene knockout. Lastly, algal engineering attempts are often plagued by low gene expression. It has been discovered that many algae, like the model alga *C. reinhardtii*, employ RNA-mediated gene silencing [109]. Numerous strategies have been applied to combat the low gene expression brought about by gene silencing in algae, including codon optimization, the use of 5' and 3' untranslated regions which may participate in regulatory functions, and the inclusion of native intron sequences [108]. Knowledge of the gene silencing mechanisms in algae has led to the development of RNA interference (RNAi) technology for gene knockdown. RNAi exploits the native cellular machinery for gene silencing to reduce the expression of target genes [109]. As we continue to expand our knowledge of algal genetics, the list of engineered algae will rapidly increase. As evidence, the biofuel-relevant alga, *Nannochloropsis* sp., was recently shown to have a high efficiency of homologous recombination in the nuclear genome [110]. This will simplify future strategies for genetic engineering in *Nannochloropsis* sp. Another promising development is the construction of a plasmid for gene expression in *C. reinhardtii* that is now commercially available through Life Technologies [111]. The greater availability and standardization of tools for the genetic manipulation of algae will move algal engineering towards the advanced stages currently seen with other industrial organisms like *E. coli* and *S. cerevisiae*.

3.2.2. Engineering Cyanobacteria For Biofuel Production

Cyanobacteria are predicted to be the first microorganisms to develop the capability of oxygenic photosynthesis, some 2.7 billion years ago [112]. Similar to algae, cyanobacteria have a great range of diverse morphologies, cellular functions, and genetics, presumably due to their long evolutionary history and their diverse habitats. As discussed previously, the ASP initially deemed cyanobacteria unfit for fuel production due to their lack of natural TAG accumulation. Since they are amenable to genetic manipulation, however, cyanobacteria can be engineered to produce a range of biofuel products (Table 1). As prokaryotes, cyanobacteria are subject to the traditional methods employed for engineering other well-developed bacterial hosts like *E. coli*. Some strains of cyanobacteria are even naturally transformable, uptaking exogenous DNA from their environment without the use of cell permeablization techniques [113]. As progenitors of the algal chloroplast, cyanobacteria also integrate DNA into their chromosomes using homologous recombination. Moreover, cyanobacteria do not possess the cellular components for gene silencing. The genetic tools for engineering some model strains of cyanobacteria are well developed and have been used to genetically modify cyanobacteria for several decades [113]. Another advantage of using cyanobacteria as the microbial host for hydrocarbon-based fuel production is that they have been shown to excrete potential fuel precursors such as FFAs [73]. Fuel excretion enables a continuous production process, eliminating the cost associated with harvesting the algal biomass and the time and nutrients needed to repeatedly grow new batches of algae for fuel production. The advantages of straightforward genetic manipulation and fuel excretion make cyanobacteria contenders for large-scale biofuel production despite the disadvantage of low natural lipid yields.

After the initial demonstration of engineering cyanobacteria for ethanol production [97], the production of hydrocarbon-based fuels in engineered cyanobacteria has expanded to include isoprene, FFAs, FAEEs, fatty alcohols, and alkanes/alkenes (Table 1). Isoprene biosynthesis was established in the model cyanobacterium, *Synechocystis* sp. PCC 6803, through expression of the isoprene synthase (*ispS*) from kudzu [78]. Codon optimization of *ispS* and the use of a strong promoter (*psbA2*) increased isoprene production. Engineering strategies targeting the upstream MEP pathway for isoprenoid biosynthesis, as described in Section 2.2 of this chapter, will likely further improve isoprene productivity. The remaining four hydrocarbon-based fuels are all derived from the fatty acid biosynthesis pathway. Common strategies for improving FFA production (see Section 2.1) have proven successful in cyanobacteria [74-76]. Eliminating non-essential, competing pathways such as polyhydroxybutyrate (PHB), cyanophycin, and acetate biosynthesis also improved FFA production [74]. Liu and colleagues engineered a more permeable peptidoglycan layer to improve FFA excretion in *Synechocystis* sp., yet this weakened cell membrane resulted in slower growth rates and may also make the engineered cyanobacterium more susceptible to external predators and toxins that may be present in large-scale cultivations. Initial engineering attempts for fatty alcohol and alkane/alkene production entail expression of a heterologous FAR and overexpression of AAR and ADC, respectively [23, 26]. Alkane/alkene synthesis was also observed with ACC overexpression and native AAR and ADC activities in cyanobacteria [23]. Despite being derived from fatty acids, the synthesis of fatty alcohols and alkanes/alkenes is up to 1000-fold lower than that observed with FFA production (Table 1), suggesting that the conversion of acyl-ACP to the final fuel product is rate limiting. These inaugural proof-of-concept reports illustrate the potential of cyanobacteria as hosts for autotrophic biofuel production, but additional metabolic engineering will be required to achieve the fuel titers necessary for large-scale synthesis.

3.3. Heterotrophic Vs. Autotrophic Biofuel Production

The selection of organic or inorganic carbon feedstock for biofuel production has downstream ramifications on host selection, product yields, and process requirements. Clearly, the feedstock choice will determine whether a heterotrophic or autotrophic host is required, and in turn, this will influence the metabolic engineering strategy. In general, heterotrophic hosts have generated higher fuel titers than autotrophic hosts, with more than 10-fold higher concentrations of FFAs, FAEEs, fatty alcohols, and alkanes/alkenes (Table 1). This does not imply that heterotrophic production is more advantageous than autotrophic production, for the entire production process must be considered (Figure 1). The sugars from lignocellulosic biomass deconstruction (heterotrophic feedstock) have a higher energy content compared to inorganic carbon (autotrophic feedstock). The overall balances for obtaining one molecule of GAP from heterotrophic and autotrophic metabolisms provide evidence for this:

$$\text{Heterotrophic: } \tfrac{1}{2}\,\text{Glc} + \text{ATP} \rightarrow \text{GAP} + \text{ADP} \tag{3}$$

$$\text{Autotrophic: } 3\,\text{CO}_2 + 9\,\text{ATP} + 6\,\text{NADPH} + 5\,\text{H}_2\text{O} \rightarrow \text{GAP} + 9\,\text{ADP} + 6\,\text{NADP}^+ + 8\,\text{P}_i \tag{4}$$

While autotrophic GAP generation requires a significant investment of energy (9 ATP) and reducingequivalents (6 NADPH), heterotrophic GAP production only requires one energy equivalent. However, if a life cycle perspective is considered, the carbon from lignocellulosic feedstocks is ultimately derived from photosynthesis, requiring the same energy and reducing equivalent input as autotrophic microorganisms. Overlooking this fact will bias a direct comparison between heterotrophic and autotrophic fuel production.

One major difference between heterotrophic and autotrophic fuel production is the design considerations for the bioreactor. Heterotrophic microbes, such as *E. coli* and *S. cerevisiae*, are traditional industrial microorganisms with well-established, large-scale cultivation practices and bioreactors. On the other hand, autotrophic hosts like algae and cyanobacteria require light as the energy source to drive photosynthesis and inorganic carbon fixation. This can have a dramatic effect on bioreactor design. Transparent materials can be used with traditional bioreactor designs to allow for light penetration. Light availability, however, will ultimately limit the cell densities of photosynthetic microalgae, and the surface area of light exposure with traditional bioreactor designs is not optimal. Some have proposed to use fiber-optics within the liquid culture to improve light availability [114], but a costly solution such as this is not feasible for a low-value, commodity product like fuel. A wide-range of photobioreactor (PBR) designs have been proposed [115], yet generally, PBRs are characterized by the use of transparent materials, high surface area to volume ratios, and a relatively short pathlength for light. Other PBR design factors include a mechanism for air/CO_2 delivery, dissipation of radiative heat, and removal of inhibitory O_2 [115]. Due to the low value of fuel products, PBRs for fuel synthesis favor low-tech designs and inexpensive materials to reduce both capital and operating costs. In fact, NASA has proposed to float plastic bags of algal cultures in wastewater to allow for nutrient exchange [116]. Alternatively, open pond systems, traditionally a raceway configuration with a paddle-wheel for mixing, have proven successful for cultivating microalgae at scale [117]. Unlike PBRs, ponds are open to the environment, allowing for evaporative water loss and pond crash due to contamination by predators and competitors. However, the low capital cost of an open pond system makes this design a contender for fuel production. Clearly, the large-scale cultivation techniques for autotrophic fuel production still require additional development and optimization compared to heterotrophic cultivation.

4. OTHER METABOLIC ENGINEERING STRATEGIES FOR INDUSTRIAL PRODUCTION OF HYDROCARBON FUELS

In addition to improving hydrocarbon-based fuel synthesis, metabolic engineering strategies can also be applied to address other factors affecting large-scale production. Two main issues will be addressed in this section: product toxicity and industrial strain robustness.

Product toxicity was shown to be a limiting factor in the production of first generation biofuels like ethanol. Since the interest in hydrocarbon-based fuels has developed only during the past decade, the toxicities of these fuels have not been fully explored, particularly with respect to autotrophic hosts. Fortunately, interest in hydrocarbon inhibition of microbial growth dates back almost a century [118], and we can capitalize on this wealth of information to engineer improved product tolerance in microbial hosts. Most fatty acid derived fuel molecules have shown some antimicrobial activity. FFAs, with a diverse range of carbon chain lengths and degrees of unsaturation, impart inhibitory effects on organisms including algae, Gram-negative and Gram-positive bacteria, fungi, protozoans, and various cell types of multicellular organisms [119]. Medium chain fatty alcohols such as pentanol, hexanol, heptanol, and octanol inhibited the biological activity of several algal and cyanobacterial strains, including fuel-relevant hosts *C. reinhardtii* and *Dunaliella salina* [120]. Interestingly, long-chain fatty alcohols (>C14) did not exhibit inhibitory effects on yeasts, suggesting that targeting longer chain fatty alcohols may eliminate the toxicity concern [121]. Similarly, medium-length alkanes (hexane, heptane, and isooctane) were toxic to microalgae while long-chain alkanes (C12-C16) elicited no effect [120,121]. Microbial TAG and FAEE toxicities have not been reported. However, the phospholipid membrane surrounding algal TAGs may mask potential inhibitory effects, and FAEE production has been linked to the toxic effects of alcohol consumption in humans [122]. Isoprenoid-based fuel molecules have also illustrated inhibitory effects. Cyclic terpenes, such as pinene and limonene (Figure 2), inhibited the growth of bacteria and *S. cerevisiae* [123, 124], while branched isoprenoids, such as farnesyl hexanoate and geranyl acetate, were shown to be toxic to *E. coli* [125]. In fact, *E. coli*'s tolerance to isoprenoid-derived biodiesels and bioaviation fuels only ranged from 0.025 – 1% (v/v) [125]. Based on these previous studies, product toxicity is a major limiting factor and should be integrated into the metabolic engineering strategy.

A variety of strategies can be adopted to address product toxicity. The easiest way to avoid complications from product toxicity is to select non-toxic fuel targets. Toxicity studies can be conducted for each potential host organism, and generally, fatty alcohols longer than C14, alkanes longer than C9, and alkenes longer than C12 have shown minimal microbial inhibition

[120, 121]. Alternatively, metabolic engineering techniques can be applied to allow for a more diverse range of hydrocarbon fuel targets. Many cellular modifications have been shown to improve microbial solvent tolerance: changes in membrane lipid composition; altered enzymatic activities of membrane repair and energy transduction enzymes; solvent expulsion via efflux pump activity; and cellular stress responses including heat shock, phage shock, and general stress responses [118, 125, 126]. These natural mechanisms offer a range of engineering targets: expression of a cis-trans isomerase to alter lipid composition; overexpression of enzymes involved in membrane repair and energy transduction; expression of efflux pumps such as *tolC*, *mar*, *rob*, *soxS*, and *acrAB*; and overexpression of stress-induced enzymes such as phage shock protein, heat shock proteins, catalases, and superoxide dismutases [125, 126]. While few metabolic engineering efforts have focused on enhancing product tolerance, a recent study explored improving hydrocarbon-based fuel tolerance in *E. coli* by testing a library of 43 efflux pumps [127]. This work identified efflux pumps that improved tolerance to five potential isoprenoid derived fuels. This preliminary success at engineering solvent tolerance should inspire additional efforts to improve the microbial production of both fatty acid and isoprenoid derived fuels.

In addition to product tolerance, other host traits are desirable for industrial biofuel production, particularly for autotrophic microorganisms. As discussed in the previous section, light availability is often a growth limiting factor in microalgal cultures. Microalgae construct light harvesting complexes (LHC) to capture the available light for use in photosynthesis, and natural species actually absorb more light than is needed for photosynthesis under light intensities > 400 μmol photons m^{-2} s^{-1} [128]. As the sun can generate light intensities as high as 2,000 μmol photons m^{-2} s^{-1} during peak hours, it is estimated that as much as 80% of light absorbed by microalgae is 'wasted' as re-emitted fluorescence and heat [129]. In addition to this loss of energy, the excess energy can also cause cellular damage, known as photoinhibition [128]. In nature, this over-absorption of light will give the microalga a competitive advantage, but from a biofuel production perspective, this excess light harvesting will lead to lower culture cell densities and therefore lower biofuel productivities. Thus, there have been many attempts to engineer microalgae to absorb only the amount of light needed for photosynthesis. These efforts target genes of the light harvesting antenna complexes. Most LHC mutants were generated using random mutagenesis techniques including chemical, UV, and transposon mutagenesis [128, 130-134]. Many of these studies focus on the model alga *C. reinhardtii*, but other microalgal species, such as the diatom *Cyclotella* sp. and the cyanobacterium *Synechocystis* sp., have been mutated to reduce the size of their photosynthetic antennae [130, 133]. Several recent works have applied RNAi technology in *C. reinhardtii* to reduce the expression of targeted LHC genes in a more controlled manner [129]. In general, the antenna mutants have shown improved photosynthetic quantum yields, reduced photoinhibition, enhanced productivity under high light conditions, and increased light penetration within the culture [128, 129, 131-134]. While these results are promising, several questions remain to be addressed: Are the photosynthetic antenna mutants genetically stable, or will they revert back to their more competitive and less efficient forms over time? And are these mutants less fit and therefore more susceptible to predators and competitors in open pond systems?

Open pond systems are subject to a variety of changing environmental conditions, and as such, the optimal autotrophic host will have the necessary cellular mechanisms to adapt to these changing conditions. Desirable host traits may include temperature tolerance, salt tolerance, and resistance to predators. Open ponds are exposed to both daily and seasonal temperature fluctuations which often exceed the normal temperature ranges for optimal cell growth and may even cause cell death. Engineering efforts have successfully altered the temperature tolerance of cyanobacteria though either gene knockout or heterologous overexpression of desaturases which influence the viscosity of both the cell and photosynthetic membranes [135]. Alternatively, microalgae with different temperature optima can be rotated seasonally in the open ponds, similar to seasonal crop rotations in agricultural practices. As mentioned previously, open pond systems are complicated by evaporative water loss, particularly for the sunny, arid regions that are ideal for microalgal biofuel production. Evaporation can lead to fluctuations in the salt concentration within the pond culture, and many have proposed to utilize marine or brackish water sources to reduce the cost associated with freshwater systems. Moreover, high salt and saturated salt systems will have lower evaporative water loss compared to freshwater cultures. Naturally salt-tolerant microalgae, such as those isolated from marine or even hypersaline environments, may be selected as host for biofuel production, or efficient fuel-producing hosts can be engineered for increased salt tolerance. For example, the cyanobacterium *Synechococcus elongatus*PCC 7942, modified with expression of a Δ12 acyl-lipid desaturase (*desA*), showed improved resistance to salt and osmotic stress compared to the wildtype [136]. Lastly, pond crash due to microalgal predators like rotifers and chytrids is a major problem for open pond biofuel production systems. While there have not been any reported attempts at engineering predator-resistant microalgae, there have been reports of natural defense mechanisms such as palmelloid formation by *C. reinhardtii*, which produces non-motile cell aggregates that are simply too large to be consumed by grazing rotifers [137]. Once the genetic mechanism responsible for palmelloid formation is deciphered, it may be possible to transfer this resistance mechanism to other microalgae using genetic engineering techniques. When devising a metabolic engineering strategy for biofuel production, it is essential to consider the entire genomic landscape and the natural diversity of genetically-driven traits to design the optimal host for the specific industrial constraints.

5. CONCLUSIONS AND FUTURE OUTLOOK

The microbial production of drop-in replacement fuels faces unprecedented challenges. The sheer quantity of hydrocarbon product required to meet the world's ever increasing demand for energy dwarfs the supply of any current microbially synthesized product. Moreover, both second (lignocellulosic feedstock) and third (microalgal feedstock) generation biofuels ultimately rely on sunlight and photosynthesis to supply the energy and carbon feedstocks necessary for production. This requires the development of new technology and infrastructure to facilitate the construction of this new supply chain. Finally, the low value of the final fuel product places additional financial restrictions on the development of large-scale biofuel production processes. For example, previous reports include the addition of exogenous metabolic precursors like mevalonate for isoprenoid production or FFA for FAEE biosynthesis [18, 50]. While these exogenous metabolites boost production of the desired hydrocarbon-based product, this practice is too expensive for large-scale biofuel applications. These challenges currently limit the industrial production of second and third generation biofuels.

Fortunately, new biological and technological tools are rapidly being developed and applied to overcome the obstacles in biofuel production. In addition to the metabolic engineering strategies previously described in this chapter, new global strategies are being applied to engineer microbes for biofuel production. With the affordability of next-generation DNA sequencing technologies, new microbial genomes are being reported at an unprecedented rate, and this information can be used to generate metabolic models for biofuel-producing hosts. In turn, these models can be leveraged to analyze proposed metabolic engineering strategies *in silico*, reducing the number of costly and time-intensive strain constructions and experiments. This technique was shown to be successful at increasing lycopene production, an isoprenoid derivative, in *E. coli* [69, 138]. The advancement of synthetic DNA technology enables new engineering approaches such as multiplex automated genome engineering (MAGE) [139]. In MAGE, synthetic oligomers, consisting of degenerate DNA sequences flanked by regions homologous to the target sequences, are simultaneously transformed into *E. coli*, and the modified strains are screened for improvements. MAGE was used to target ribosome binding sites, for optimization of protein translation, and to inactivate genes by inserting nonsense mutations; this technique can also be applied to target promoters for improved gene transcription and enzyme active sites for enhanced activities. The technique does have some limitations, however. MAGE will likely require modification of the host organism to allow for efficient integration of the single-stranded oligonucleotides, and a high-throughput screening method is essential for screening the billions of genetic variants that are generated with MAGE. Global or systems-level technologies can also be applied to advance our fundamental understanding of genetic and regulatory mechanisms within a microbial host; this is vital to host development of non-model organisms and newly isolated strains. Omics technologies including genomics, transcriptomics, metabolomics, and proteomics provide global insight at the cellular level, which can be compared across different conditions or time points to identify the native mechanisms that control the cell metabolism. Integration of omics data can identify bottlenecks at the transcriptional, translational, and protein levels, and as such, can be applied to inform the metabolic engineering strategy for biofuel production [34]. Systems-level tools for engineering microbial hosts, including metabolic modeling, MAGE, and omics technologies, will be integral to the successful development of hosts for biofuel production.

Commercial interest in the production of second and third generation biofuels has developed rapidly in the past decade. As evidence of this, there has been a flurry of activity in patent applications regarding microbial hydrocarbon production. Companies invested in heterotrophic hydrocarbon-based fuel production include LS9 [27, 59, 65, 66, 140, 141] and Amyris Biotechnologies [72, 142], which focus mainly on *E. coli* as the host, and Solazyme [143, 144], which initially focused on fuels derived from algae but has since moved toward more high-value markets, such as cosmetics and nutraceuticals. Most companies interested in algae and cyanobacteria are focused on autotrophically-produced hydrocarbon fuels. Notable companies in this industry include Sapphire Energy [145, 146], Joule Unlimited [26,77, 147], and Synthetic Genomics [68, 75]. The hydrocarbon-based fuels targeted by these companies span the entire gamut of fatty acid and isoprenoid derived fuel products. Despite this commercial interest, hydrocarbon biofuel production still remains to be demonstrated at scale and in a sustainable manner.

This chapter has described the challenges in microbial hydrocarbon production and presented metabolic engineering strategies to resolve these issues. As is evident from this discussion, microbial-based fuel production is only in the initial stages of exploration, and additional research and innovation is necessary to enable large-scale biofuel production. New metabolic engineering tools and techniques are currently being developed for engineering untraditional hosts like eukaryotic algae and cyanobacteria, and as our understanding of these new hosts matures, significant improvement in hydrocarbon yields is anticipated.

NAN. ACKNOWLEDGEMENTS

This work was supported by the Harry S. Truman Fellowship in National Security Science and Engineering and the Laboratory Directed Research and Development program at Sandia National Laboratories. Sandia National Laboratories is a multi-program laboratory managed and operated by Sandia Corporation, a wholly owned subsidiary of Lockheed Martin Corporation, for the U.S. Department of Energy's National Nuclear Security Administration under contract DE-AC04-94AL85000.

REFERENCES

1. Williams JL. (2011) Oil Price History and Analysis. WTRG Economics. Available: http://www.wtrg.com/prices.htm. Accessed 2012 April 16.

2. Sawin JL, Martinot E, Barnes D, McCrone A, Roussell J, Sims R, et al. (2011) Renewables 2011 Global Status Report. Renewable Energy Policy Network for the 21st Century.

3. Rittmann BE. (2008) Opportunities for Renewable Bioenergy Using Microorganisms. Biotechnol. Bioeng. 100(2):203-212.

4. Christi Y. (2008) Biodiesel from Microalgae Beats Bioethanol. Trends Biotechnol. 26(3):126-131.

5. Davis MS, Solbiati J, Cronan JE. (2000) Overproduction of Acetyl-CoA Carboxylase Activity Increases the Rate of Fatty Acid Biosynthesis in*Escherichia coli*. J. Biol. Chem. 275(37):28593-28598.

6. Lee S, Jeon E, Yun H, Lee J. (2011) Improvement of Fatty Acid Biosynthesis by Engineered Recombinant *Escherichia coli*. Biotech. Biopro. Eng. 16(4):706-713.

7. Ramos MJ, Fernández CM, Casas A, Rodríguez L, Pérez Á. (2009) Influence of Fatty Acid Composition of Raw Materials on Biodiesel Properties. Bioresour. Technol. 100(1):261-268.

8. Dehesh K, Jones A, Knutzon DS, Voelker TA. (1996) Production of High Levels of 8:0 and 10:0 Fatty Acids in Transgenic Canola by Overexpression of Ch FatB2, a Thioesterase cDNA from *Cuphea hookeriana*. Plant J. 9(2):167-172.

9. Jiang P, Cronan JE. (1994) Inhibition of Fatty Acid Synthesis in *Escherichia coli* in the Absence of Phospholipid Synthesis and Release of Inhibition by Thioesterase Action. J. Bacteriol. 176(10):2814-2821.

10. Voelker TA, Davies HM. (1994) Alteration of the Specificity and Regulation of Fatty Acid Synthesis of *Escherichia coli* by Expression of a Plant Medium-Chain Acyl-Acyl Carrier Protein Thioesterase. J. Bacteriol. 176(23):7320-7327.

11. Lotero E, Liu Y, Lopez DE, Suwannakarn K, Bruce DA, Goodwin JG. (2005) Synthesis of Biodiesel via Acid Catalysis. Ind. Eng. Chem. Res. 44(14):5353-5363.

12. Liu T, Khosla C. (2010) Genetic Engineering of *Escherichia coli* for Biofuel Production. Annu. Rev. Genet. 44(1):53-69.

13. Dellomonaco C, Clomburg JM, Miller EN, Gonzalez R. (2011) Engineered Reversal of the β-Oxidation Cycle for the Synthesis of Fuels and Chemicals. Nature. 476(7360):355-359.

14. Jako C, Kumar A, Wei Y, Zou J, Barton DL, Giblin EM, et al. (2001) Seed-Specific Over-Expression of an *Arabidopsis* cDNA Encoding a Diacylglycerol Acyltransferase Enhances Seed Oil Content and Seed Weight. Plant Physiol. 126(2):861-874.

15. Dunahay T, Jarvis E, Dais S, Roessler P. (1996) Manipulation of Microalgal Lipid Production Using Genetic Engineering. Appl. Biochem. Biotechnol. 57-58(1):223-231.

16. Nguyen HTT, Dieterich A, Athenstaedt K, Truong NH, Stahl U, Nevoigt E. (2004) Engineering of *Saccharomyces cerevisiae* for the Production of L-Glycerol 3-Phosphate. Metab. Eng. 6(2):155-163.

17. Kalscheuer R, Luftmann H, Steinbüchel A. (2004) Synthesis of Novel Lipids in *Saccharomyces cerevisiae* by Heterologous Expression of an Unspecific Bacterial Acyltransferase. Appl. Environ. Microbiol. 70(12):7119-7125.

18. Kalscheuer R, Stölting T, Steinbüchel A. (2006) Microdiesel: *Escherichia coli* Engineered for Fuel Production. Microbiol. 152(9):2529-2536.

19. Steen EJ, Kang Y, Bokinsky G, Hu Z, Schirmer A, McClure A, et al. (2010) Microbial Production of Fatty-Acid-Derived Fuels and Chemicals from Plant Biomass. Nature. 463(7280):559-562.

20. Ohta K, Beall DS, Mejia JP, Shanmugam KT, Ingram LO. (1991) Genetic Improvement of *Escherichia coli* for Ethanol Production: Chromosomal Integration of *Zymomonas mobilis* Genes Encoding Pyruvate Decarboxylase and Alcohol Dehydrogenase II. Appl. Environ. Microbiol. 57(4):893-900.

21. Hofvander P, Doan TTP, Hamberg M. (2011) A Prokaryotic Acyl-CoA Reductase Performing Reduction of Fatty Acyl-CoA to Fatty Alcohol. FEBS Lett. 585(22):3538-3543.

22. Doan TTP, Carlsson AS, Hamberg M, Bülow L, Stymne S, Olsson P. (2009) Functional Expression of Five *Arabidopsis* Fatty Acyl-CoA Reductase Genes in *Escherichia coli*. J. Plant Physiol. 166(8):787-796.

23. Tan X, Yao L, Gao Q, Wang W, Qi F, Lu X. (2011) Photosynthesis Driven Conversion of Carbon Dioxide to Fatty Alcohols and Hydrocarbons in Cyanobacteria. Metab. Eng. 13(2):169-176.

24. Willis RM, Wahlen BD, Seefeldt LC, Barney BM. (2011) Characterization of a Fatty Acyl-CoA Reductase from *Marinobacter aquaeolei* VT8: A Bacterial Enzyme Catalyzing the Reduction of Fatty Acyl-CoA to Fatty Alcohol. Biochem. 50(48):10550-10558.

25. Schirmer A, Rude MA, Li X, Popova E, del Cardayre SB. (2010) Microbial Biosynthesis of Alkanes. Science. 329(5991):559-562.

26. Reppas NB, Ridley CP, inventors; Joule Unlimited, Inc., assignee. (2010) Methods and Compositions for the Recombinant Biosynthesis of *n*-Alkanes. patent US 7794969.

27. Schirmer A, Rude MA, Brubaker S, inventors; LS9, Inc., assignee. (2010) Methods and Compositions for Producing Hydrocarbons. patent US 2010/0249470.

28. Lee SK, Chou H, Ham TS, Lee TS, Keasling JD. (2008) Metabolic Engineering of Microorganisms for Biofuels Production: From Bugs to Synthetic Biology to Fuels. Curr. Opin. Biotechnol. 19(6):556-563.

29. Chandran SS, Kealey JT, Reeves CD. (2011) Microbial Production of Isoprenoids. Process Biochem. 46(9):1703-1710.

30. Fortman JL, Chhabra S, Mukhopadhyay A, Chou H, Lee TS, Steen E, et al. (2008) Biofuel Alternatives to Ethanol: Pumping the Microbial Well. Trends Biotechnol. 26(7):375-381.

31. Rude MA, Schirmer A. (2009) New Microbial Fuels: A Biotech Perspective. Curr. Opin. Microbiol. 12(3):274-281.

32. Rilling H, K B. (1959) On the Mechanism of Sqalene Biogenesis from Mevalonic Acid. J. Biol. Chem. 234(6):1424-1432.

33. Rohmer M, Knani M, Simonin P, Sutter B, Sahm H. (1993) Isoprenoid Biosynthesis in Bacteria: A Novel Pathway for the Early Steps Leading to Isopentenyl Diphosphate. Biochem. J. 295:517-524.

34. Muntendam R, Melillo E, Ryden A, Kayser O. (2009) Perspectives and Limits of Engineering the Isoprenoid Metabolism in Heterologous Hosts. Appl. Microbiol. Biotechnol. 84(6):1003-1019.

35. Yoon S-H, Lee S-H, Das A, Ryu H-K, Jang H-J, Kim J-Y, et al. (2009) Combinatorial Expression of Bacterial Whole Mevalonate Pathway for the Production of β-Carotene in *E. coli.* J. Biotechnol. 140(3–4):218-226.

36. Pitera DJ, Paddon CJ, Newman JD, Keasling JD. (2007) Balancing a Heterologous Mevalonate Pathway for Improved Isoprenoid Production in*Escherichia coli.* Metab. Eng. 9(2):193-207.

37. Asadollahi MA, Maury J, Schalk M, Clark A, Nielsen J. (2010) Enhancement of Farnesyl Diphosphate Pool as Direct Precursor of Sesquiterpenes Through Metabolic Engineering of the Mevalonate Pathway in *Saccharomyces cerevisiae.* Biotechnol. Bioeng. 106(1):86-96.

38. Ohto C, Muramatsu M, Obata S, Sakuradani E, Shimizu S. (2009) Overexpression of the Gene Encoding HMG-CoA Reductase in *Saccharomyces cerevisiae* for Production of Prenyl Alcohols. Appl. Microbiol. Biotechnol. 82(5):837-845.

39. Kim S-W, Keasling JD. (2001) Metabolic Engineering of the Nonmevalonate Isopentenyl Diphosphate Synthesis Pathway in *Escherichia coli*Enhances Lycopene Production. Biotechnol. Bioeng. 72(4):408-415.

40. Matthews PD, Wurtzel ET. (2000) Metabolic Engineering of Carotenoid Accumulation in *Escherichia coli* by Modulation of the Isoprenoid Precursor Pool with Expression of Deoxyxylulose Phosphate Synthase. Appl. Microbiol. Biotechnol. 53(4):396-400.

41. Zhao Y, Yang J, Qin B, Li Y, Sun Y, Su S, et al. (2011) Biosynthesis of Isoprene in *Escherichia coli* via Methylerythritol Phosphate (MEP) Pathway. Appl. Microbiol. Biotechnol. 90(6):1915-1922.

42. Kajiwara S, Fraser P, Kondo K, Misawa N. (1997) Expression of an Exogenous Isopentenyl Diphosphate Isomerase Gene Enhances Isoprenoid Biosynthesis in *Escherichia coli.* Biochem. J. 324:421-426.

43. Ghimire GP, Lee HC, Sohng JK. (2009) Improved Squalene Production via Modulation of the Methylerythritol 4-Phosphate Pathway and Heterologous Expression of Genes from *Streptomyces peucetius* ATCC 27952 in *Escherichia coli.* Appl. Environ. Microbiol. 75(22):7291-7293.

44. Morrone D, Lowry L, Determan M, Hershey D, Xu M, Peters R. (2010) Increasing Diterpene Yield with a Modular Metabolic Engineering System in*E. coli* Comparison of MEV and MEP Isoprenoid Precursor Pathway Engineering. Appl. Microbiol. Biotechnol. 85(6):1893-1906.

45. Leonard E, Ajikumar PK, Thayer K, Xiao W-H, Mo JD, Tidor B, et al. (2010) Combining Metabolic and Protein Engineering of a Terpenoid Biosynthetic Pathway for Overproduction and Selectivity Control. Proceedings of the National Academy of Sciences.

46. Martin VJJ, Pitera DJ, Withers ST, Newman JD, Keasling JD. (2003) Engineering a Mevalonate Pathway in *Escherichia coli* for Production of Terpenoids. Nat Biotech. 21(7):796-802.

47. Wang C, Yoon S-H, Jang H-J, Chung Y-R, Kim J-Y, Choi E-S, et al. (2011) Metabolic Engineering of *Escherichia coli* for α-Farnesene Production. Metab. Eng. 13(6):648-655.

48. Wang C, Yoon S-H, Shah AA, Chung Y-R, Kim J-Y, Choi E-S, et al. (2010) Farnesol Production from *Escherichia coli* by Harnessing the Exogenous Mevalonate Pathway. Biotechnol. Bioeng. 107(3):421-429.

49. Yang J, Zhao G, Sun Y, Zheng Y, Jiang X, Liu W, et al. (2012) Bio-Isoprene Production Using Exogenous MVA Pathway and Isoprene Synthase in*Escherichia coli*. Bioresour. Technol. 104(0):642-647.

50. Yoon S-H, Lee Y-M, Kim J-E, Lee S-H, Lee J-H, Kim J-Y, et al. (2006) Enhanced Lycopene Production in *Escherichia coli* Engineered to Synthesize Isopentenyl Diphosphate and Dimethylallyl Diphosphate from Mevalonate. Biotechnol. Bioeng. 94(6):1025-1032.

51. Farmer WR, Liao JC. (2000) Improving Lycopene Production in *Escherichia coli* by Engineering Metabolic Control. Nat Biotech. 18(5):533-537.

52. Shiba Y, Paradise EM, Kirby J, Ro D-K, Keasling JD. (2007) Engineering of the Pyruvate Dehydrogenase Bypass in *Saccharomyces cerevisiae* for High-Level Production of Isoprenoids. Metab. Eng. 9(2):160-168.

53. Farmer WR, Liao JC. (2001) Precursor Balancing for Metabolic Engineering of Lycopene Production in *Escherichia coli*. Biotechnol. Prog. 17(1):57-61.

54. Asadollahi MA, Maury J, Patil KR, Schalk M, Clark A, Nielsen J. (2009) Enhancing Sesquiterpene Production in *Saccharomyces cerevisiae* Through in Silico Driven Metabolic Engineering. Metab. Eng. 11(6):328-334.

55. Xue J, Ahring BK. (2011) Enhancing Isoprene Production by Genetic Modification of the 1-Deoxy-d-Xylulose-5-Phosphate Pathway in *Bacillus subtilis*. Appl. Environ. Microbiol. 77(7):2399-2405.

56. Li Y, Han D, Hu G, Dauvillee D, Sommerfeld M, Ball S, et al. (2010) *Chlamydomonas* Starchless Mutant Defective in ADP-Glucose Pyrophosphorylase Hyper-Accumulates Triacylglycerol. Metab. Eng. 12(4):387-391.

57. Li Y, Han D, Hu G, Sommerfeld M, Hu Q. (2010) Inhibition of Starch Synthesis Results in Overproduction of Lipids in *Chlamydomonas reinhardtii*. Biotechnol. Bioeng. 107(2):258-268.

58. Work VH, Radakovits R, Jinkerson RE, Meuser JE, Elliott LG, Vinyard DJ, et al. (2010) Increased Lipid Accumulation in the *Chlamydomonas reinhardtii* sta7-10 Starchless Isoamylase Mutant and Increased Carbohydrate Synthesis in Complemented Strains. Eukaryot. Cell. 9(8):1251-1261.

59. Hu Z, Valle F, inventors; LS9, Inc, assignee. (2011) Enhanced Production of Fatty Acid Derivatives. patent US 2011/0256599.

60. Lu X, Vora H, Khosla C. (2008) Overproduction of Free Fatty Acids in *E. coli*: Implications for Biodiesel Production. Metab. Eng. 10(6):333-339.

61. Michinaka Y, Shimauchi T, Aki T, Nakajima T, Kawamoto S, Shigeta S, et al. (2003) Extracellular Secretion of Free Fatty Acids by Disruption of a Fatty Acyl-CoA Synthetase Gene in *Saccharomyces cerevisiae*. J. Biosci. Bioeng. 95(5):435-440.

62. Scharnewski M, Pongdontri P, Mora G, Hoppert M, Fulda M. (2008) Mutants of *Saccharomyces cerevisiae* Deficient in Acyl-CoA Synthetases Secrete Fatty Acids due to Interrupted Fatty Acid Recycling. FEBS J. 275(11):2765-2778.

63. Kamisaka Y, Tomita N, Kimura K, Kainou K, Uemura H. (2007) *DGA1* (Diacylglycerol Acyltransferase Gene) Overexpression and Leucine Biosynthesis Significantly Increase Lipid Accumulation in the Δ*snf2* disruptant of *Saccharomyces cerevisiae*. Biochem. J. 408:61-68.

64. Nojima Y, Kibayashi A, Matsuzaki H, Hatano T, Fukui S. (1999) Isolation and Characterization of Triacylglycerol-Secreting Mutant Strain from Yeast, *Saccharomyces cerevisiae*. The Journal of General and Applied Microbiology. 45(1):1-6.

65. del Cardayre SB, Milner Cockrem MC, inventors; LS9, Inc., assignee. (2009) Systems and Methods for the Production of Fatty Esters. patent WO/2009/009391.

66. Keasling JD, Hu Z, Somerville C, Church G, Berry D, Friedman L, et al., inventors; LS9, Inc., assignee. (2007) Production of Fatty Acids and Derivatives Thereof. patent WO/2007/136762.

67. Zhang F, Carothers JM, Keasling JD. (2012) Design of a Dynamic Sensor-Regulator System for Production of Chemicals and Fuels Derived from Fatty Acids. Nat Biotech. 30(4):354-359.

68. Roessler PG, Watts K, Liu B, inventors; Synthetic Genomics, Inc., assignee. (2011) Microbial Production of Fatty Alcohols. patent US 2011/0195469.

69. Alper H, Miyaoku K, Stephanopoulos G. (2005) Construction of Lycopene-Overproducing *E. coli* Strains by Combining Systematic and Combinatorial Gene Knockout Targets. Nat Biotech. 23(5):612-616.

70. Muramatsu M, Ohto C, Obata S, Sakuradani E, Shimizu S. (2008) Various Oils and Detergents Enhance the Microbial Production of Farnesol and Related Prenyl Alcohols. J. Biosci. Bioeng. 106(3):263-267.

71. Takahashi S, Yeo Y, Greenhagen BT, McMullin T, Song L, Maurina-Brunker J, et al. (2007) Metabolic Engineering of Sesquiterpene Metabolism in Yeast. Biotechnol. Bioeng. 97(1):170-181.

72. Renninger NS, McPhee DJ, inventors; Amyris Biotechnologies, Inc., assignee. (2008) Fuel Compositions Comprising Farnesane and Farnesane Derivatives and Methods of Making and Using Same. patent WO/2008/045555.

73. Kaczmarzyk D, Fulda M. (2010) Fatty Acid Activation in Cyanobacteria Mediated by Acyl-Acyl Carrier Protein Synthetase Enables Fatty Acid Recycling. Plant Physiol. 152(3):1598-1610.

74. Liu X, Sheng J, Curtiss III R. (2011) Fatty Acid Production in Genetically Modified Cyanobacteria. Proceedings of the National Academy of Sciences. Epub April 11, 2011.

75. Roessler PG, Chen Y, Liu B, Dodge CN, inventors; Synthetic Genomics, Inc., assignee. (2009) Secretion of Fatty Acids by Photosynthetic Microorganisms. patent US 2009/0298143.

76. Ruffing AM, Jones HDT. (2012) Physiological Effects of Free Fatty Acid Production in Genetically Engineered *Synechococcus elongatus* PCC 7942. Biotechnol. Bioeng. 109(9):2190-2199.

77. Berry DA, Afeyan NB, Skraly FA, Ridley CP, Robertson DE, Wilpiszeski R, et al., inventors; Joule Unlimited, Inc., assignee. (2011)

Methods and Compositions for the Recombinant Biosynthesis of Fatty Acids and Esters. patent US 2011/0111470.

78. Lindberg P, Park S, Melis A. (2010) Engineering a Platform for Photosynthetic Isoprene Production in Cyanobacteria, Using *Synechocystis* as the Model Organism. Metab. Eng. 12(1):70-79.

79. Hahn-Hägerdal B. (1996) Ethanolic Fermentation of Lignocellulose Hydrolysates. Appl. Biochem. Biotechnol. 57-58(1):195-199.

80. Dellomonaco C, Fava F, Gonzalez R. (2010) The Path to Next Generation Biofuels: Successes and Challenges in the Era of Synthetic Biology. Microb. Cell Fact. 9(1):3.

81. Jin Y, Lee T, Choi Y, Ryu Y, Seo J. (2000) Conversion of Xylose to Ethanol by Recombinant *Saccharomyces cerevisiae* Containing Genes for Xylose Reductase and Xylitol Dehydrogenase from *Pichia stipitis*. J. Microbiol. Biotech. 10(4):564-567.

82. van Maris A, Winkler A, Kuyper M, de Laat W, van Dijken J, Pronk J. (2007) Development of Efficient Xylose Fermentation in *Saccharomyces cerevisiae:* Xylose Isomerase as a Key Component Biofuels. Adv Biochem Eng Biotechnol. 108:179-204.

83. Walfridsson M, Anderlund M, Bao X, Hahn-Hägerdal B. (1997) Expression of Different Levels of Enzymes from the *Pichia stipitis* XYL1 and XYL2 Genes in *Saccharomyces cerevisiae* and its Effects on Product Formation During Xylose Utilisation. Appl. Microbiol. Biotechnol. 48(2):218-224.

84. Kuyper M, Hartog MMP, Toirkens MJ, Almering MJH, Winkler AA, van Dijken JP, et al. (2005) Metabolic Engineering of a Xylose-Isomerase-Expressing *Saccharomyces cerevisiae* Strain for Rapid Anaerobic Xylose Fermentation. FEMS Yeast Res. 5(4-5):399-409.

85. Walfridsson M, Hallborn J, Penttilä M, Keränen S, Hahn-Hägerdal B. (1995) Xylose-Metabolizing *Saccharomyces cerevisiae* Strains Overexpressing the TKL1 and TAL1 Genes Encoding the Pentose Phosphate Pathway Enzymes Transketolase and Transaldolase. Appl. Environ. Microbiol. 61(12):4184-4190.

86. Becker J, Boles E. (2003) A Modified *Saccharomyces cerevisiae* Strain That Consumes l-Arabinose and Produces Ethanol. Appl. Environ. Microbiol. 69(7):4144-4150.

87. Richard P, Verho R, Putkonen M, Londesborough J, Penttilä M. (2003) Production of Ethanol from L-Arabinose by *Saccharomyces cerevisiae*Containing a Fungal L-Arabinose Pathway. FEMS Yeast Res. 3(2):185-189.

88. Wisselink HW, Toirkens MJ, del Rosario Franco Berriel M, Winkler AA, van Dijken JP, Pronk JT, et al. (2007) Engineering of *Saccharomyces cerevisiae* for Efficient Anaerobic Alcoholic Fermentation of L-Arabinose. Appl. Environ. Microbiol. 73(15):4881-4891.

89. Stülke J, Hillen W. (1999) Carbon Catabolite Repression in Bacteria. Curr. Opin. Microbiol. 2(2):195-201.

90. Nichols NN, Dien BS, Bothast RJ. (2001) Use of Catabolite Repression Mutants for Fermentation of Sugar Mixtures to Ethanol. Appl. Microbiol. Biotechnol. 56(1):120-125.

91. Hernández-Montalvo VH-M, Valle FV, Bolivar FB, Gosset GG. (2001) Characterization of Sugar Mixtures Utilization by an *Escherichia coli* Mutant Devoid of the Phosphotransferase System. Appl. Microbiol. Biotechnol. 57(1):186-191.

92. Yomano L, York S, Shanmugam K, Ingram L. (2009) Deletion of Methylglyoxal Synthase Gene (*mgsA*) Increased Sugar Co-Metabolism in Ethanol-Producing *Escherichia coli*. Biotechnol. Lett. 31(9):1389-1398.

93. da Silva GP, Mack M, Contiero J. (2009) Glycerol: A Promising and Abundant Carbon Source for Industrial Microbiology. Biotechnol. Adv. 27(1):30-39.

94. Houghton J, Weatherwax S, Ferrell J. (2006) Breaking the Biological Barriers to Cellulosic Ethanol. Rockville, Maryland: Office of Energy Efficiency and Renewable Energy, Office of the Biomass Program, U.S. Department of Energy.

95. NSF. (2008) Breaking the Chemical and Engineering Barriers to Lignocellulosic Biofuels: Next Generation Hydrocarbon Biorefineries. Washington, D.C.: University of Massachusetts Amerst. National Science Foundation. Chemical, Bioengineering, Environmental, and Transport Systems Division.

96. Sheehan J, Dunahay T, Benemann J, Roessler P. (1998) A Look Back at the U.S. Department of Energy's Aquatic Species Program - Biodiesel from Algae. National Renewable Energy Laboratory.

97. Deng M-D, Coleman JR. (1999) Ethanol Synthesis by Genetic Engineering in Cyanobacteria. Appl. Environ. Microbiol. 65(2):523-528.

98. Falkowski PG, Katz ME, Knoll AH, Quigg A, Raven JA, Schofield O, et al. (2004) The Evolution of Modern Eukaryotic Phytoplankton. Science. 305(5682):354-360.

99. Seckbach J, Oren A, Pattanaik B, Reisser W, Klaveness D, Lovhoiden F, et al. (2007) Algae and Cyanobacteria in Extreme Environments. Seckbach J, editor. The Netherlands: Springer.

100. Singh J, Gu S. (2010) Commercialization Potential of Microalgae for Biofuels Production. Renew. Sustain. Ener. Rev. 14(9):2596-2610.

101. León R, Fernández E. (2007) Nuclear Transformation of Eukaryotic Microalgae: Historical Overview, Achievements and Problems. In: León R, Galván A, Fernández E, editors.: Springer New York; p. 1-11.

102. León-Bañares R, González-Ballester D, Galván A, Fernández E. (2004) Transgenic Microalgae as Green Cell-Factories. Trends Biotechnol. 22(1):45-52.

103. Wang ZT, Ullrich N, Joo S, Waffenschmidt S, Goodenough U. (2009) Algal Lipid Bodies: Stress Induction, Purification, and Biochemical Characterization in Wild-Type and Starchless *Chlamydomonas reinhardtii*. Eukaryot. Cell. 8(12):1856-1868.

104. Radakovits R, Eduafo PM, Posewitz MC. (2011) Genetic Engineering of Fatty Acid Chain Length in *Phaeodactylum tricornutum*. Metab. Eng. 13(1):89-95.

105. Yu W-L, Ansari W, Schoepp N, Hannon M, Mayfield S, Burkart M. (2011) Modifications of the Metabolic Pathways of Lipid and Triacylglycerol Production in Microalgae. Microb. Cell Fact. 10(1):91.

106. Stevens DR, Purton S. (1997) Genetic Engineering of Eukaryotic Algae: Progress and Prospects. J. Phycol. 33(5):713-722.

107. Rosenberg JN, Oyler GA, Wilkinson L, Betenbaugh MJ. (2008) A Green Light for Engineered Algae: Redirecting Metabolism to Fuel a Biotechnology Revolution. Curr. Opin. Biotechnol. 19(5):430-436.

108. Radakovits R, Jinkerson RE, Darzins A, Posewitz MC. (2010) Genetic Engineering of Algae for Enhanced Biofuel Production. Eukaryot. Cell. 9(4):486-501.

109. Cerutti H, Ma X, Msanne J, Repas T. (2011) RNA-Mediated Silencing in Algae: Biological Roles and Tools for Analysis of Gene Function. Eukaryot. Cell. 10(9):1164-1172.

110. Kilian O, Benemann CSE, Niyogi KK, Vick B. (2011) High-Efficiency Homologous Recombination in the Oil-Producing Alga *Nannochloropsis* sp. Proceedings of the National Academy of Sciences. 108(52):21265-21269.

111. (2012) *Chlamydomonas* Engineering Kits. Life Technologies Corporation. Available: https://www.invitrogen.com/site/us/en/home/Products-and-Services/Applications/Protein-Expression-and-Analysis/Protein-Expression/algae-engineering-kits/chlamydomonas-engineering-kits.html. Accessed 2012 April 16.

112. Rasmussen B, Fletcher IR, Brocks JJ, Kilburn MR. (2008) Reassessing the First Appearance of Eukaryotes and Cyanobacteria. Nature. 455(7216):1101-1104.

113. Koksharova OA, Wolk CP. (2002) Genetic Tools for Cyanobacteria. Appl. Microbiol. Biotechnol. 58(2):123-137.

114. Ogbonna JC, Soejima T, Tanaka H. (1999) An Integrated Solar and Artificial Light System for Internal Illumination of Photobioreactors. J. Biotechnol. 70(1–3):289-297.

115. Kunjapur AM, Eldridge RB. (2010) Photobioreactor Design for Commercial Biofuel Production from Microalgae. Ind. Eng. Chem. Res. 49(8):3516-3526.

116. Bullis K. (2012) NASA Wants to Launch Floating Algae Farms. Technology Review.

117. Chaumont D. (1993) Biotechnology of Algal Biomass Production: A Review of Systems for Outdoor Mass Culture. J. Appl. Phycol. 5(6):593-604.

118. Sikkema J, de Bont JA, Poolman B. (1995) Mechanisms of Membrane Toxicity of Hydrocarbons. Microbiol. Rev. 59(2):201-222.

119. Desbois A, Smith V. (2010) Antibacterial Free Fatty Acids: Activities, Mechanisms of Action and Biotechnological Potential. Appl. Microbiol. Biotechnol. 85(6):1629-1642.

120. León R, Garbayo I, Hernández R, Vigara J, Vilchcz C. (2001) Organic Solvent Toxicity in Photoautotrophic Unicellular Microorganisms. Enzyme Microb. Technol. 29(2–3):173-180.

121. Gill C, Ratledge C. (1972) Toxicity of *n*-Alkanes, *n*-Alk-1-enes, *n*-Alkan-1-ols and *n*-Alkyl-1-bromides Towards Yeasts. J. Gen. Microbiol. 72(1):165-172.

122. Best CA, Laposata M. (2003) Fatty Acid Ethyl Esters: Toxic Non-Oxidative Metabolites of Ethanol and Markers of Ethanol Intake. Front. Biosci. 8:202-217.

123. Andrews RE, Parks LW, Spence KD. (1980) Some Effects of Douglas Fir Terpenes on Certain Microorganisms. Appl. Environ. Microbiol. 40(2):301-304.

124. Uribe S, Ramirez J, Peña A. (1985) Effects of β-Pinene on Yeast Membrane Functions. J. Bacteriol. 161(3):1195-1200.

125. Dunlop MJ. (2011) Engineering Microbes for Tolerance to Next-Generation Biofuels. Biotech. Biofuels. 4(32).

126. Sardessai Y, Bhosle S. (2002) Tolerance of Bacteria to Organic Solvents. Res. Microbiol. 153(5):263-268.

127. Dunlop MJ, Dossani ZY, Szmidt HL, Chu HC, Lee TS, Keasling JD, et al. (2011) Engineering Microbial Biofuel Tolerance and Export Using Efflux Pumps. Mol Syst Biol. 7.

128. Melis A. (2009) Solar Energy Conversion Efficiencies in Photosynthesis: Minimizing the Chlorophyll Antennae to Maximize Efficiency. Plant Sci. 177(4):272-280.

129. Mussgnug JH, Thomas-Hall S, Rupprecht J, Foo A, Klassen V, McDowall A, et al. (2007) Engineering Photosynthetic Light Capture: Impacts on Improved Solar Energy to Biomass Conversion. Plant Biotechnol. J. 5(6):802-814.

130. Huesemann M, Hausmann T, Bartha R, Aksoy M, Weissman J, Benemann J. (2009) Biomass Productivities in Wild Type and Pigment Mutant of*Cyclotella* sp. (Diatom). Appl. Biochem. Biotechnol. 157(3):507-526.

131. Lee J, Mets L, Greenbaum E. (2002) Improvement of Photosynthetic CO2 Fixation at High Light Intensity Through Reduction of Chlorophyll Antenna Size. Appl. Biochem. Biotechnol. 98-100(1):37-48.

132. Nakajima Y, Tsuzuki M, Ueda R. (2001) Improved Productivity by Reduction of the Content of Light-Harvesting Pigment in *Chlamydomonas perigranulata*. J. Appl. Phycol. 13(2):95-101.

133. Nakajima Y, Ueda R. (1997) Improvement of Photosynthesis in Dense Microalgal Suspension by Reduction of Light Harvesting Pigments. J. Appl. Phycol. 9(6):503-510.

134. Polle JEW, Kanakagiri S-D, Melis A. (2003) *tla1*; a DNA Insertional Transformant of the Green Alga *Chlamydomonas reinhardtii* with a Truncated Light-Harvesting Chlorophyll Antenna Size. Planta. 217(1):49-59.

135. Gombos Z, Murata N. (2004) Genetic Engineering of the Unsaturation of Membrane Glycerolipid: Effects on the Ability of the Photosynthetic Machinery to Tolerate Temperature Stress Lipids in Photosynthesis: Structure, Function and Genetics. In: Paul-André S, Norio M, editors.: Springer Netherlands; p. 249-262.

136. Allakhverdiev SI, Kinoshita M, Inaba M, Suzuki I, Murata N. (2001) Unsaturated Fatty Acids in Membrane Lipids Protect the Photosynthetic Machinery against Salt-Induced Damage in *Synechococcus*. Plant Physiol. 125(4):1842-1853.

137. Lurling M, Beekman W. (2006) Palmelloids Formation in *Chlamydomonas reinhardtii*: Defence Against Rotifer Predators? Ann. Limnol. - Int. J. Limnol. 42:65-72.

138. Jin Y-S, Stephanopoulos G. (2007) Multi-Dimensional Gene Target Search for Improving Lycopene Biosynthesis in *Escherichia coli*. Metab. Eng. 9(4):337-347.

139. Wang HH, Isaacs FJ, Carr PA, Sun ZZ, Xu G, Forest CR, et al. (2009) Programming Cells by Multiplex Genome Engineering and Accelerated Evolution. Nature. 460(7257):894-898.

140. Friedman L, Rude MA, inventors; LS9, Inc., assignee. (2008) Process for Producing Low Molecular Weight Hydrocarbons from Renewable Resources. patent WO/2008/113041.

141. Friedman L, da Costa B, inventors; LS9, Inc., assignee. (2008) Hydrocarbon-Producing Genes and Methods of Their Use. patent WO/2008/147781.

142. Renninger NS, Ryder JA, Fisher KJ, inventors; Amyris Biotechnologies, Inc., assignee. (2008) Jet Fuel Compositions and Methods of Making and Using Same. patent WO/2008/140492.

143. Trimbur DE, Im C-S, Dillon HF, Day AG, Franklin S, Coragliotti A, inventors; Solazyme, Inc., assignee. (2009) Lipid Pathway Modification in Oil-Bearing Microorganisms. patent US 2009/0061493.

144. Trimbur D, Im C-S, Dillon HF, Day AG, Franklin S, Coragliotti A, inventors; Solazyme, Inc., assignee. (2009) Use of Cellulosic Materials for Cultivation of Microorganisms. patent US 2009/0011480.

145. Mendez M, Fang S-C, Richard S, inventors; Sapphire Energy, Inc., assignee. (2010) Engineering Salt Tolerance in Photosynthetic Microorganisms. patent WO 2010/105095.

146. Heaps NA, Behnke CA, Molina D, inventors; Sapphire Energy, Inc., assignee. (2010) Biofuel Production in Prokaryotes and Eukaryotes. patent WO/2010/104763.

147. Berry DA, Robertson DE, Skraly FA, Green BD, Ridley CP, Kosuri S, et al., inventors; Joule Unlimited, Inc., assignee. (2011) Engineered CO2 Fixing Microorganisms Producing Carbon-Based Products of Interest. patent US 2011/0008861.

CHAPTER 7

Valorisation of food waste to biofuel: current trends and technological challenges

Sanjib Kumar Karmee and Carol Sze Ki Lin

School of Energy and Environment, City University of Hong Kong,
Tat Chee, Avenue, Kowloon, Hong Kong

ABSTRACT

Demand for biofuels is rapidly growing worldwide as petroleum based fuels are finite reserves. In this context, biodiesel and bioethanol are popular biofuels that are commercially available in various countries. Biofuels can be prepared from edible biomass. However, this is already generating food versus fuel debate among the members of civil societies. Therefore, there are needs to synthesize biofuels from nonedible waste materials. Food wastes can be utilized as resources for the production of biodiesel and bioethanol since they contain significant amount of lipids and carbohydrates. In future, industrial production of biodiesel and bioethanol from food waste can contribute to resolve the waste disposal, energy scarcity and energy security problems.

KEYWORDS

Food waste Biofuel Biodiesel Bioethanol Sustainable chemistry

1. INTRODUCTION

Food waste is creating serious environmental and social problems in Hong Kong and across the world. According to a recent report released by the Hong Kong Environment Bureau, among the 9,000 tonnes of municipal solid waste that is thrown away everyday at landfills, 40% of which is composed of "putrescibles" [1]. These putrescibles are organic wastes that are known to create odour upon decomposition. Approximately, 90% of putrescibles are food wastes. Food waste can be raw, cooked, edible and inedible parts generated during production, storage distribution, and consumption of food stuffs. In 2011, Hong Kongers threw away approximately 3,600 tonnes of food waste everyday [1,2]. Two third of the food waste was obtained from household; whereas, one third of food waste came from commercial and industrial sources. Hong Kong is not the only country generating large quantities of food waste. For instance, other developed cities like Taipei and Seoul are producing 182,000 tonnes/per year and 767,000 tonnes/per year of food wastes respectively [1,2].

The food wastes produced in Hong Kong includes rotten fruits, vegetables, fish, poultry organs, fruits and vegetable peelings, meat, fish, shellfish shells, bones, food fats, sauces, condiments, soup pulp, Chinease medicinal pulp, egg shells, cheeses, ice cream, yogurts, tea leaves, teabags, coffee grounds, breads, cakes, biscuits, desserts, jam, different cereals, leftover of cooked food, BBQ raw or cooked leftovers, and pet food [1]. The Hong Kong Government plans to cut down the food waste that goes to landfill to approximately 40% by 2022. Landfills are the most common place for garbage deposition. Landfills spread offensive smell and are known to cause hazardous effects on people, animals, and the environment. Landfills are unsustainable as they produce methane which is a common green house gas. Furthermore, landfills also generate large amount of harmful leachate when rainwater falls on the garbage. This leachate can contaminate water and soil. Nevertheless, anaerobic digestion of food wastes that occurs naturally in the absence of oxygen employing bacteria can be used to produce biogas. Biogas is used as an energy source. Alternatively, food wastes can be valorized for the production of energy by using different common techniques such as composting, recycling and incineration. Although, these processes are capable of converting food waste into fuels and value-added products development of greener and advanced technologies are required [3].

Biofuel production is rapidly growing as the world encounters pollution problems due to burning of petroleum and coal based fuels. In addition, petroleum fuels are finite reserves and most of the petroleum reserves are geographically located in politically unstable countries. This reinforces the fact that alternative fuels are important from both environmental and energy security point of view. Along this line, many countries are formulating energy policies for the production of renewable energy.

At present, biofuels such as biodiesel and bioethanol are largely produced from edible food materials [4]. Various edible plant oils from soybean, rapeseed and canola oils are used for the preparation of biodiesel. Whereas, ethanol can be produced from a variety of feedstocks such as sugar cane, bagasse, sugar beet, grain, switchgrass, barley, potatoes, molasses, corn, stover, wheat and many other sources rich in carbohydrate [4,5]. Chemical biodiesel production process is called transesterification [4,5]. During transesterification, the tri-,di-and mono-glycerides react with methanol in the presence of a catalyst to produce biodiesel. On the other hand, the production of bioethanol process involves pretreatment, enzymatic hydrolysis, fermentation and distillation steps. Preparation of biofuels from edible food materials is attributed for the reason of food scarcity and a food vs fuel debate is already raging [6]. Alternatively, nonedible feedstocks can be used for the production of biofuels. Jatropha, Pongamia and other nonedible plant oils are already used for the preparation of biodiesel [5]. Along this line, nonedible lignocellulosic biomass is also employed for the production of bioethanol [7].

Food waste is a well-known nonedible source of lipids, carbohydrates, amino acids and phosphates [8,9]. Reserach in our laboratory reveals that bakery and mixed food wastes contain significant amount of lipids and carbohydrates [8,9]. Depending on the source of food waste the average lipid content was around 30% and the average carbohydrate content was around 50% [8,9]. Different types of food wastes can be hydrolysed enzymatically to produce food hydrolysate and lipids [8,9]. The food hydrolysate was rich in carbohydrate and can be used for the production of bioethanol; whereas, the obtained lipid can be converted to biodiesel (Figure 1).

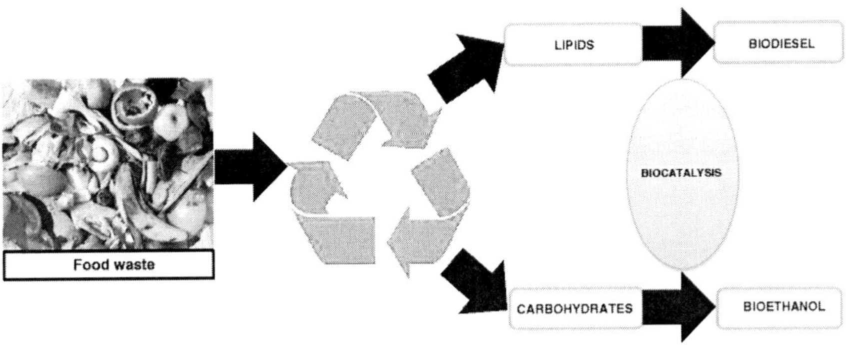

Figure 1. Recycling of food waste into biodiesel and bioethanol.

Along this line, noodle is a common starch based food material. In South Korea around 3 billion packages of instant noodle were consumed in 2011 and more than 2,100 tons of instant noodle residues were disposed as waste. Kim et al. have used instant noodle waste for the production of biofuels [10,11]. Kim et al. recovered the oil from noodle waste by extraction using nonpolar hexane as a solvent. From 100 g of noodle waste 83 g of purified starch and 5 g of oil was obtained. The obtained oil free starch residue was used for simultaneous saccharification and fermentation process for the production of bioethanol [10,11]. The hexane extract was evaporated and the obtained oil was reacted with methanol in the presence of acid and alkali catalysts for the preparation of biodiesel. One major limitation of this process is that the excess use of hexane during the extraction of oil from food waste [10,11]. The Centers for Disease Control classifies n-hexane as a neurotoxin. It is also listed as a "hazardous air pollutant" by the Environmental Protection Agency as it helps in the formation of ozone at the ground level which is primarily responsible for smog. Nevertheless, these experiments demonstrate the potential utilization of waste noodle waste as a resource for biofuel production.

Research on the use of food wastes for the production of biofuel is becoming attractive in different countries. Sulaiman et al. have conceptualized a halal biorefinery for the production of fuels and value added products in Malaysia [12]. Yao et al. from the Chinese Academy of Sciences investigated the application of food waste to generate hydrolysates for the production of bioethanol [13]. In this context, potato peel is a well-known waste generated by the potato industries in Europe [14]. As reported by Christakopoulos.et al. this "no value" potato peel waste was converted to bioethanol using environmentally benign biocatalytic methods [13]. Researchers have also used household food waste for the production of bioethanol. During this process liquefaction and saccharification methods were employed to increase both ethanol production and productivity of the process [15]. Subsequently, fermentation of the remaining solids obtained from this process was performed to increase the overall yield of ethanol [15].

It is clear from the above discussion that environmental pollution and upcoming shortage of fossil fuels have turned the attention of researchers largely on the utilization of renewable feedstocks. Additionally, scientists and policy makers are devoting much efforts to use nonedible and zero cost food wastes for fuel and energy production to reduce the direct competition between fuel and food. It is known that food wastes are generated in large quantities and their handling is a challenge. As discussed earlier, these food wastes are potential resources as they contain substantial amount of carbohydrates and lipids. Thus, zero value food waste can be used as a resource for the production of low-cost biofuels. Research by various groups is currently underway for the production of biodiesel and bioethanol from food waste [16-22]. So far, the "proof of concept" for the synthesis and characterization of biofuels from different food wastes has been established [16-22].

Currently, technologies are available for the production of biodiesel and bioethanol in industrial scale. In this regard, a pilot scale production of ethanol from food waste using *Saccharomyces cerevisiae* H058 is already reported [22]. Nevertheless, low cost, greener and advanced technologies are needed for the production of fuels from food wastes [3,23]. The industrial production of biofuel from food waste is largely depended on i) availability of food waste, ii) efficiency of hydrolyis process, iii) the amount of lipid and carbohydrate obtained from food waste, and iv) efficiency of fermentation and transesterification methods. In Hong Kong, Taiwan, Korea, US and in many other European countries plenty of food wastes are available. Thus, the future work should be primarily focused on large scale pretreatment and hydrolysis of food waste for production of lipid and sugar enriched hydrolysate. Several microorganisms and enzymes are known to hydrolyze food waste to carbohydrate, lipid, amino acids and phosphates. The catalytic efficiencies of the existing biocatalysts can be tested for the large scale hydrolysis of food wastes. Afterwards, the commercially available technologies can be employed for the production of biodiesel and bioethanol.

To make biofuels economically viable, the business and scientific communities and policy makers should come together to start a joint venture into the business of converting food waste into fuels. With proper financial and policy based supports from government, food waste biorefineries can be realized. In this context, it is particularly important to overcome the existing technological challenges of conventional food waste valorization methods. Simultaneously, it is crucial to develop environmental friendly and cost effective recycling methods that can convert food wastes into biofuels and chemicals [24-26]. Recently, combi-protein coated microcrystals of lipases are used for the production of biodiesel from oil of spent coffee grounds [27]. In this regard, both chemo- and biocatalytic methods can be explored for the preparation of biofuels from food waste [27-31].

2. CONCLUSIONS

Since food waste is considered zero cost material it is possible to develop cost-effective commercial methods for the production of biofuel using lipid and carbohydrate generated from food waste. Research on the conversion of food waste into biofuel will provide an innovative food valorisation strategy; which could contribute substantially to bio-based economy. The future goal of this research should focus on investigation of the feasibility of utilizing food waste into biofuel through simple and clean methods. Utilisation of lipid and carbohydrates obtained from food waste for biofuel production is an excellent example to demonstrate the enormous potential of waste valorisation for building a sustainable society.

ACKNOWLEDGEMENTS

SKK is grateful to the School of Energy and Environment, City University of Hong Kong for a postdoctoral fellowship under the Graduate Teaching Assistant (GTA) scheme.

AUTHORS' CONTRIBUTIONS

SKK contributed to the writing of the manuscript. CSKL assisted in the revision of manuscript. The final version was read and approved by both authors.

REFERENCES

1. **A food waste & yard waste plan for Hong Kong.** 2014–2022. http://www.enb.gov.hk/en/files/FoodWastePolicyEng.pdf, accessed on 30.06.2014.

2. **Monitoring of solid waste in Hong Kong.** https://www.wastereduction.gov.hk/en/assistancewizard/waste_red_sat.htm, accessed on 30.06.2014.

3. Luque R, Clark JH: **Valorisation of food residues: waste to wealth using green chemical technologies.** *Sustainable Chem Process*2013, **1:**10.

4. Pimentel D, Patzek TW: **Ethanol production using corn, switchgrass, and wood;biodiesel production using soybean and sunflower.** *Nat Resour Res* 2005, **14:**65–76.

5. Karmee SK, Chadha A: **Preparation of biodiesel from crude oil of *Pongamia pinnata* .** *Bioresour Technol* 2005, **96:**1425–1429.

6. Mathews JA: **Is growing biofuel crops a crime against humanity?** *Biofuels Bioprod Bioref* 2008, **2**(2):97–99.

7. Sun Y, Cheng J: **Hydrolysis of lignocellulosic materials for ethanol production: a review.** *Bioresour Technol* 2002, **83:**1–11.

8. Pleissner D, Lam WC, Sun Z, Lin CSK: **Food waste as nutrient source in heterotrophic microalgae cultivation.** *Bioresour Technol* 2013, **137:**139–146.

9. Pleissner D, Kwan TH, Lin CSK: **Fungal hydrolysis in submerged fermentation for food waste treatment and fermentation feedstock preparation.** *Bioresour Technol* 2014, **158:**48–54.

10. Yang X, Lee SJ, Yoo HY, Choi HS, Park C, Kim SW: **Biorefinery of instant noodle waste to biofuels.** *Bioresour Technol* 2014, **159:**17–23.

11. Yang X, Lee JH, Yoo HY, Shin HY, Thapa LP, Park C, Kim SW: **Production of bioethanol and biodiesel using instant noodle waste.***Bioprocess Biosyst Eng* 2014,

12. Sulaiman A, Othman N, Baharuddin AS, Mokhtar MN, Tabatabaei M: **Enhancing the halal food industry by utilizing food wastes to produce value-added bioproducts.** *Procedia Soc Behav Sci* 2014, **121:**35–43.

13. Yan S, Li J, Chen X, Wu J, Wang P, Ye J, Yao J: **Enzymatical hydrolysis of food waste and ethanol production from the hydrolysate.***Renew Energ* 2011, **36:**1259–1265.

14. Arapoglou D, Varzakas T, Vlyssides A, Israilides C: **Ethanol production from potato peel waste (PPW).** *Waste Manage* 2010, **30:**1898–1902.

15. Matsakas L, Kekos D, Loizidou M, Christakopoulos P: **Utilization of household food waste for the production of ethanol at high dry material content.** *Biotechnol Biofuels* 2014, **7:**4.

16. Kim JH, Lee JC, Pak D: **Feasibility of producing ethanol from food waste.** *Waste Manage* 2011, **31:**2121–2125.

17. Kim JK, Oh BR, Shin H-J, Eom C-Y, Kim SW: **Statistical optimization of enzymatic saccharification and ethanol fermentation using food waste.** *Process Biochem* 2008, **43:**1308–1312

18. Lin CSK, Pfaltzgraff LA, Herrero-Davila L, Mubofu EB, Abderrahim S, Clark JH, Koutinas AA, Kopsahelis N, Stamatelatou K, Dickson F, Thankappan S, Mohamed Z, Brocklesby R, Luque R: **Food waste as a valuable resource for the production of chemicals, materials and fuels. Current situation and global perspective.** *Energy Environ Sci* 2013, **6**(2):426–464.

19. Kondamundi N, Mohapatra SK, Misra M: **Spent coffee grounds as a versatile source of green energy.** *J Agric Food Chem* 2008,**56:**11757–11760.

20. Garcia J: **Food wasterecycled as biofuel.** http://large.stanford.edu/courses/2011/ph240/garcia1/, accessed on 30.06.2014.

21. Akpan UG, Alhakim AA, Ijah UJJ: **Production of ethanol fuel from organic and food wastes.** *Leonardo Electron J Pract Technol* 2008,**13:**1–11.

22. Yan S, Chen X, Wu J, Wang P: **Pilot scale production of fuel ethanol from concentrated food waste hydrolysates using Saccharomyces cerevisiae H058.** *Bioprocess Biosyst Eng* 2013, **36:**937–946.

23. Arancon RAD, Lin CSK, Chan KM, Kwan TH, Luque R: **Advances on wastevalorization: new horizons for a more sustainable society.***Energy Sci Eng* 2013, **1:**53–71

24. Zhang AYZ, Sun Z, Leung CCJ, Han W, Lau KY, Li M, Lin CSK: **Valorisation of bakery waste for succinicacid production.** *Green Chem*2013, **15:**690–695.

25. Leung CC, Cheung SY, Zhang YZ, Lam KF, Lin CSK: **Utilisation of waste bread for the sustainable production of succinic acid.***Biochem Eng J* 2012, **65:**10–15.

26. Koutinas AA, Pleissner D, Vlysidis A, Kopsahelis N, Lopez Garciac I, Kookosd I, Papanikolaou S, Lin CSK: **Valorization of industrial waste and by-product streams via fermentation for the production of chemicals and biopolymers.** *Chem Soc Rev* 2014,**43**(8):2587–2627.

27. Banerjee A, Singh V, Solanki K, Mukherjee J, Gupta MN: **Combi-protein coated microcrystals of lipases for production of biodiesel from oil from spent coffee grounds.** *Sustainable Chemical Processes* 2013, **1:**14.

28. Karmee SK: **Biocatalytic synthesis of ascorbyl esters and their biotechnological applications.** *Appl Microbiol Biotechnol* 2009,**81**(6):1013–1022.

29. Karmee SK: **Lipase catalyzed synthesis of ester-based surfactants from biomass derivatives.** *Biofuels Bioprod Bioref* 2008, **2:**144–154.

30. Kiss AA, Dimian AC, Rothenberg G: **Solid acid catalysts for biodiesel production- towards sustainable energy.** *Adv Synth Catal*2006, **348:**75–81.

31. Lin CSK, Koutinas AA, Stamatelatou K, Mubofu EB, Matharu AS, Kopsahelis N, Pfaltzgraff LA, Clark JH, Papanikolaou S, Kwan TH, Luque R: **Current and future trends in food waste valorization for the production of chemicals, materials and fuels: a global perspective.** *Biofuels Bioprod Bioref* 2014, **8**(5):686–715.

CHAPTER 8

Gas Fermentation for Commercial Biofuels Production

Fung Min Liew[1], Michael Köpke[1] and Séan Dennis Simpson[1]

[1] LanzaTech NZ Ltd., Parnell, Auckland, New Zealand

1. INTRODUCTION

With diminishing global reserves of crude oil and increasing demand, especially from developing countries, the pressure on oil supply will grow. Although the 2007-2010 financial crisis brought down the price of crude oil (per barrel) from a record peak of US$145 in July 2008, factors such as recovering global economies and political instability in the Middle East have restored the price of crude oil to the US$100 mark. At current rate of consumption, the global reserves of petroleum are predicted to be exhausted within 50 years [1, 2]. This, coupled with the deleterious environmental impacts that result from accumulating atmospheric CO_2 from the burning of fossil fuels, the development of affordable, and environmentally sustainable fuels is urgently required. Many countries have responded to this challenge by legislating mandates and introducing policies to stimulate research and development (R&D) and commercialization of technologies that allow the production of low cost, low fossil carbon emitting fuels. For instance, the European Union (EU) has mandated member countries to a target of deriving 10% of all transportation fuel from renewable sources by 2020 [3]. Between 2005 and 2010, renewable energies such as solar, wind, and biofuels have been increasing at an average annual rate of 15-50% [4]. Renewable energy accounted for an estimated 16% of global final energy consumption in 2009 [4].

Biofuels have been defined as solid (bio-char), liquid (bioethanol, biobutanol, and biodiesel) and gaseous (biogas, biosyngas, and biohydrogen) fuels that are mainly derived from biomass [5]. Liquid biofuels provided a small but growing contribution towards worldwide fuel usage, accounting for 2.7% of global road transport fuels in 2010 [4]. The world's largest producer of biofuels is the United States (US), followed by Brazil and the EU [4]. In 2009, US and Brazil accounted for approximately 85% of global bioethanol production while Europe generated about 85% of the world's biodiesel [6]. The global market for liquid biofuels (bioethanol and biodiesel) increased dramatically in recent years, reaching US$83 billion in 2011 and is projected to US$139 billion by 2021 [7].

The use and production of biofuels has a long history, starting with the inventors Nikolaus August Otto and Rudolph Diesel, who already envisioned the use of biofuels such as ethanol and natural oils when developing the first Otto cycle combustion and diesel engines [6]. While fermentative production of ethanol has been used for thousands of years, mainly for brewing beer starting in Mesopotamia 5000 B.C., fermentative production of another potential biofuel butanol, has only been discovered over the last century, but had significant impact. During the World War 1, Chaim Weizmann successfully applied a process called ABE (acetone-butanol-ethanol) fermentation using *Clostridium acetobutylicum*to generate industrial scale acetone (for cordites, the propellant of cartridges and shells) from starchy materials [6, 8]. His contribution was later recognised in the Balfour declaration in 1917 and he became the first President of the newly founded State of Israel [6, 8]. Intriguingly, the enormous potential of butanol produced at that time was not realized and the substance was simply stored in huge containers [6]. ABE fermentation became the second biggest ever biotechnological process (after the ethanol fermentation process) ever performed, but the low demand of acetone following the conclusion of the war led to closure of all the plants [8]. Although ABE fermentation briefly made a comeback during the Second World War, increasing substrate costs and increasing stable supply of low cost crude oil from the Middle East rendered the technology economically unviable. Recently, a resurgence of the technology is underway as some old plants are reopened and new plants are being built or planned in China, the US, the United Kingdom (UK), Brazil, France and Austria [6, 8].

Traditionally sugar substrates derived from food crops such as sugar cane, corn (maize) and sugar beet have been the preferred feedstocks for the production of biofuels. However, world raw sugar prices have witnessed significant volatility over the last decade or so, ranging from US$216/ton in year 2000 to a 30 year high of US$795/ton in February 2011 due to global sugar deficits and crop shortfall [9]. This has created uncertainty and raised sustainability issues about its use as a feedstock for large scale biofuel production. This review aims to shed light on the use of syngas and industrial waste gas as feedstocks, and the emerging field of gas fermentation to generate not only biofuels, but also other high-value added products. The advantages of gas fermentation over conventional sugar-based fermentation and thermochemical conversions, and their flexibility in utilizing a spectrum of feedstocks to generate syngas will be discussed. The biochemistry, genetic and energetic background of the microorganisms that perform this bioconversion process will be critically examined, together with recent advances in systems biology and synthetic biology that offer growing opportunities to improve biocatalysts in terms of both the potential products that can be produced and their process performance. The key processes such as gasification, bioreactor designs, media formulation, and product recovery will be analysed. Finally, the state of commercialization of gas fermentation will be highlighted and an outlook will be provided.

2. ADVANTAGES OF GAS FERMENTATION

The production of first generation biofuels relies on food crops such as sugar beet, sugar cane, corn, wheat and cassava as substrates for bioethanol; and vegetable oils and animal fats for biodiesel. Although years of intense R&D have made methods of bioethanol production (typically using the yeast*Saccharomyces cerevisiae*) technologically mature, there remain some serious questions regarding its sustainability. The use of food crops as a source of carbohydrate feedstocks by these processes requires high-quality agricultural land. The inevitable conflict between the increasing diversion of crops or land for fuel rather than food production has been highlighted as one of the prime causes of rising global food prices. Furthermore, corn ethanol producers in the US, have historically enjoyed a 45-cent-a-gallon federal tax credit for years (which ended in early 2012), costing the government US$30.5 billion between 2005 to 2011, raising questions about its economic competitiveness with gasoline [10,11].

These arguments have stimulated the search for so-called second generation biofuels, which utilize non-food lignocellulose biomass such as wood, dedicated energy crops, agricultural residues and municipal solid wastes as feedstocks. Biomass consists of cellulose, hemicellulose and lignin, and the latter of which is extremely resistant to degradation. One approach to unlocking the potential in this abundant feedstock is to separate the lignin from the carbohydrate fraction of the biomass via extensive pre-treatment of the lignocellulose involving, for example, steam-explosion and/or acid hydrolysis. These pre-treatments are designed to allow the carbohydrate portion of the biomass to be broken down into simple sugars, for example by enzymatic hydrolysis using exogenously added cellulases to release fermentable sugars [12]. Such approaches have been found to be expensive and rate limiting [6, 12,13]. Alternatively, processes using cellulolytic microorganisms (such as *C. cellulolyticum*, *C. thermocellum*, and *C.*

phytofermentans) to carry out both the hydrolysis of lignocelluloses and sugar fermentation in a single step, termed 'Consolidated Bioprocessing Process (CBP)' [12] have been proposed, however the development of these is still at an early stage, and again low conversion rates seem to be a major limitation that needs to be overcome.

Microorganisms such as acetogens, carboxytrophs and methanogens are able to utilize the $CO_2 + H_2$, and/or CO available in such syngas as their sole source of carbon and energy for growth as well as the production of biofuels and other valuable products. However, only acetogens are described to synthesize metabolic end products that have potentials as liquid transportation fuels. While biological processes are generally considered slower than chemical reactions, the use of these microbes to carry out syngas fermentation offers several key advantages over alternative thermo-chemical approaches such as the Fischer-Tropsch' process (FTP). First, microbial processes operate at ambient temperatures and low pressures which offer significant energy and cost savings. Second, the ambient conditions and irreversible nature of biological reactions also avoid thermodynamic equilibrium relationships and allow near complete conversion efficiencies [14, 15]. Third, biological conversions are commonly more specific due to high enzymatic specificities, resulting in higher product yield with the formation of fewer by-products. Fourth, unlike traditional chemical catalysts which require a set feed gas composition to yield desired product ratios or suite, microbial processes have freedom to operate for the production ofthe same suite of products across a wider range of $CO:H_2$ ratios in the feed gas [16]. Fifth, biocatalysts exhibit a much higher tolerance to poisoning by tars, sulphur and chlorine than inorganic catalysts [6, 16]. However, some challenges have been identified for syngas fermentation to be commercialized, including gas mass transfer limitations, long retention times due to slow cell growth, and lower alcohol production rates and broth concentrations. Recent progress and development to remedy these issues will be highlighted in this review.

3. FEEDSTOCK AND GASIFICATION

Due to the flexibility of the microbes to ferment syngas with diverse composition, virtually any carbonaceous materials can be used as feedstock for gasification. Non-food biomass that can be employed as feedstock for gasification includes agricultural wastes, dedicated energy crops, forest residues, and municipal organic wastes, or even glycerol and feathers [16-20]. Biomass is available on a renewable basis, either through natural processes or anthropogenic activities (e.g. organic wastes). It has been estimated that out of a global energy potential from modern biomass of 250 EJ per year in 2005, only 9 EJ (3.6%) was used for energy generation [18].

The use of existing waste streams such as municipal organic waste also differentiate itself from other feedstocks such as dedicated energy crops because these wastes are available today at economically attractive prices, and they are often already aggregated and require less indirect land use. Alternatively, gasification of non-biomass sources such as coal, cokes, oil shale, tar sands, sewage sludge and heavy residues from oil refining, as well as reformed natural gas are commonly applied as feedstocks for the FTP and can also be used for syngas fermentation [15, 21]. Furthermore, some industries such as steel manufacturing, oil refining and chemical production generate large volume of CO and/or CO_2 rich gas streams as wastes. Tapping into these sources using microbial fermentation process essentially convert existing toxic waste gas streams into valuable commodities such as biofuels. The overall process of gas fermentation is outlined in Figure 1.

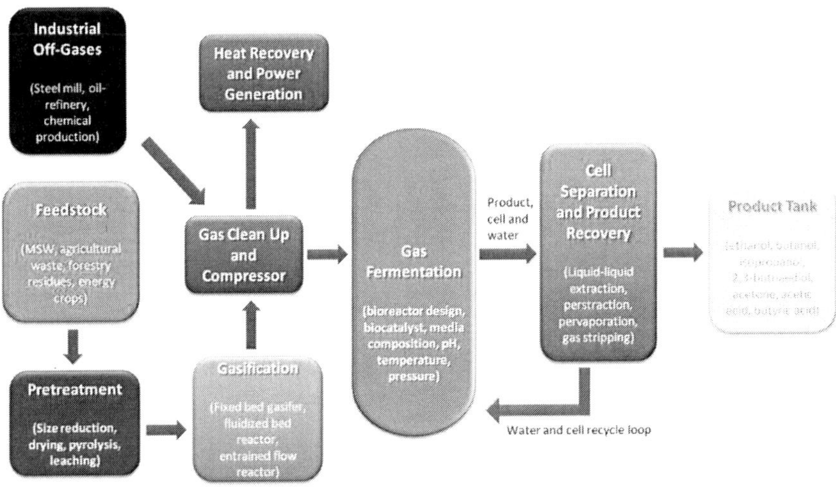

Figure 1. Overview of gas fermentation process

Prior to gasification, biomass generally needs to go through a pre-treatment process encompassing drying, size reduction (e.g. chipping, grinding and chopping), pyrolysis, fractionation and leaching depending on the gasifier configuration [22, 23]. This upstream pre-treatment process can incur significant capital expense and add to the overall biomass feedstock cost, ranging from US$16-70 per dry ton [22]. Gasification is a thermo-chemical process that converts carbonaceous materials to gaseous intermediates at elevated temperature (600-1000°C), in the presence of an oxidizing agent such as air, steam or oxygen [16, 22]. The resulting syngas contains mainly CO, CO_2, H_2 and N_2, with varying amounts of CH_4, water vapour and trace amount of impurities such as H_2S, COS, NH_3, HCl, HCN, NO_x, phenol, light hydrocarbons and tar [17, 22, 24]. The composition and amount of impurities of syngas depends on the feedstock properties (e.g. moisture, dust and particle size), gasifier type and operational conditions (e.g. temperature, pressure, and oxidant) [17, 22]. Table 1 summarizes typical composition of syngas and other potential gas streams derived from various sources.

4. FIXED BED GASIFIER

Depending on the direction of the flows of carbonaceous fuel and oxidant (air or steam), fixed bed gasifier can be further categorized into updraft or downdraft reactor. In the updraft (counter-current) version of the fixed bed gasifier, biomass enters from the top while gasifying agent from the bottom. The biomass moves down the reactor through zones of drying (100°C), pyrolysis (300°C), gasification (900°C) and finally oxidation zone (1400°C) [18]. Although this mode of gasifier is often associated with high tar content in the exit gas, recent advances in tar cracking demonstrated that very low tar level is achievable [31]. The direct heat exchange of the oxidizing agent with the entering fuel feed results in low gas exit temperature and hence high thermal efficiency [18, 23]. The downdraft (co-current) gasifier has very similar design as the updraft reactor, except the carbonaceous fuel and oxidizing agent flow in the same direction. In comparison to the updraft gasifier, the downdraft reactor has lower tar content in the exit gas but exhibit lower thermal efficiency [23]. Due to the size limitation in the constriction (where most of the gasification occurs) of the reactor, this mode of gasifier is considered unsuitable for large scale operation [18].

5. FLUIDIZED BED REACTOR

In fluidized bed reactor, the carbonaceous fuel is mixed together with inert bed material (e.g. silica sand) by forcing fluidization medium (e.g. air and/or steam) through the reactor. The inert bed facilitates better heat exchange between the fuel materials, resulting in nearly isothermal operation conditions and high feedstock conversion efficiencies [18, 22]. The maximum operating temperature of the gasifier is typically around 800 - 900°C, which is limited by the melting point of the bed material [18]. Furthermore, the geometry of the reactor and excellent mixing properties also means that fluidized bed reactors are suitable for up-scaling [18, 22]. Due to these properties, fluidized bed reactor is currently the most commonly used gasifier for biomass feedstock [32]. However, this mode of gasifier is not suitable for feedstocks with high levels of ash and alkali metals because the melting of these components causes stickiness and formation of bigger lumps, which ultimately negatively affect the hydrodynamics of the reactor [18].

Table 1. Typical composition of syngas and other potential gas streams from various sources

	Composition vol%, dry basis						Ref.
	CO	CO_2	H_2	N_2	CH_4	Other	
Non-biomass source							
Coal gasification	59.4	10	29.4	0.6	0	0.6	[25]
Coke oven gas	5.6	1.4	55.4	4.3	28.4	4.9	[25]
Partial oxidation of heavy fuel oil	47	4.3	46	1.4	0.3	1	[26]
Hardwood chips + 20 wt%liquid crude glycerol	19.73	11.67	19.38	NR*	3.82	NR*	[19]
Steam reforming of natural gas	15.5	8.1	75.7	0.2	0.5	0	[25]
Steam reforming of Naphtha	6.7	15.8	65.9	2.6	6.3	2.7	[25]
Water gas	30	3.4	31.7	13.1	12.2	9.6	[25]
Steel Mill	44	22	2	32	0	0	[27]
Biomass and organic waste source							
Biomass and organic waste source							
Demolition wood + sewage sludge	10.53	15.02	8.02	60.46	3.19	2.78	[28]
Cacao shell	8	16.02	9.02	61.45	2.34	3.17	[28]
Dairy biomass	8.7	15.7	18.6	56	0.6	0.4	[29]
Switchgrass	14.7	16.5	4.4	56.8	4.2	3.4	[13]
Kentucky bluegrass straw	12.9	17.4	2.6	64.2	2.1	0.8	[30]
Willow	9.4	17.1	7.2	60.42	3.3	2.58	[28]

Note: NR, not reported

The factors that determine which type of gasifier to employ are scale of operation, feedstock size and composition, tar yield and sensitivity towards ash [18]. Currently, three main types of gasifier are commercially employed: fixed bed, fluidized bed and entrained flow reactors [18].

6. ENTRAINED FLOW REACTOR

Entrained flow reactor is the preferred route for large scale gasification of coal, petcoke and refinery residues because of high carbon conversion efficiencies and low tar production [22]. This mode of gasifier does not require inert bed material but relies on feeding the feedstocks co-currently with oxidizing agent at high velocity to achieve a pneumatic transport regime [18]. At operating temperature of 1200-1500°C, this method is able to convert tars and methane, resulting in better syngas quality [18]. Importantly this technology requires the feedstocks to be pulverised into fine particles of ~50 μm before feeding, which is not a major issue for coal but very difficult and costly for biomass sources [18, 22].

7. MICROBES AND BIOCHEMISTRY OF GAS FERMENTATION

Acetogens are defined as obligate anaerobes that utilize the reductive acetyl-CoA pathway for the reduction of CO_2 to the acetyl moiety of acetyl-coenzyme A (CoA), for the conservation of energy, and for the assimilation of CO_2 into cell carbon [33]. In addition to the reductive acetyl-CoA pathway, four other biological pathways are known for complete autotrophic CO_2 fixation: the Calvin cycle, the reductive tricarboxylic acid (TCA) cycle, the 3-hydroxypropionate/malyl-CoA cycle and the 3-hydroxypropionate/4-hydroxybutyrate cycle [34]. Since the earlier atmosphere of earth was anoxic and the acetyl-CoA pathway is biochemically the simplest among the autotrophic pathways (the only linear pathway, whereas the other four pathways are cyclic), it has been postulated to be the first autotrophic process on earth [35, 36]. The reductive acetyl-CoA pathway is also known as the 'Wood-Ljungdahl' pathway, in recognition of the two pioneers, Lars G. Ljungdahl and Harland G. Wood, who elucidated the chemical and enzymology of the pathway using *Moorella thermoacetica* (formerly: *Clostridium thermoaceticum*) [35] or CODH/ACS pathway after the key enzyme of the pathway Carbon Monoxide dehydrogenase/Acetyl-CoA synthase. This ancient pathway is diversely distributed among at least 23 different bacterial genera: *Acetitomaculum*, *Acetoanaerobium*, *Acetobacterium*, *Acetohalobium*, *Acetonema*, *Alkalibaculum*, "*Bryantella*", "*Butyribacterium*",*Caloramator*, *Clostridium*, *Eubacterium*,*Holophaga*, *Mo orella*, *Natroniella*, *Natronincola*, *Oxobacter*, *Ruminococcus*, *Sporomusa*,*Sy ntrophococcus*, *Tindallia*, *Thermoacetogenium*, *Thermoanaerobacter*, and *Treponema* [33]. A selection of mesophilic and thermophilic acetogens are presented in Table 2. Acetogens are able to utilize gases $CO_2 + H_2$, and/or CO to produce acetic acid and ethanol according to the following stoichiometries:

$$2\,CO_2 + 4\,H_2 \Rightarrow CH_3COOH + 2\,H_2O \qquad \Delta H = -75.3\ kJ/mol \tag{1}$$

$$2\,CO_2 + 6\,H_2 \Rightarrow C_2H_5OH + 3\,H_2O \qquad \Delta H = -97.3\ kJ/mol \tag{2}$$

$$4\,CO + 2\,H_2O \Rightarrow CH_3COOH + 2\,CO_2 \qquad \Delta H = -154.9\ kJ/mol \tag{3}$$

$$6CO + 3\,H_2O \Rightarrow C_2H_5OH + 4\,CO_2 \qquad \Delta H = -217.9\ kJ/mol \tag{4}$$

Table 2. Acetogens

Species	Substrate	Product(s)	T_{opt} (°C)	pH_{opt}	Genome Status	Ref.
Mesophilic Microorganisms						
Acetobacterium woodii	H_2/CO_2, CO	Acetate	30	6.8	Available	[41, 42]
Acetonema longum	H_2/CO_2	Acetate, butyrate	30-33	7.8		[43]
Alkalibaculum bacchi	H_2/CO_2, CO	Acetate, ethanol	37	8.0-8.5		[44, 45]
Blautia producta	H_2/CO_2, CO	Acetate	37	7		[46]
Butyribacterium methylotrophicum	H_2/CO_2, CO	Acetate, ethanol, butyrate, butanol	37	6		[47-49]
Clostridium aceticum	H_2/CO_2, CO	Acetate	30	8.3	Under construction	[50-52]
Clostridium autoethanogenum	H_2/CO_2, CO	Acetate, ethanol, 2,3-butanediol, lactate	37	5.8-6.0		[27, 53]
Clostridium carboxidivorans or "P7"	H_2/CO_2, CO	Acetate, ethanol, butyrate, butanol, lactate	38	6.2	Draft	[54, 55]
Clostridium drakei	H_2/CO_2, CO	Acetate, ethanol, butyrate	25-30	5.8- 6.9		[55-57]
Clostridium formicoaceticum	CO	Acetate, formate	37	NR		[50, 58, 59]
Clostridium glycolicum	H_2/CO_2	Acetate	37-40	7.0-7.5		[60, 61]
Clostridium ljungdahlii	H_2/CO_2, CO	Acetate, ethanol, 2,3-butanediol, lactate	37	6	Available	[27, 62, 63]
Clostridium magnum	H_2/CO_2	Acetate	30-32	7.0		[64, 65]
Clostridium mayombei	H_2/CO_2	Acetate	33	7.3		[66]
Clostridium methoxybenzovorans	H_2/CO_2	Acetate, formate	37	7.4		[67]
"Clostridium ragsdalei" or "P11"	H_2/CO_2, CO	Acetate, ethanol, 2,3-butanediol, lactate	37	6.3		[68]
Clostridium scatologenes	H_2/CO_2, CO	Acetate, ethanol, butyrate	37-40	5.4-7.5		[55, 56]
Eubacterium limosum	H_2/CO_2, CO	Acetate	38-39	7.0-7.2	Available	[41, 69]
Oxobacter pfennigii	H_2/CO_2, CO	Acetate, butyrate	36-38	7.3		[70]
Oxobacter pfennigii	H_2/CO_2, CO	Acetate, butyrate	36-38	7.3		[70]
Thermophilic Microorganisms						
Moorella thermoacetica	H_2/CO_2, CO	Acetate	55	6.5-6.8	Available	[71-73]
Moorella thermoautotrophica	H_2/CO_2, CO	Acetate	58	6.1		[74]
Thermoanaerobacter kiuvi	H_2/CO_2	Acetate	66	6.4		[72]

Notes: NR, not reported

The Acetyl-CoA pathway is essentially a terminal electron-accepting process that assimilates CO_2 into biomass [35]. It constitutes an *Eastern* (or Carbonyl) branch and a *Western* (or Methyl) branch (Figure 2.). The *Western* branch employs a series of enzymes to carry out a six-electron reduction of CO_2 to the methyl group of acetyl-CoA, starting from the conversion of CO_2 to formate by formate dehydrogenase. Formyl-H_4folate synthase then condenses formate with H_4folate to form 10-formyl-H_4folate, which is then converted to 5,10-methenyl-H_4folate by a cyclohydrolase. This is followed by a dehydrogenase that reducesmethenyl- to 5,10-methylene-H_4hydrofolate, before (6S)-5-CH_3-H_4folate is formed by methylene-H_4folate reductase [37]. A B12-depedent methyltransferase (MeTr) then transfer the methyl group of (6S)-5-CH_3-H_4folate to corrinoid iron-sulphur protein (CoFeSP) of the bi-functional carbon monoxide dehydrogenase/acetyl-CoA synthase (CODH/ACS) complex [37]. The bi-functional CODH/ACS enzyme complex is formed by two autonomous proteins, an $\alpha_2\beta_2$ tetramer (CODH/ACS) and a $\gamma\delta$ heterodimer (CoFeSP), and the genes are often arranged in an operon, together with MeTr [37, 38]. In the *Eastern* branch, the CODH component catalyzes the reduction of CO_2 to CO. The central molecule, acetyl-CoA, is finally generated when CO, methyl group (bound to CoFeSP) and CoASH are condensed by ACS. Given the pivotal role of CODH/ACS, it is unsurprising that this complex was found to be the most highly expressed transcripts under autotrophic conditions in*C. autoethanogenum* [27], and can represent up to 2% of the soluble cell protein of an acetogen [39]. CODH/ACS is not unique to acetogenic bacteria, as it is also present in sulphate-reducing bacteria, desulfitobacteria, and Archaea (methanogens and Archaeoglobus) [38, 40].

The reducing equivalents required for fixation of CO_2 carbon into acetyl-CoA come from the oxidation of molecular hydrogen under chemolithoautotrophic growth, or NADH and reduced ferredoxin under heterotrophic growth [75]. An extensive review by Calusinska *et al.* (2010) highlighted the diversity of ubiquitous hydrogenases that Clostridia possess although only one acetogen*C. carboxidivorans* was included in this study [76], which catalyze the reversible oxidation of hydrogen:

$$H_2 \Leftrightarrow 2\,H^+ + 2\,e^- \tag{5}$$

The direction of the hydrogenase reaction is directed by the redox potential of the components able to interact with the enzyme. Hydrogen evolution occurs when electron donor is available, whereas the presence of electron acceptor results in hydrogen oxidation [77]. Hydrogenases can be classified into three phylogenetically distinct classes of metalloenzymes: [NiFe]-, [FeFe]-, and [Fe]-hydrogenases [76]. In *Methanosarcina barkeri*, the Ech hydrogenase, a [NiFe]-hydrogenase, was demonstrated to oxidize H_2 to reduce ferredoxin [78]. During acetoclastic methanogenesis, Ech hydrogenase oxidize ferredoxin to generate H_2 [78]. Although genome analysis revealed the presence of Ech-like hydrogenase in *C. thermocellum, C. phytofermentans*, *C. papyrosolvens*, and *C. cellulolyticum*, their physiological roles remained unknown [76]. Clostridia harbour multiple distinct [FeFe]-hydrogenases, perhaps reflecting their ability to respond swiftly to changing environmental conditions [76]. The monomeric, soluble [FeFe]-hydrogenase of *C. pasteurianum* is one of the best studied. It transfer electrons from reduced ferredoxins or flavodoxins to protons, forming H_2 [79]. A trimeric [FeFe]-hydrogenase found in *C. difficile*, *C. beijerinckii*, and *C. carboxidivorans* were hypothesized to couple formate oxidation to reduce protons into H_2 [76]. In *Thermotoga maritima*, an electron bifurcating, trimeric [FeFe]-hydrogenase was identified, that was shown to simultaneously oxidize reduced ferredoxin and NADH to evolve hydrogen under low H_2 partial pressure [80]. Under high H_2 partial pressure, the authors hypothesized that the NADH is oxidized to produce ethanol. *In silico* analysis revealed homologs of this bifurcating hydrogenase in a few Clostridia including *C. beijerinckii* and *C. thermocellum* [80]. In addition to classical hydrogenases, CODH/ACS and pyruvate:ferredoxin oxidoreductase (PFOR) from *M. thermoacetica* were shown to have hydrogen evolving capability, possibly as a mean of disposing excess reducing equivalents when electron carriers are limited and/or CO concentration is sufficient to inhibit conventional hydrogenases [81].

Most acetogens are also able to utilize another gas carbon monoxide (CO). In contrast to CO_2, CO can serve as both a source of carbon btut also as source of electrons such that hydrogen is not necessarily required. With a CO_2/CO reduction potential of -524 to -558mV, CO is approximately 1000-fold more capable of generating extremely low potential electrons than NADH, capable of reducing cellular electron carriers such as ferredoxin and flavodoxin [38, 82]. The reducing equivalents generated from CO oxidation can be coupled to reduction of CO_2 into acetate, butyrate and/or methane, evolution of molecular hydrogen from protons, reduction of nitrate/nitrite, reduction of sulfur species and reduction of aldehydes into alcohols [35, 83]. However, relatively few microorganisms are able to utilize CO as sole carbon and energy source, probably due to growth inhibition from sensitivity of their metalloproteins and hydrogenases towards CO [38, 83]. During exponential growth of *Pseudomonas carboxydovorans* (an aerobic carboxydotroph), it was demonstrated via immunological localization studies that 87% of the key enzyme CODH is associated with the inner cytoplasmic membrane, but this association was lost at the end of the exponential growth phase and a reduction in CO-dependent respiration rate was observed [84, 85]. It should be mentioned that aerobic and anaerobic CODH enzymes are structurally very different. CODH has been reported to be a very rapid and efficient CO oxidizer at rates between 4,000 and 40,000 s^{-1}, and reduces CO_2 at $11s^{-1}$ [86, 87]. Other electron donors commonly used by acetogens include formate, CH_3Cl, lactate, pyruvate, alcohols, betaine, carbohydrate, acetoin, oxalate and citrate [88]. CODH is able to split water in a biological water-gas shift reaction into hydrogen and electron according to the stoichiometry:

$$CO + H_2O \Rightarrow CO_2 + 2H^+ + 2e^- \quad (6)$$

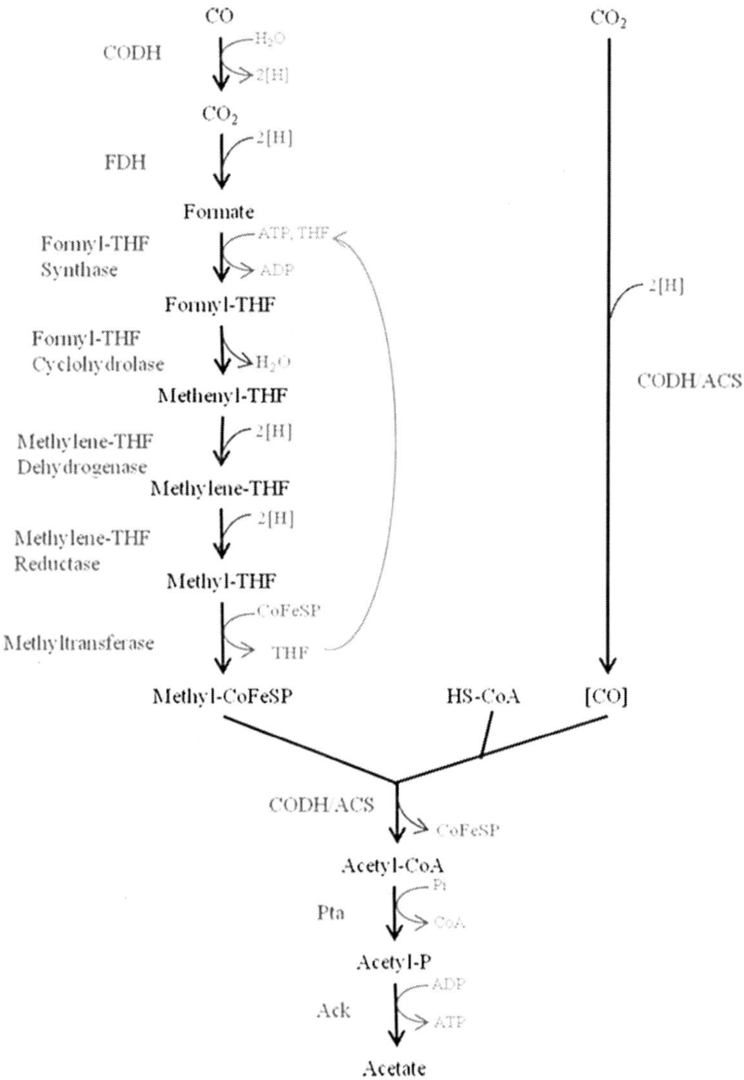

Figure 2. Wood-Ljungdahl Pathway. Ack, acetate kinase; ACS, acetyl-CoA synthase; CODH, carbon monoxide dehydrogenase; CoFeSP, corrinoid iron sulfur protein; FDH, formate dehydrogenase; Pta, phosphotransacetylase; THF, tetrahydrofolate.

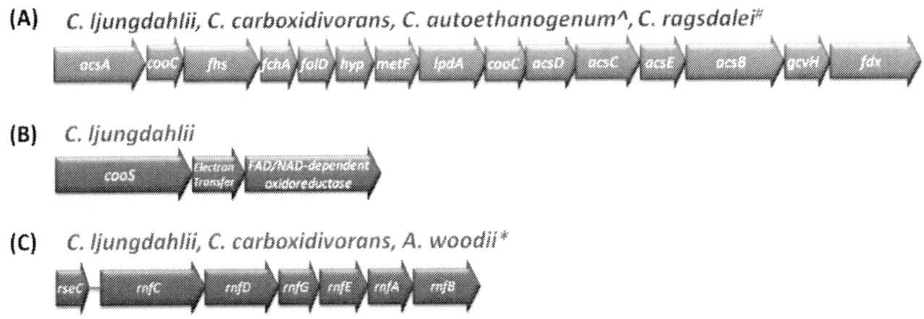

Figure 3. The organization of genes involved in acetogenesis and energy conservation from sequenced key acetogens. (A) Wood-Ljungdahl cluster; (B) carbon monoxide dehydrogenase (CODH) cluster; (C) Rnf complex cluster. *acsA*, CODH subunit; *acsB*, ACS subunit; *acsC*, corrinoid iron-sulfur protein large subunit; *acsD*, corrinoid iron-sulfur protein small subunit; *acsE*, methyltransferase subunit;*cooC*, gene for CODH accessory protein; *cooS*, CODH; *fchA*, formimido-tetrahydrofolate cyclodeaminase; *fdx*, ferredoxin; *fhs*, formyl-tetrahydrofolate synthase; *folD*, bifunctional methylene-tetrahydrofolate dehydrogenase/formyl-tetrahydrofolate cyclohydrolase; *gcvH*, gene for glycine cleavage system H protein; *hyp*, hypothetical protein; *lpdA*, dihydrolipoamide dehydrogenase; *metF*, methylene-tetrahydrofolate reductase; *rnfA, rnfB, rnfC, rnfD, rnfE, rnfG,*electron transport complex protein subunits; *rseC,* sigma E positive regulator., truncated *acsA*. #, truncated fdx. *, lack *rseC*.

The operation of this water gas shift reaction is the biochemical basis for the tremendous flexibility that acetogens have in terms of input gas composition. Via this reaction these organisms can flexibly use CO or H_2 as a source of electrons.Recently, some acetogens such as *C. ljungdahlii*, *C. aceticum*,*M. thermoacetica*, *Sporomusa ovata*, and *S. sphaeroides* have additionally been shown to utilize electrons derived from electrodes to reduce CO_2 into organic compounds such as acetate, formate, fumarate, caffeine, and 2-oxo-butyrate [89]. Termed microbial electrosynthesis, this nascent concept offers another route for acetogens to harvest the electrons generated from sustainable sources (e.g. solar and wind) to reduce CO_2 into useful multi-carbon products such as biofuels [90].

Under chemolithoautotrophic conditions, acetogenesis must not only fix carbon but also conserve energy. Approximately 0.1 mol of ATP is required for generation of 1g of dry biomass in anaerobes [82]. Acetyl-CoA is an energy rich molecule that through the combined actions of Pta (phosphotransacetylase) and Ack (acetate kinase), one ATP can be generated via substrate level phosphorylation (SLP). However, the activation of formate to 10-formyl-H4folate in the methyl-branch of Acetyl-CoA pathway consumes one ATP so no net gain in ATP is achieved via this mechanism [35, 75]. Furthermore, the reduction of CO_2 to the carbonyl group also requires energy, estimated at one third of ATP equivalent [35]. Recent advances indicated that other modes of energy conservation such as electron transport phosphorylation (ETP) or chemiosmotic processes that are coupled to the translocation of protons or sodium ions are implicated in acetogens. Acetogens such as *M. thermoacetica* harbour membrane-associated electron transport system containing cytochrome, menaquinones, and oxidoreductases that translocate H^+ out of the cell [33]. For acetogens that lack such membranous electron transport system, such as *Acetobacterium woodii* and *C. ljungdahlii*, a membrane-bound corrinoid protein is hypothesized to facilitate extrusion of Na^+ or protons during the transfer of methyl group from methyl-H_4F to CODH/ACS [75]. However, all enzymes involved are predicted to be soluble rather than membrane bound. Recent evidence suggested coupling to an Rnf complex in *A. woodii*, and *C. ljungdahlii* (Figure 3) which acts as ferredoxin:NAD^+-oxidoreductase [62, 91-93]. The Rnf complex is also found in other Clostridia (but not in ABE model organism *C. acetobutylicum*) and bacteria, and was originally discovered in *Rhodobacter capsulatus* where it is involved in nitrogen fixation [93]. Using reduced ferredoxin (Fd^{2-}) generated from CO oxidation, carbohydrate utilization and/or hydrogenase reactions, this membrane-bound electron transfer complex is predicted to reduce NAD^+ with concomitant translocation of Na^+/ H^+. The ion gradient generated from the above processes is harvested by H^+- or Na^+- ATP synthase to generate ATP [33, 93]. The recent genome sequencing of *A. woodii* revealed that Rnf complex is likely to be the only ion-pumping enzyme active during autotrophic growth and the organism's entire catabolic metabolism is optimized to maximize the Fd^{2-}/NAD^+ ratio [42]. Recently, a third mechanism of energy conservation which involves bifurcation of electrons by hydrogenases was proposed for anaerobes [94] and demonstrated for enzymes hydrogenase (see above; [80]), butyryl-CoA dehydrogenase [94, 95], or an iron-sulfur flavoprotein Nfn [96]. A similar mechanism has also been proposed for the methylene-THF reductase of the reductive acetyl-CoA pathway, which would enable this highly exergonic reduction step ($\Delta G^{0'} = -22$ kJ/mol) to be coupled with the Rnf complex for additional energy conservation [62]. However, no experimental proof to support this hypothesis has been published to date.

In an attempt to generate an autotrophic *E. coli*, the genes encoding MeTr, the two subunits of CODH/ACS, and the two subunits of CoFeSP from *M. thermoacetica* were cloned and heterologously expressed in *E. coli* [97]. Although the MeTr was found to be active, the other subunits misassembled hence no active enzymes were found [97]. Autotrophic capability is clearly a very complex process that involves many genes other than the CODH/ACS complex and tetrahydrofolate pathway, including compatible cofactors, electron carriers, specific chaperones and energy conservation mechanisms. For instance, more than 200 genes are predicted to be involved in methanogenesis and energy conservation from CO_2 and H_2 in methanogens [98]. A recent patent application described the introduction of three Wood-Ljungdahl pathway genes encoding MeTr, CoFeSP subunit α and β from *C. difficile* into *C. acetobutylicum* [99]. The recombinant strain was shown to incorporate more CO_2 into extracellular products than wild-type [99].

8. PRODUCTS OF GAS FERMENTATION

Acetyl-CoA generated via the Wood-Ljungdahl pathway serves as key intermediate for synthesis of cell mass as well as products. All acetogens are described to produce acetate, in order to gain energy via SLP to compensate for the energy invested in activating formate in the *Western* branch of the reductive acetyl-CoA pathway. Acetate and ATP are formed via acetyl-phosphate through the successive actions of Pta and Ack. *pta* and *ack* are arranged in the same operon and they were reported to be constitutively expressed [100]. With CO_2 and H_2 as substrate, only acetate has been observed as major product [44], with minor amounts of ethanol produced in rare cases with *C. ljungdahlii* [101], *C. autoethanogenum* [53], or "*Moorella sp.*" [102, 103]. Using the more reduced substrate CO, production of a range of other products have been reported, such as ethanol, butanol, butyrate, 2,3-butanediol [104], and lactate (Figure 4.) [105]. From a biofuel perspective, ethanol and butanol are of particular interest. Ethanol and butanol have even been described as the main fermentation products over acetate in some acetogens under specific conditions. Ethanol producers include *C. ljungdahlii* [62,63], *C. autoethanogenum* [53], "*C. ragsdalei*" ("*Clostridium* strain P11") [106, 107], "*Moorella sp.*" [102, 103], *Alkalibaculum bacchii* [44], *C. carboxidivorans* ("*Clostridium* strain P7") [54, 55], and *B. methylotrophicum* [49, 108]. The latter two have also been described to produce butanol.

Due to historical roles in ABE fermentation, organisms like *C. acetobutylicum*, *C. beijerinckii*, *C. saccharobutylicum*, and *C. saccharoperbutylacetonicum* have been much more extensively characterized than acetogenic Clostridia [95]. Since *C. acetobutylicum* was the first *Clostridium* to be fully sequenced [109] and it remains the most commonly used species for industrial production of solvents to date [110], it provides a model for study of solventogenesis. Although sugar- and starch-utilizing ABE Clostridia and acetogens exhibit clear distinctions in substrate utilization and thus metabolism, they share some similarities in the biochemical pathway and genetic organization of product synthesis and can be used as model for comparison. Structure of key genes and operons (except for the absence of acetone biosynthetic genes) have been found to be very similar in sequenced acetogen *C. carboxidivorans* [54], and in respect of acetate and ethanol genes to some extent also in *C. ljungdahlii* [62]. For instance, the operon structure of *pta-ack*, *ptb-buk* and the *bcs* cluster of acetogen *C. carboxidivorans* are highly similar to starch-utilizing *C. acetobutylicum* and *C. beijerinckii* [54,109] (Figure 5). Due to these reasons, solventogenic genes from starch-utilizing Clostridia are ideal targets for heterologous expression in acetogens for improvement of product yield and expansion of product range.

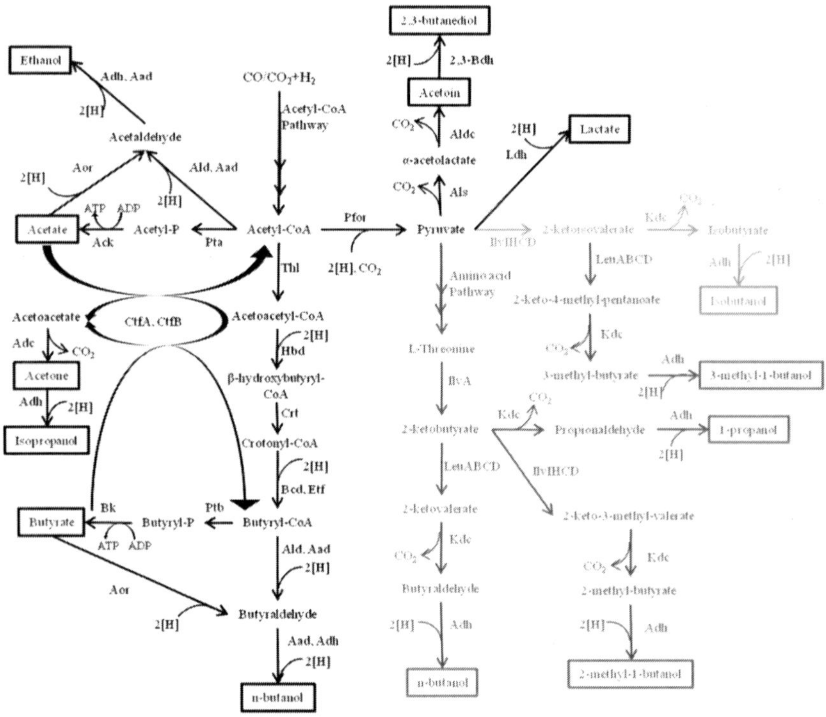

Figure 4. Scheme of metabolite production from gas fermentation using native and genetically modified Clostridia. Black denotes well-characterized pathways in Clostridia. Blue shows demonstrated heterologous pathways that have been engineered into Clostridia. Purple designates hypothetical pathways that can be engineered into Clostridia. Products are highlighted in boxes. Aad, aldehyde/alcohol dehydrogenase; Ack, acetate kinase; Adc, acetoacetate decarboxylase; Adh, alcohol dehydrogenase; Ald, aldehyde dehydrogenase; Aldc, acetolactate decarboxylase; Aor, aldehyde oxidoreductase; Bcd, butyryl-CoA dehydrogenase; Bk, butyrate kinase; Crt, crotonase; CtfA & CtfB, CoA transferase A & B; Etf, electron-transferring flavoprotein; Hbd, hydroxybutyryl-CoA dehydrogenase; IlvA, threonine deaminase; IlvIHCD, valine and isoleucine biosynthesis; Kdc, 2-ketoacid decarboxylase; Ldh, lactate dehydrogenase; LeuABCD, leucine and norvaline biosynthesis; Pfor, Pyruvate ferredoxin oxidoreductase; Pta, phosphotransacetylase; Ptb, phosphotransbutyrylase; Thl, thiolase; 2,3-Bdh, 2,3-butanediol dehydrogenase.

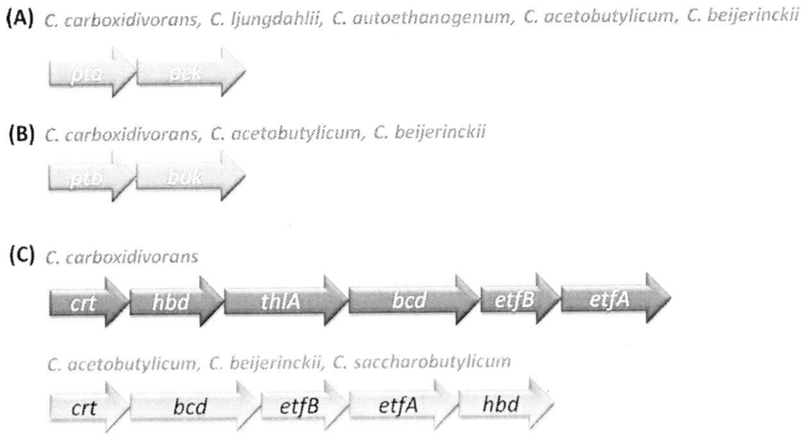

Figure 5. Similarity of acidogenesis and butanol formation gene clusters of acetogens and sugar-utilizing Clostridia. (A) Acetate-forming operon; (B) butyrate-forming operon; (C) butanol-forming operon.*ack*, acetate kinase; *buk,* butyrate kinase; *bcd*, butyryl-CoA dehydrogenase; *crt*, crotonase; *etfA,*electron-transferring flavoprotein subunit A; *etfB*, electron-transferring flavoprotein subunit B; *hbd,*3-hydroxybutyryl-CoA dehydrogenase; *ptb*, phosphotransbutyrylase; *thlA*, thiolase.

Similar to sugar- and starch-utilizing ABE Clostridia, acetogens such as *C. carboxidivorans* [111, 112],*C. ljungdahlii* [113], and *C. autoethanogenum* [27] also typically undergo biphasic fermentation under autotrophic conditions. The first phase involves the production of carboxylic acids (acidogenic), $H_2$and CO_2 during exponential growth. This is followed by the solventogenic phase in which part of the produced acids are reassimilated or reduced into solvents, which usually occurs during stationary growth phase [114]. This shift from acidogenesis to solventogenesis is of industrial importance and several transcriptional analysis on *C. acetobutylicum* [100, 115], and *C. beijerinckii* [116] have been performed to shed light on this process. In both organisms, the onset of solventogenesis coincides with an increase in expression of master sporulation/solventogenesis regulator gene *spo0A*, solventogenic genes such as *ald*, *ctfA-ctfB*, and *adc*, as well as down-regulation of chemotaxis/motility genes [100,115, 116]. Physiologically, the signals that induce solventogenesis were hypothesized to involve temperature, low pH, high concentrations of undissociated acetic and butyric acids, limiting concentrations of sulphate or phosphate, ATP/ADP ratio and/or NAD(P)H levels [117].

For Clostridia such as acetogen *C. carboxidivorans* [54], which harbour the genes thiolase (*thlA*), 3-hydroxybutyryl-CoA dehydrogenase (*hbd*), crotonase (*crt*) and butyryl-CoA dehydrogenase (*bcd*), the two carbon acetyl-CoA can be converted to four carbon butyryl-CoA [95]. ThlA compete with the activities of Pta, Ald (aldehyde dehydrogenase), and PFOR to condense two acetyl-CoA into one acetoacetyl-CoA, and plays a key role in regulating the C2:C4 acid ratio [110, 118]. Since the formation of acetate yields twice as much ATP per mole of acetyl-CoA relative to butyrate formation, thiolase activity indirectly affects ATP yield [118]. Under physiological conditions, Crt catalyzes dehydration of β-hydroxybutyryl-CoA to crotonyl-CoA [119]. Bcd was shown to require a pair of electron transfer flavoproteins (EtfA and EtfB) to convert crotonyl-CoA to butyryl-CoA [120]. Furthermore, the Bcd was demonstrated to form a stable complex with EtfA and EtfB, and they were shown to couple the reduction of crotonyl-CoA to butyryl-CoA with concomitant generation of reduced ferredoxins, which can be used for energy conservation via Rnf complex [94, 119]. Subsequent actions of phosphotransbutyrylase (*ptb*) and butyrate kinase (*buk*) then generate ATP and butyrate from butyryl-CoA [118].

Under low extracellular pH of 4-4.5, the secreted undissociated acetic acid (pK_a 4.79) and/or butyric acid (pK_a 4.82) diffuse back into cell cytoplasm and then dissociate into the respective salts and protons because of the more alkaline intracellular conditions. Without further interventions, the result of this is abolishment of the proton gradient and inevitable cell death [95]. The conversion of acetate and butyrate into solvents increase the pH, thus provide some time for the organism to sporulate and secure long term survival. However, the solvents produced are toxic because they increase membrane fluidity and disrupt critical membrane-associated functions such as ATP synthesis, glucose uptake and other transport processes [114, 121]. In *C. acetobutylicum*, it has been demonstrated that the addition of 7-13 g/l of butanol, or up to 40 g/l of acetone and ethanol resulted in 50% growth inhibition [122]. The bacterium is likely to experience a different cytotoxic effect from endogenously produced solvents because the organism has time to adapt to increasing amount of solvents.

The reassimilation of acetate and butyrate into the respective acyl-CoA and acetoacetate is catalyzed by acetoacetyl-CoA:acetate/butyrate CoA transferase (CtfA and CtfB) [110, 117, 118]. Acetoacetate is deconstructed by acetoacetate decarboxylase (Adc) into acetone and CO_2. This enzyme is missing in acetogenic *C. carboxidivorans* compared to the ABE strains [54, 123]. Some ABE strains such as *C. beijerinckii* NRRL B593 also possess a primary/secondary alcohol dehydrogenase that converts acetone to isopropanol [124]. In acetogenic "*C. ragsdalei*", reduction of acetone to isopropanol was also observed although the mechanism of this reduction is as yet unknown [124, 125]. Again, *C. carboxidivorans* lacks this activity [125]. The recycled acetyl-CoA and butyryl-CoA can be converted to ethanol and butanol through the actions of coenzyme A-acylating aldehyde dehydrogenase (Ald) and alcohol dehydrogenase (Adh) [110, 118]. Ald converts acyl-CoA into aldehydes, and the enzyme has been purified from *C.*

beijerinckii NRRL B593 and was shown to be NADH-specific, exhibit higher affinity with butyraldehyde than acetaldehyde, but possess no Adh activity [126]. In *C. ljungdahlii*, two variants of aldehyde:ferredoxin oxidoreductases (AOR) are present in the genome, and they are hypothesized to couple reduced ferredoxin from CO oxidation via the CODH (see above) to perform the reversible reduction of acetate into acetaldehyde, which can be further reduced into ethanol [62].

The final step of solventogenesis utilizes Adh to reduce acetaldehyde and butyraldehyde into ethanol and butanol, respectively. For ethanol synthesis, transposon mutagenesis and enzymatic assay in *C. acetobutylicum* showed the involvement of a specific Ald that does not interact with butyryl-CoA, and a NAD(P)H-dependent Adh [127, 128]. The production of butanol by *C. acetobutylicum* is mainly due to the action of butanol dehydrogenase A and B (BdhA and BdhB), and bifunctional butyraldehyde/butanol dehydrogenase 1 and 2 (AdhE1 and AdhE2) [95]. In *C. carboxidivorans* [54] and *C. ljungdahlii* [62] both *adhE1* and *adhE2* are arranged in tandem and separated by a 200bp gap which contains a putative terminator [62, 111]. This is likely the result of gene duplication [62]. qRT-PCR analysis from *C. carboxidivorans* fed with syngas showed that the two *adhE* showed differential expression, and the more abundant *adhE2* was significantly upregulated over 1000 fold in a time span that coincided with the greatest rate of butanol production [111].

Pyruvate is a central molecule for anabolism and it is predominantly generated from glycolysis during heterotrophic growth. But under autotrophic growth, this four carbon molecule can be synthesized by PFOR and potentially also the pyruvate-formate lyase (PFL). Two variants of PFOR were reported in *C. autoethanogenum*, and transcriptional analysis showed that they were differentially expressed when grown using industrial waste gases (containing CO, CO_2 and H_2) [104]. Unlike PFL from most other microorganisms that only catalyze the lysis of pyruvate into formate and acetyl-CoA, clostridial PFL (*C. kluyveri, C. butylicum*, and *C. butyricum*) were reported to readily catalyze the reverse reaction (i.e. pyruvate formation) [129]. Apart from roles in anabolism, pyruvate is also a precursor to other products such as lactic acid and 2,3-butanediol. Small amounts of lactic acid are converted from pyruvate in acetogens, a reaction which is catalyzed by lactate dehydrogenase (Ldh) [104, 118]. Recently, Köpke *et al.* (2011) reported the production of 2mM 2,3-butanediol from acetogenic bacteria (*C. autoethanogenum*, *C. ljungdahlii*, and *C. ragsdalei*) using industrial waste gases (containing CO, CO_2 and H_2) as feedstock [104]. Pyruvate is first converted into α-acetolactate by the enzyme acetolactate synthase, followed by acetolactate decarboxylase which split acetolactate into acetoin and CO_2, before a final reduction of acetoin into 2,3-butanediol by 2,3-butanediol dehydrogenase [104] (Figure 4).

9. STRAIN IMPROVEMENT AND METABOLIC ENGINEERING

The genomes of several solventogenic Clostridia, including gas fermenting species, have been sequenced since 2001 [54, 62, 109, 119, 123, 130], and an array of transcriptomic [100, 116, 121,131, 132], proteomic [132] and systems analysis [133, 134] are being made increasingly available. However, the generation of stable recombinant Clostridia has been severely hindered by the difficulties encountered introducing foreign DNA into cells and a lack of established genetic tools for this genera of bacteria. In comparison to starch-utilizing Clostridia, very little information is available for metabolic engineering of acetogens. Although this section describes recent advances in the development of genetic tools for mostly sugar-utilizing Clostridia, these techniques are highly relevant and applicable to the closely related acetogenic Clostridia for biofuels or chemical production via gas fermentation.

The ideal microbial catalyst for industrial scale gas fermentation might exhibit the following traits: high product yield and selectivity, low product inhibition, no strain degeneration, asporogenous, prolonged cell viability, strong aero-tolerance, high biomass density and efficient utilization of gas substrates. These can be achieved by directed evolution, random mutagenesis and/or targeted genetic engineering. Traditionally, chemical mutagenesis [135-137] and adaption strategies [138, 139] have been deployed to select for these traits. However, these strategies are limited and often come with the expense of unwanted events. First attempts of targeted genetic modification of Clostridia were made in the early 1990s by the laboratory of Prof. Terry Papoutsakis [140-142]. While these pioneering efforts relied on use of plasmids for (over)expression of genes in *C. acetobutylicum*, more sophisticated tools were later developed for a range of solventogenic and pathogenic Clostridia.

Antisense RNA (asRNA) has been employed to down-regulate genes. Here, single stranded RNA binds to a complementary target mRNA and prevents translation by hindering ribosome-binding site interactions [143]. For instance, this method has been used to knockdown *ctfB* resulting in production of 30 g/l solvents with significantly suppressed acetone yield in *C. acetobutylicum* ATCC 824 [144,145].

Several homologous recombination methods have been developed for integration or knock-out of genes in a range of sugar-utilizing Clostridia. In early stage, knockout mutants were almost exclusively generated from single crossover events that could revert back to wild-type [146-152], with stable double crossovers only observed in rare cases [153, 154]. For *C. acetobutylicum* [155] and cellulolytic *C. thermocellum* [156] counter selectable markers have been developed to allow more efficient screening for the rare second recombination event.

ClosTron utilizes the specificity of mobile group II intron Ll.*ltrB* from *Lactoccocus lactis* to propagate into a specified site in the genome via a RNA-mediated, retro-homing mechanism which can be used to disrupt genes [157]. This technique has initially been developed by InGex and Sigma-Aldrich under the name 'TargeTron™' and successfully adapted to a range of solventogenic and pathogenic Clostridia including *C. acetobutylicum*, *C. difficile*, *C. sporogenes*, *C. perfringens*, and *C. botulinum*[158-160] by the laboratory of Prof. Nigel Minton.

The same laboratory recently also developed another method for integration of DNA into the genome. Termed Allele-Coupled Exchange (ACE), this approach does not employ a counter selective marker to select for the rare second recombination event. Rather, it utilizes the activation or inactivation of gene(s) that result in a selectable phenotype, and asymmetrical homology arms to direct the order of recombination events [161]. Remarkably, the whole genome of phage lambda (48.5kb minus a 6kb region) was successfully inserted into the genome of *C. acetobutylicum* ATCC 824 in three successive steps using this genetic tool. This technique was also demonstrated in *C. difficile* and *C. sporogenes*[161].

For reverse engineering, mainly transposon mutagenesis has been utilized. Earlier efforts of transposon mutagenesis were demonstrated in *C. acetobutylicum* P262 (now: *C. saccharobutylicum* [162]), *C. acetobutylicum* DSM792, *C. acetobutylicum* DSM1732, and *C. beijerinckii* NCIM 8052, but issues with multiple transposon insertions per mutant, and non-random distribution of insertion were reported [163, 164]. Recent developments have seen the successful generation of mono-copy random insertion of transposon *Tn1545* into cellulolytic *C. cellulolyticum* [165] and mariner transposon *Himar1* into pathogenic *C. difficile* [166].

While there is still a lack of some other essential metabolic engineering tools such as efficient inducible promoters, the array of available tools that enabled significant improvements to the ABE process and cellulolytic Clostridia fermentations as summarized in Table 3.

In contrast, to date only a limited number of acetogenic Clostridia have been successfully modified. Pioneering work in this area has been undertaken in the laboratory of Prof. Peter Dürre. *C. ljungdahlii*, a species that does not naturally produce butanol, was modified with butanol biosynthetic genes (*thlA*, *hbd*, *crt*, *bcd*, *adhE* and *bdhA*) from *C. acetobutylicum* ATCC 824 resulting in production of up to 2 mM of butanol using synthesis gas as sole energy and carbon source [62]. By delivering a plasmid with acetone biosynthesis genes *ctfA*, *ctfB*, *adc*, and *thlA* in *C. aceticum,* production of up to 140 µM acetone was demonstrated from various gas mixes (80% H_2/20% CO_2 and 67% H_2/33% CO_2) [169, 170]. Recent patent filings by LanzaTech describe the production of butanol as main fermentation product and increased alcohol tolerance in genetically engineered acetogens. Up to 26 mM butanol were produced with genetically modified *C. ljungdahlii* and *C. autoethanogenum* using steel mill gas (composition 44% CO, 32% N_2, 22% CO_2, and 2% H_2) as the only source of carbon and energy when the butanol biosynthetic genes *thlA, hbd, crt, bcd, etfA,* and *etfB* were heterologously expressed [167]. Overexpression of native *groESL* operon in *C. autoethanogenum* resulted in a strain that displayed higher alcohol tolerance relative to wild-type when challenged with ethanol [168].

Table 3. Genetically modified solventogenic Clostridia

Organism	Genetic modification	Phenotypes/Effects	Ref
Acetogens			
C. ljungdahlii	Plasmid overexpression of butanol biosynthetic genes from *C. acetobutylicum* (*thlA, crt, hbd, bcd, adhE* and *bdhA*)	Produced 2 mM butanol from syngas	[62]
C. autoethanogenum	Plasmid overexpression of butanol biosynthetic genes from *C. acetobutylicum* (*thlA, crt, hbd, bcd, etfA, & etfB*)	Produced 26 mM butanol using steel mill gas	[167]
C. autoethanogenum	Plasmid expression of native *groES* and *groEL*	Increased alcohol tolerance	[168]
C. aceticum	Plasmid overexpression of acetone operon from *C. acetobutylicum* (*adc, ctfAB, thlA*)	Produced up to 140 μM acetone using gas	[169, 170]
Acidogenesis and Solventogenesis			
C. acetobutylicum	Inactivation of *buk* and overexpression of *aad*	Produced same amount of butanol as control but relatively more ethanol, corresponding to a total alcohol tolerance of 21.2 g/l	[171]
C. acetobutylicum	Inactivation of *hbd* using ClosTron	Produced 716 mM ethanol by diverting C4 products	[172]
C. acetobutylicum	Inactivation of *ack* using ClosTron	Reduction in acetate kinase activity by more than 97% resulted in 80% less acetate produced but similar final solvent amount	[173]
C. tyrobutylicum	Inactivation of *ack* and plasmid overexpression of *adhE2* from *C. acetobutylicum*	Produced 216 mM butanol	[174]

C. thermocellum	Inactivation of *ldh* and *pta* via homologous recombination	Showed 4 fold increase in ethanol yield (122 mM instead of 28 mM)	[156]
C. cellulolyticum	Inactivation of*ldh* and *mdh* (malate dehydrogenase) using ClosTron	Generated 8.5 times higher ethanol yield (56.4 mM) than wild type (6.5 mM)	[175]
C. acetobutylicum	Plasmid overexpression of a synthetic acetoneoperon (*adc, ctfA, ctfB*) and primary/secondary *adh* from *C. beijerinckii* NRRL B593	Produced 85 mM isopropanol	[176]
C. acetobutylicum	Genome insertion of *adh* gene from *C beijerinckii* NRRL B593 using Allele-Coupled Exchange	Converted acetone into 28 mM isopropanol without affecting the yield of other fermentation products	[161]
Biosynthesis of New Products			
C. cellulolyticum	Plasmid overexpression of *kivD, yqhD, alsS, ilvC* and *ilvD*	Produced 8.9 mM isobutanol by diverting 2-ketoacid intermediates	[177]
C. acetobutylicum	Plasmid expression of native *ribGBAH* operon and mutated PRPP amidotransferase	Produced 70 mg/l riboflavin and 190 mM butanol	[178]
Solvent- and Aero-tolerance			
C. acetobutylicum	Plasmid overexpression of glutathione *gshA* and *gshB* from *E. coli*	Improved aero- and solvent-tolerance	[179]
C. acetobutylicum	Plasmid overexpression of chaperone *groESL*	Showed 85% decrease in butanol inhibition and 33% increase in solvent yield	[180]
Substrate Utilization			
C. acetobutylicum	Plasmid expression of *acsC, acsD* and *acsE* from *C. difficile*	Increased incorporation of CO_2 into extracellular products	[99]
C. saccharoperbutylacetonicum strain N1-4	Knockdown hydrogenase *hupCBA* expression using siRNA delivered from plasmid	Significantly reduced hydrogen uptake activity to 13% (relative to control strain)	[181]

Besides the classical Clostridial butanol pathway (which constitutes genes *thlA, crt, hbd, bcd, etfA and etfB*; see earlier section), a non-fermentative approach has been described and demonstrated in *E. coli*for branched chain higher alcohol production [182]. This alternative approach requires a combination of highly active amino acid biosynthetic pathway and artificial diversion of 2-keto acid intermediates into alcohols by introduction of two additional genes: broad substrate range 2-keto-acid decarboxylase (*kdc*) which converts 2-keto acids into aldehydes, followed by *Adh* to form alcohols [182]. Engineered strains of *E. coli* have been shown to produce alcohols such as isobutanol, n-butanol, 2-methyl-1-butanol, 3-methyl-1-butanol and 2-phenylethanol via this strategy [182]. For instance, the overexpression of *kivD* (KDC from *Lactococcus lactis*), *adh2, ilvA,* and *leuABCD* operon, coupled with deletion of *ilvD* gene and supplementation of L-threonine, increased n-butanol yield to 9 mM while producing 10 mM of 1-propanol [182].An even more remarkable yield of 300 mM isobutanol was achieved through introduction of k*ivD, adh2, alsS* (from *B. subtilis*), and *ilvCD* into *E. coli* [182]. Like butanol, isobutanol exhibits superior properties as a transportation fuel when compared to ethanol [177]. By applying similar strategy into *C. cellulolyticum,* 8.9 mM isobutanol was produced from cellulose when k*ivD, yqhD, alsS, ilvC,* and *ilvD* were overexpressed [177]. This result suggests that such non-fermentative pathway is suitable target for metabolic engineering of acetogens for the biosynthesis of branched chain higher alcohols. Via synthetic biology and metabolic engineering, production of additional potential liquid transportation fuels like farnesese or fatty acid based fuels has successfully been demonstrated in *E. coli* or yeast from sugar [183, 184]. Given the unsolved energetics in acetogens, it is unclear if production of such energy dense liquid fuels could be viable via gas fermentation.

10. FERMENTATION AND PRODUCT RECOVERY

10.1. Bioreactor Design

An optimum gas fermentation system requires efficient mass transfer of gaseous substrates to the culture medium (liquid phase) and microbial catalysts (solid phase). Gas-to-liquid mass transfer has been identified as the rate-limiting step and bottleneck for gas fermentation because of the low aqueous solubility of CO and H_2, respectively at only 77% and 68% of that of oxygen (on molar basis) at 35°C [185]. Hence, a bioreactor design that delivers sufficient gas-to-liquid mass transfer in an energy-efficient manner at commercial scale for gas fermentation represents a significant engineering challenge. A brief overview of reactor configurations reported in gas fermentation operations is given below.

10.2. Continuous Stirred Tank Reactor (Cstr)

In continuous stirred tank reactor (CSTR), gas substrates are continuously fed into the reactor and mechanically sheared by baffled impellers into smaller bubbles, which has greater interfacial surface area for mass transfer [16]. In addition, finer bubbles have a slower rising velocity and a longer retention time in the aqueous medium, resulting in higher gas-to-liquid mass transfer [24]. Fermentation reactions using *C. ljungdahlii* have been successfully maintained in a 2 litre CSTR under autotrophic conditions for more than a month, while achieving peak ethanol level of 6.5 g/l and CO conversion rate of 93% [186]. The production of 49 g/l of ethanol from gas substrates using *C. ljungdahlii* was demonstrated using CSTR [113]. In another example, a 100 litre stirred tank reactor was demonstrated to produce up to 24.57 g/l ethanol, 9.25 g/l isopropanol and 0.47 g/l n-butanol during a 59-day semi-batch gas fermentation using "*C. ragsdalei*" strain P11 as biocatalysts [112]. An improved version of CSTR incorporates microbubble sparger to generate finer bubbles to achieve higher mass transfer coefficient [187]. Although CSTR offers complete mixing and uniform distribution of gas substrates to the microbes, the high power per unit volume required to drive the stirrer are thought to make this approach economically unviable for commercial scale gas fermentation systems [187].

10.3. Bubble Column Reactor

In contrast to CSTR, gas mixing in bubble column reactor is achievable by gas sparging, without mechanical agitation. This reactor configuration has fewer moving parts, and consequently has a lower associated capital and operational costs while exhibiting good heat and mass transfer efficiencies, making it a good candidate for large scale gas fermentation [17]. However, excessive level of gas inflow for enhanced mixing have been cited as an issue that leads to heterogeneous flow and back-mixing of the gas substrates [16, 17]. *C. carboxidivorans* strain P7 was cultured in a 4 litre bubble column reactor for 20 days using a combination of producer gas and synthetic syngas, generating a peak ethanol concentration of 6 g/l [13].

10.4. Immobilized Cell Column Reactor

One of the key challenges of gas fermentation is cell density. Immobilization of microbes through crosslinking or adsorption to insoluble biosupport materials and the subsequent packing within the column offers a range of benefits [14]. These include high cell densities, plug flow operation, high mass transfer rate via direct contact between microbe and gas, reduction of retention time, and operation without mechanical agitation [14, 16]. However, channelling issues may arise when the microbe overgrows and completely fill the interstitial space. Due to limitations in column dimensions and packing, this reactor configuration lacks flexibility to operate in various gas fermentation conditions [14, 16].

10.5. Trickle-Bed Reactor

Trickle-bed reactor is a gas- or liquid- continuous reactor consisting of packed bed, which liquid culture trickles down through packing media containing suspended or immobilized cells [16, 24, 187]. The gas substrate is delivered either co-currently or counter-currently to the liquid flow, and no mechanical agitation is required [187].In this reactor format, low gas and liquid flow rates are typically applied, generating relatively low pressure drops [187]. Trickle-bed reactor was found to exhibit excellent gas conversion rates and higher productivities than CSTR and bubble column reactor [15].

11. GAS FERMENTATION PARAMETERS

11.1. Gas Composition

The gas composition and its impurities can have an impact on the productivity of the gas fermentation process. Greater molar ratio of H_2:CO allows greater efficiency in the conversion of the carbon from CO into products such as ethanol, because reducing equivalents are generated from oxidation of H_2(rather than CO). However, CO is also a known inhibitor of hydrogenase which can affect utilization of H_2 during fermentation. In *B. methylotrophicum*, H_2 utilization was inhibited until CO was exhausted [108]. When CO is consumed, acetogens are able to grow using CO_2 and H_2. Common impurities from biomass gasification or other waste gases are tar, ash, char, ethane, ethylene, acetylene, H_2S, NH_3 and NO [17, 22, 24, 188].These have been shown to cause cell dormancy, inhibition of hydrogen uptake, low cell growth and shift between acidogenesis and solventogenesis in acetogens [13,188]. For instance, NH_3 from the feed gas readily convert into NH_4^+ in the culture media and these ions were recently shown to inhibit hydrogenase and cell growth of acetogen "*C. ragsdalei*" [189]. A number of strategies to mitigate the impact of such impurities have been proposed, for example installing 0.025 mm filters, or the use of gas scrubbers or cyclones, and improvement in gasification efficiency and scavenging for contaminants in the gas stream using agents such as potassium permanganate, sodium hydroxide or sodium hypochlorite [24, 190-192]. H_2S does not have a negative effect on acetogens such as *C. ljungdahlii* up to 5.2% (v/v) [193].

11.2. Substrate Pressure

The partial pressure of syngas components have a major influence on microbial growth and product profiles because the enzymes involved are sensitive to substrate exposure [194]. Due to the low solubility of CO and H_2 in water, the growth of dense bacterial cell cultures can face mass transfer limitations, so increasing the partial pressure of gaseous substrates can help alleviate this problem. For instance, studies in which the CO partial pressure (P_{CO}) increased from 0.35 to 2.0 atm showed that this resulted in a 440% increase in maximum cell density, a significant increase in ethanol productivity and a decrease in acetate production in *C. carboxidivorans* strain P7 [195]. In another study involving*C. ljungdahlii*, the increase of P_{CO} from 0.8 to 1.8 atm had a positive effect on ethanol production, and the microbe did not exhibit any substrate inhibition at high P_{CO} [196].In less CO-tolerant microorganisms, the effect of increasing P_{CO} partial pressure range from non-appreciable in the case of *Rhodospirillum rubrum* [197], to negative impact on doubling time of *Peptostreptococcus productus*(now: *Blautia product*) [194] and *Eubacterium limosum* [198]. Similar to CO, the increase in partial pressure of H_2 (pH_2) to 1700 mbar enhanced acetate productivity of *A. woodii* to 7.4g acetate/ l/day [199].

11.3. Medium Formulation

Although acetogens are able of utilizing CO and CO_2/H_2 as carbon and energy source, other constituents such as vitamins, trace metal elements, minerals and reducing agents are also required for maintenance of high metabolic activity [16, 113]. Studies indicated that formation of ethanol in solventogenic Clostridiais non-growth associated and limitation of growth by reducing availability of carbon-, nitrogen- and phosphate- nutrients shift the balance from acidogenesis to solventogenesis [113, 200, 201]. Optimization of medium formulation for *C. ljungdahlii* through reduction of B-vitamin concentrations and elimination of yeast extract significantly enhanced the final ethanol yield to 48 g/l in a CSTR with cell recycling (23 g/l without cell recycling) [113]. Another study by Klasson *et al.* showed thatthe replacement of yeast extract with cellobiose not only increased maximum cell concentration, but also enhanced ethanol yield by 4-fold [14]. Media formulation for *C. autoethanogenum* was investigated using Plackett-Burman and central composite designs, but only low ethanol yield was recorded overall [202]. In an attempt to reduce the cost of fermentation medium and improve process economics, 0.5 g/l of cotton seed extract without other nutrient supplementation was shown to be a superior medium for *C. carboxidivorans* strain P7 in producing ethanol from syngas fermentation [203]. A recent study showed that increasing concentrations of trace metal ions such as Ni^{2+}, Zn^{2+}, SeO_4^-, WO_4^-, Fe^{2+} and elimination of Cu^{2+} from medium improved enzymatic activities (FDH, CODH, and hydrogenase), growth and ethanol production in "*C. ragsdalei*" under autotrophic conditions [107].

A low redox potential is necessary for strict anaerobes to grow, hence reducing agents such as sodium thioglycolate, ascorbic acid, methyl viologen, benzyl viologen, titanium (III)–citrate, potassium ferricyanide, cysteine-HCl and sodium sulfide are commonly added to fermentation medium [14, 16,204]. Furthermore, the addition of reducing agent directs the electron and carbon flow towards solventogenesis by enhancing the availability of reducing equivalents to form NADH for alcohol production [16, 205]. Excessive addition of reducing agents can cause slower microbial growth due to reduced ATP formation from acetogenesis so it is important to determine the optimum concentration of reducing agents [14, 16]. The sulfur containing gases (e.g. H_2S) present in syngas are toxic to chemical catalysts but can be beneficial for microbial catalysts by reducing medium redox potential, stimulate redox sensitive enzymes such as CODH, and promote alcohol formation [206, 207].

11.4. Medium Ph

Like other organisms, acetogens have a limited range of pH for optimal growth so the pH of the fermentation medium needs to be closely controlled. The extracellular pH directly influences the intracellular pH, membrane potential, proton motive force, and consequently substrate utilization and product profile [208, 209]. In most studies, lowering pH medium divert carbon and electron flow from cell and acid formation towards alcohol production [113, 209-211]. By applying this knowledge, Gaddy and Clausen performed a two-stage CSTR syngas fermentation systems using *C. ljungdahlii*where they set the first reactor at pH 5 to promote cell growth, and pH 4 - 4.5 in the second reactor to induce ethanol production [212]. One recent study with *C. ljungdahlii* showed conflicting results in which cell density and ethanol production were both higher at pH 6.8 when compared to pH 5.5 [213].

11.5. Temperature

The optimum temperature for mesophilic acetogens are between 30-40°C, while thermophilic acetogens grow best between 55 and 58°C. The fermentation temperature not only affects substrate utilization, growth rate and membrane lipid composition of the acetogens, but also gas substrate availability because gas solubility increases with decreasing temperature [24, 211]. "*C. ragsdalei*"was reported to produce more ethanol at 32°C than at the optimum growth temperature of 37°C [211].

12. CELL SEPARATION AND PRODUCT RECOVERY

To retain high cell densities in reactor, microbes can be grown as biofilm attached to carrier material. Planktonic cells can be retained in the fermentation broth by installing solid/liquid separators such as membranous ultra-filtration units, spiral wound filtration systems, hollow fibres, cell-recycling membranes and centrifuges [214-216]. The concentrations of solvents from gas fermentation rarely exceed 6% [w/v] so a cost- and energy- efficient product recovery process is required. Furthermore, acetogens also exhibit low resistance towards solvents like ethanol [217, 218] and butanol [219, 220] so an *in situ*/online product recovery system can enhance solvent productivity by decreasing solvent concentrations (and hence toxicity) in the fermentation broth. Distillation has been the traditional method of product recovery but the associated high energy costs have led to the development of alternative methods such as liquid-liquid extraction, pervaporation, perstraction, and gas stripping [24,221].

12.1. Liquid-Liquid Extraction

In liquid-liquid extraction, a water-insoluble organic extractant is mixed with the fermentation broth [222]. Because solvents are more soluble in the organic phase than in the aqueous phase, they get selectively concentrated in the extractant.Although this technique does not remove water or nutrients from the fermentation broth, some gaseous substrates might be removed because CO and H_2 have much higher solubility in organic solvents than water [222, 223]. Oleyl alcohol has been the extractant of choice due to its relatively non-toxicity [224].

12.2. Perstraction

Liquid-liquid extraction is associated with several problems including toxicity to the microbes, formation of emulsion, and the accumulation of microbes at the extractant and fermentation broth interphase [222]. In an attempt to remediate these problems, perstraction was developed and this technique employs membrane to separate the extractant from the fermentation broth. This physical barrier prevent direct contact between the microbe and the toxicity of extractant, but it can also limit the rate of solvent extraction and is susceptible to fouling [219, 221]

12.3. Pervaporation

In a product recovery technique termed pervaporation, a membrane that directly comes in contact with fermentation broth is used to selectively remove volatile compounds such as ethanol and butanol [219,222]. The volatile compounds diffuse through the membrane as vapour and are then collected by condensation. To facilitate volatilization of permeates into vapour, a partial pressure difference across the membrane is usually maintained by applying a vacuum or inert gas (e.g. N_2) across the permeate side of the membrane [219]. Polydimethylsiloxane (PDMS) is the current material of choice for the membrane, but other materials such as poly(1-trimethylsilyl-1-propyne) (PTMSP), hydrophobic zeolite membranes, and composite membranes have also been investigated [225].

12.4. Gas Stripping

Gas stripping is an attractive product recovery method for gas fermentation because the exit gas stream from the bioreactor can be used for *in situ*/online product recovery [219]. Following product recovery via condensation, the effluent and gas can be recycled back into the bioreactor. In sugar-based fermentation using *C. beijerinckii* mutant strain BA101, *in situ* gas stripping was shown to improve ABE productivity by 200%, complete substrate utilization and also complete acid conversion into solvents, when compared to non-integrated process [226].

13. COMMERCIALIZATION

The growing commercial interests in using gas fermentation as a platform for biofuels production is evident in the recent spike in patent fillings within the field [105]. A 2009 report compared mass and energy conversion efficiencies from a process engineering standpoint between enzymatic hydrolysis fermentation of lignocellulose, syngas fermentation and FTP [227]. The authors concluded that while syngas fermentation offers a range of advantages such as low pretreatment requirement and low energy requirement for bioconversion, the technology is severely limited by low ethanol productivity [227]. Another report documented the techno-economic analysis of gas fermentation and concluded that the selling price of ethanol using this technology would still be significantly higher than gasoline in 2009 [228]. In contrast, Griffin and Schultz recently compared the production of ethanol from CO-rich gas using thermo-chemical route and biological gas fermentation route [22]. The authors concluded that gas fermentation offers superior fuel yield per volume of biomass feed, carbon conversion to fuel, energy efficiency and lower carbon emissions relative to the thermo-chemical approach to bioethanol production.

Ethanol and butanol are the most attractive fuel products from current gas fermentation but other by-products such as 2,3-butanediol, acetic acid and butyric acid are also valuable commodities that have the potential to provide significant additional revenue streams, setting off costs for biofuel production. 2,3-butanediol is a high value commodity which can be used to synthesize chemical products such as 1,3-butanediane, methyl ethyl ketone, and gamma butyrolactone, with a combined potential market value of $43 billion [104]. Acetic acid is an important precursor for synthesis of polymers while butyric acid can be used as a flavouring agent in the food industry [229, 230]. With the development of advanced genetic tools for expansion of product range, the industry might witness an increasing emphasis on the production of high-value commodities in addition to biofuels.

Several companies are actively engaged in the development of the gas fermentation technology and some are approaching commercialization. Bioengineering Resources Inc (BRI) founded by Prof. James Gaddy of University of Arkensas, Fayetteville, an early pioneer in the investigation of gas fermentation at scale, was the first company to explore the potential of gas fermentation for industrial bioethanol production. BRI was acquired by chemical company INEOS and rebranded as INEOS Bio (www.ineosbio.com). A pilot-scale facility in Arkansas has been operated since 2003 using several isolates of *C. ljungdahlii* [231] and is building a US$130 million commercial facility in Florida with its joint venture partner New Planet Energy Florida [232]. The commercial facility is expected to start operation in the second quarter of 2012 and is aiming to generate 8 million gallon of cellulosic ethanol per annum and 6 MW of power to the local communities [232]. INEOS Bio also announced design of a second plant, the Seal Sands Biorefinery in Teeside, UK [233].

Founded in 2006, Coskata Inc. (www.coskata.com) is a US-based company that has reported achieving ethanol yields of 100 gallons per dry ton of wood biomass in a semi-commercial facility in Pennsylvania [234]. The company licensed several microbial strains from the University of Oklahoma [235], which has filed patents and journal publications for acetogens such as "*C. ragsdalei*" [211, 236,237] and *C. carboxidivorans* [55, 112]. A patent documenting a new ethanologenic species, "*C. coskatii*" was also recently filed by Coskata [238]. Backed by a conditional US$250 million loan guarantee from the US Department of Agriculture (USDA), Coskata has announced that it is planning to build a commercial plant with the capacity to produce 55 million gallon fuel grade ethanol per annum in Alabama [234, 239]. While the initial strategy saw biomass as feedstock, the company recently announced its first commercial plant will be switched to 100% natural gas as feedstock [240]. A planned IPO with the aim to tap into private investors to finance the plant was put on hold [241]. In 2012,

Coskata and INEOS Bio were involved in a trade secret dispute which culminated in a settlement that see INEOS Bio receiving US$2.5 million cash payment, shares and right to receive 2.5% of future ethanol royalties from Coskata [242].

LanzaTech is a NZ/US based company that has developed a gas fermentation technology to utilize industrial off-gases from steel making and other sources, as well as syngas produced from biomass as feedstocks. The company has reported the development of a proprietary Clostridial biocatalyst that is able to convert the CO-rich waste gas with minimal gas conditioning into bioethanol and the platform chemical 2,3-butanediol. The use of industrial off-gases as feedstock not only helps to reduce the carbon footprint of the steel-making operations but also allows the production of valuable commodities without the costs associated with feedstock gasification. The company has estimated that up to 30 billion gallon of bioethanol per year can be produced from the CO-rich off gases produced through steel manufacturers globally [243]. Founded in 2005, LanzaTech has successfully demonstrated bioethanol production at a pilot plant at BlueScope Steel in Glenbrook, NZ, since 2008 and the company has recently started operating its 100,000 gallon bioethanol per year demonstration facility in Shanghai, China, using waste gas collected from an adjacent steel mill plant owned by its partner Baosteel Group [243, 244]. LanzaTech is planning to build a commercial facility with the capacity to produce 50 million gallon of bioethanol per annum in China by 2013 [243]. The recent acquisition of a biorefinery facility developed by the US-based gasification technology company Range Fuels in Georgia, and a milestone signing of its first commercial customer, Concord Enviro Systems (India), highlighted LanzaTech's intention to utilize MSW and lignocellulosic waste as feedstocks for biofuel and chemical production [243, 244].

14. CONCLUSION

One of the fundamental factors that govern the environmental and economical sustainability of biofuel production is feedstock. Through gasification, a spectrum of renewable non-food feedstock such as agricultural wastes, dedicated energy crops, forest residues, and MSW can be converted into syngas. This article presents a detailed examination of gas fermentation technology in capturing the carbon and energy from syngas and produce biofuels and chemicals. In comparison to indirect fermentation of lignocellulose via enzymatic hydrolysis, and thermo-chemical FTP, gas fermentation offers several advantages such as good product yield and selectivity, operation in ambient conditions, high tolerance to gas impurities, and elimination of expensive pre-treatment steps and costly enzymes. Furthermore, some industries such as steel mill, natural gas steam reforming, oil refining and chemical production generate large volumes of CO-rich off-gas. Gas fermentation can access these existing feedstocks and generate valuable products from these while reducing carbon emissions. Pivotal to gas fermentation is acetogens such as *C. ljungdahlii*, *C. carboxidivorans*, "*C. ragsdalei*" and *C. autoethanogenum,* which are able to metabolize CO, and CO_2/H_2 into a range of products such as ethanol, butanol, isopropanol, acetone, 2,3-butanediol, acetic acid and butyric acid. Sustained effort in studying the physiology and biochemistry using advanced molecular techniques such as genomics, transcriptomics, proteomics, metabolomics and systems biology are essential to further the understanding of these microbes. Furthermore, recent advances in Clostridial genetic tools offer endless opportunities to engineer strains that have improved product yield, substrate utilization, no strain degeneration, and synthesis of new products.

The main challenges associated with commercialization of gas fermentation have been identified as gas-to-liquid mass transfer limitation, product yield, substrate utilization efficiency, low biomass density and product recovery. Further development of bioreactor is necessary to improve the availability of gas substrates and maintain high cell density for higher productivity. Improvement in integrated product recovery technology is also essential to lower the costs of product recovery and alleviate product inhibition. Gas fermentation appears to be mature enough for commercialization since several companies have already demonstrated their technologies at pilot scale and are moving towards commercialization in the near future.

REFERENCES

1. Vasudevan PT, Gagnon MD, Briggs MS. Environmentally sustainable biofuels – The case for biodiesel, biobutanol and cellulosic ethanol. In: Singh OV, Harvey SP, editors. Sustainable Biotechnology: Sources of Renewable Energy: Springer Science+Business Media B.V.; 2010. p. 43-62.

2. Appenzeller T. The end of cheap oil. National Geographic Magazine. 2004.

3. European_Union. Directive 2009/28/EC of the European Parliament and of the Council of 23 April 2009 on the promotion of the use of energy from renewable sources and amending and subsequently repealing Directives 2001/77/EC and 2003/30/EC. 2009.

4. REN21. Renewables 2011 Global Status Report. Paris: REN21, 2011.

5. Demirbas A. Political, economic and environmental impacts of biofuels: A review. Applied Energy. 2009 Nov;86:S108-S17.

6. Köpke M, Noack S, Dürre P. The past, present, and future of biofuels – Biobutanol as Promising Alternative. In: dos Santos Bernades MA, editor.: InTech; 2011. p. 451-86.

7. Pernick R, Wilder C, Winnie T. Clean Energy Trends 2012. Clean Edge Inc., 2012.

8. Dürre P. Biobutanol: an attractive biofuel. Biotechnology journal. 2007;2(12):1525-34.

9. OECD-FAO. OECD-FAO Agricultural Outlook 2011-2020. 2011.

10. Murse T. Understanding the ethanol subsidy 2011. Available from: http://usgovinfo.about.com/od/moneymatters/a/The-Federal-Ethanol-Subsidy.htm.

11. Pear R. After three decades, tax credit for ethanol expires. The New York Times. 2nd January 2012.

12. Carere CR, Sparling R, Cicek N, Levin DB. Third generation biofuels via direct cellulose fermentation. International Journal of Molecular Sciences. 2008 Jul;9(7):1342-60.

13. Datar RP, Shenkman RM, Cateni BG, Huhnke RL, Lewis RS. Fermentation of biomass-generated producer gas to ethanol. Biotechnology and bioengineering. 2004;86(5):587-94.

14. Klasson KT, Ackerson MD, Clausen EC, Gaddy JL. Bioreactor design for synthesis gas fermentations. Fuel. 1991 May;70(5):605-14.

15. Klasson KT, Ackerson MD, Clausen EC, Gaddy JL. Bioconversion of synthesis gas into liquid or gasesous fuels. enzyme and microbial technology. 1992 Aug;14(8):602-8.

16. Mohammadi M, Najafpour GD, Younesi F, Lahijani P, Uzir MH, Mohamed AR. Bioconversion of synthesis gas to second generation biofuels: A review. Renewable & Sustainable Energy Reviews. 2011 Dec;15(9):4255-73.

17. Abubackar HN, Veiga MC, Kennes C, Coruña L. Biological conversion of carbon monoxide: rich syngas or waste gases to bioethanol. Biofuels, Bioproducts & Biorefining. 2011:93-114.

18. Siedlecki M, de Jong W, Verkooijen AHM. Fluidized bed gasification as a mature and reliable technology for the production of bio-syngas and applied in the production of liquid transportation fuels-A review. Energies. 2011 Mar;4(3):389-434.

19. Wei L, Pordesimo LO, Haryanto A, Wooten J. Co-gasification of hardwood chips and crude glycerol in a pilot scale downdraft gasifier. Bioresource Technology. 2011 May;102(10):6266-72.

20. Dudynski M, Kwiatkowski K, Bajer K. From feathers to syngas - Technologies and devices. Waste Management. 2012 Apr;32(4):685-91.

21. Sipma J, Henstra AM, Parshina SM, Lens PN, Lettinga G, Stams AJM. Microbial CO conversions with applications in synthesis gas purification and bio-desulfurization. Critical reviews in biotechnology. 2006;26(1):41-65.

22. Griffin DW, Schultz MA. Fuel and chemical products from biomass syngas: A comparison of gas fermentation to thermochemical conversion routes. Environmental Progress & Sustainable Energy. 2012 Jul;31(2):219-24.

23. McKendry P. Energy production from biomass (part 3): gasification technologies. Bioresource Technology. 2002 May;83(1):55-63.

24. Munasinghe PC, Khanal SK. Biomass-derived syngas fermentation into biofuels: Opportunities and challenges. Bioresour Technol. 2010;101(13):5013-22.

25. Bartish CM, Drissel GM. Carbon monoxide. In: Othmer K, editor. Encyclopedia of Chemical Technology. New York: John Wiley and Sons; 1978.

26. van Houten RT, Lettinga G. Biological sulphate reduction with synthesis gas: Microbiology and technology. Immobilized Cells: Basics and Applications. 1996;11:793-9.

27. Köpke M, Mihalcea C, Liew FM, Tizard JH, Ali MS, Conolly JJ, et al. 2,3-butanediol production by acetogenic bacteria, an alternative route to chemical synthesis, using industrial waste gas. Applied and Environmental Microbiology. 2011 Aug;77(15):5467-75.

28. van der Drift A, van Doorn J, Vermeulen JW. Ten residual biomass fuels for circulating fluidized-bed gasification. Biomass & Bioenergy. 2001;20(1):45-56.

29. Gordillo G, Annamalai K. Adiabatic fixed bed gasification of dairy biomass with air and steam. Fuel. 2010 Feb;89(2):384-91.

30. Boateng AA, Banowetz GM, Steiner JJ, Barton TF, Taylor DG, Hicks KB, et al. Gasification of Kentucky bluegrass (Poa pratensis l.) straw in a farm-scale reactor. Biomass & Bioenergy. 2007 Feb-Mar;31(2-3):153-61.

31. Bridgwater T. Biomass for energy. Journal of the Science of Food and Agriculture. 2006 Sep;86(12):1755-68.

32. Swanson RM, Platon A, Satrio JA, Brown RC. Techno-economic analysis of biomass-to-liquids production based on gasification. Fuel. 2010 Nov;89:S2-S10.

33. Drake HL, Gössner AS, Daniel SL. Old acetogens, new light. Annals of the New York Academy of Sciences. 2008;1125:100-28.

34. Thauer RK. Microbiology - A fifth pathway of carbon fixation. Science. 2007 Dec;318(5857):1732-3.

35. Drake HL, Küsel K, Matthies C. Acetogenic prokaryotes. In: Dworkin M, Falkow S, Rosenberg E, Schleifer K-H, Stackebrandt E, editors.: Springer New York; 2006. p. 354-420.

36. Lindahl PA, Chang B. The evolution of acetyl-CoA synthase. Origins of Life and Evolution of the Biosphere. 2001;31(4-5):403-34. PubMed PMID: WOS:000170975700006.

37. Ragsdale SW. Enzymology of the Wood-Ljungdahl pathway of acetogenesis. Incredible Anaerobes: from Physiology to Genomics to Fuels. 2008;1125:129-36.

38. Ragsdale SW. Life with carbon monoxide. Critical Reviews in Biochemistry and Molecular Biology. 2004 May-Jun;39(3):165-95.

39. Ragsdale SW, Clark JE, Ljungdahl LG, Lundie LL, Drake HL. Properties of purified carbon monoxide dehydrogenase from Clostridium thermoaceticum, a nickel, iron-sulfur protein. Journal of Biological Chemistry. 1983;258(4):2364-9.

40. Morsdorf G, Frunzke K, Gadkari D, Meyer O. Microbial growth oncarbon monoxide. Biodegradation. 1992;3:61-82.

41. Genthner BRS, Bryant MP. Additional characteristics of one-carbon-compound utilization by Eubacterium limosum and Acetobacterium woodi. Applied and Environmental Microbiology. 1987 Mar;53(3):471-6..

42. Poehlein A, Schmidt S, Kaster AK, Goenrich M, Vollmers J, Thurmer A, et al. An ancient pathway combining carbon dioxide fixation with the generation and utilization of a sodium ion gradient for ATP synthesis. Plos One. 2012 Mar;7(3).

43. Kane MD, Breznak JA. Acetonema longum gen nov SP-nov, an H2/CO2 acetogenic bacterium from the termite, Pterotermes occidentis. Archives of Microbiology. 1991 Aug;156(2):91-8.

44. Allen TD, Caldwell ME, Lawson Pa, Huhnke RL, Tanner RS. Alkalibaculum bacchi gen. nov., sp. nov., a CO-oxidizing, ethanol-producing acetogen isolated from livestock-impacted soil. International journal of systematic and evolutionary microbiology. 2010;60(Pt 10):2483-9.

45. Liu K, Atiyeh HK, Tanner RS, Wilkins MR, Huhnke RL. Fermentative production of ethanol from syngas using novel moderately alkaliphilic strains of Alkalibaculum bacchi. Bioresource Technology. 2012 Jan;104:336-41.

46. Lorowitz WH, Bryant MP. Peptostreptococcus productus strain that grows rapidly with CO as the energy source. Applied and Environmental Microbiology. 1984;47(5):961-4.

47. Lynd L, Kerby R, Zeikus JG. Carbon monoxide metabolism of the methylotrophic acidogen Butyribacterium methylotrophicum. Journal of bacteriology. 1982;149(1):255-63.

48. Zeikus JG, Lynd LH, Thompson TE, Krzycki Ja, Weimer PJ, Hegge PW. Isolation and characterization of a new, methylotrophic, acidogenic anaerobe, the marburg strain. Current Microbiology. 1980;3(6):381-6.

49. Grethlein AJ, Worden RM, Jain MK, Datta R. Evidence for production of n-butanol from carbon monoxide by Butyribacterium methylotrophicum. Journal of Fermentation and Bioengineering. 1991;72(1):58-60.

50. Lux MF, Drake HL. Reexamination of the metabolic potentials of the acetogens Clostridium aceticum and Clostridium formicoaceticum - chemolithoautotrophic and aromatic-dependent growth. Fems Microbiology Letters. 1992 Aug;95(1):49-56.

51. Adamse AD. New isolation of Clostridium aceticum (Wieringa). Antonie Van Leeuwenhoek Journal of Microbiology. 1980;46(6):523-31.

52. Braun M, Mayer F, Gottschalk G. Clostridium aceticum (Wieringa), A microorganism producing acetic acid from molecular hydrogen and carbon dioxide. Archives of Microbiology. 1981;128(3):288-93.

53. Abrini J, Naveau H, Nyns EJ. Clostridium autoethanogenum, sp. nov., an anaerobic bacterium that produces ethanol from carbon monoxide. Archives of microbiology. 1994;161(4):345-51.

54. Bruant G, Lévesque M-J, Peter C, Guiot SR, Masson L. Genomic analysis of carbon monoxide utilization and butanol production by Clostridium carboxidivorans strain P7. PloS one. 2010;5(9):e13033.

55. Liou JSC, Balkwill DL, Drake GR, Tanner RS. Clostridium carboxidivorans sp nov., a solvent-producing clostridium isolated from an agricultural settling lagoon, and reclassification of the acetogen Clostridium scatologenes strain SL1 as Clostridium drakei sp nov. International Journal of Systematic and Evolutionary Microbiology. 2005 Sep;55:2085-91.

56. Küsel K, Dorsch T, Acker G, Stackebrandt E, Drake HL. Clostridium scatologenes strain SL1 isolated as an acetogenic bacterium from acidic sediments. International journal of systematic and evolutionary microbiology. 2000;50 Pt 2:537-46.

57. Gössner AS, Picardal F, Tanner RS, Drake HL. Carbon metabolism of the moderately acid-tolerant acetogen Clostridium drakei isolated from peat. FEMS microbiology letters. 2008;287(2):236-42.

58. Andreese.Jr, Gottscha.G, Schlegel HG. Clostridium formicoaceticumnov spec - isolation, description and distinction from C aceticum and C thermoaceticum. Archiv Fur Mikrobiologie. 1970;72(2):154.

59. Diekert GB, Thauer RK. Carbon monoxide oxidation by Clostridium thermoaceticum and Clostridium formicoaceticum. Journal of bacteriology. 1978;136(2):597-606.

60. Kusel K, Karnholz A, Trinkwalter T, Devereux R, Acker G, Drake HL. Physiological ecology of Clostridium glycolicum RD-1, an aerotolerant acetogen isolated from sea grass roots. Applied and Environmental Microbiology. 2001 Oct;67(10):4734-41.

61. Ohwaki K, Hungate RE. Hydrogen utilization by Clostridia in sewage sludge. Applied and Environmental Microbiology. 1977;33(6):1270-4.

62. Köpke M, Held C, Hujer S, Liesegang H, Wiezer A, Wollherr A, et al. Clostridium ljungdahlii represents a microbial production platform based on syngas. Proceedings of the National Academy of Sciences of the United States of America. 2010;107(29):13087-92.

63. Tanner RS, Miller LM, Yang D. Clostridium ljungdahlii sp. nov., an acetogenic species in clostridial rRNA homology group I. International journal of systematic bacteriology. 1993;43(2):232-.

64. Bomar M, Hippe H, Schink B. Lithotrophic growth and hydrogen metabolism by Clostridium magnum. Fems Microbiology Letters. 1991;83(3):347-50.

65. Schink B. Clostridium magnum sp nov, a non-autotrophic homoacetogenic bacterium. Archives of Microbiology. 1984; 137(3):250-5.

66. Kane MD, Brauman A, Breznak JA. Clostridium mayombei sp nov, an H2/CO2 acetogenic bacterium from the gut of the african soil-feeding termite, Cubitermes speciosus. Archives of Microbiology. 1991;156(2):99-104.

67. Mechichi T, Labat M, Patel BKC, Woo THS, Thomas P, Garcia JL. Clostridium methoxybenzovorans sp nov., a new aromatic o-demethylating homoacetogen from an olive mill wastewater treatment digester. International Journal of Systematic Bacteriology. 1999 Jul;49:1201-9.

68. Huhnke R, Lewis RS, Tanner RS, inventors. Isolation and characterization of novel Clostridial species. 2008.

69. Genthner BRS, Davis CL, Bryant MP. Features of rumen and sewage sludge strains of Eubacterium limosum, a methanol-utilizing and H2-CO2-utilizing species. Applied and Environmental Microbiology. 1981;42(1):12-9.

70. Krumholz LR, Bryant MP. Clostridium pfennigii sp nov uses methoxyl groups of monobenzenoids and produces butyrate. International Journal of Systematic Bacteriology. 1985;35(4):454-6..

71. Pierce E, Xie G, Barabote RD, Saunders E, Han CS, Detter JC, et al. The complete genome sequence of Moorella thermoacetica (f. Clostridium thermoaceticum). Environmental microbiology. 2008;10(10):2550-73.

72. Daniel SL, Hsu T, Dean SI, Drake HL. Characterization of the H2-dependent and CO-dependent chemolithotrophic potentials of the acetogens Clostridium thermoaceticum and Acetogenium kivui. Journal of Bacteriology. 1990 Aug;172(8):4464-71

73. Kerby R, Zeikus JG. Growth of Clostridium thermoaceticum on H2/CO2 or CO as energy source. Current Microbiology. 1983;8(1):27-30.

74. Savage MD, Wu ZR, Daniel SL, Lundie LL, Drake HL. Carbon monoxide-dependent chemolithotrophic growth of Clostridium thermoautotrophicum. Applied and Environmental Microbiology. 1987 Aug;53(8):1902-6..

75. Müller V. Energy conservation in acetogenic bacteria. Applied and environmental microbiology. 2003;69(11):6345-53.

76. Calusinska M, Happe T, Joris B, Wilmotte A. The surprising diversity of clostridial hydrogenases: a comparative genomic perspective. Microbiology-Sgm. 2010 Jun;156:1575-88.

77. Vignais PM, Colbeau A. Molecular biology of microbial hydrogenases. Current Issues in Molecular Biology. 2004;6:159-88.

78. Vignais PM. Hydrogenases and H+-reduction in primary energy conservation. In: Schafer G, Penefsky HS, editors. Bioenergetics: Energy Conservation and Conversion. 45: Springer Berlin / Heidelberg; 2007. p. 223-52.

79. Demuez M, Cournac L, Guerrini O, Soucaille P, Girbal L. Complete activity profile of Clostridium acetobutylicum FeFe -hydrogenase and kinetic parameters for endogenous redox partners. Fems Microbiology Letters. 2007 Oct;275(1):113-21.

80. Schut GJ, Adams MWW. The iron-hydrogenase of Thermotoga maritima utilizes ferredoxin and NADH synergistically: a new perspective on anaerobic hydrogen production. Journal of bacteriology. 2009;191(13):4451-7.

81. Menon S. Unleashing hydrogenase activity in carbon monoxide dehydrogenase/acetyl-CoA synthase and pyruvate: ferredoxin oxidoreductase. Biochemistry. 1996;2960(96):15814-21.

82. Thauer RK, Jungermann K, Decker K. Energy conservation in chemotrophic anerobic bacteria. Bacteriological Reviews. 1977;41(1):100-80.

83. Oelgeschläger E, Rother M. Carbon monoxide-dependent energy metabolism in anaerobic bacteria and archaea. Archives of microbiology. 2008;190(3):257-69.

84. Rohde M, Mayer F, Meyer O. Immunocytochemical localization of carbon-monoxide oxidase in Pseudomonas carboxydovorans - The enzyme is attached to the inner aspect of the cytoplasmic membrane. Journal of Biological Chemistry. 1984;259(23):4788-92.

85. Rohde M, Mayer F, Jacobitz S, Meyer O. Attachement of CO dehydrogenase to the cytoplasmic membrane is limiting the respiratory rate of Pseudomonas carboxydovorans. Fems Microbiology Letters. 1985;28(2):141-4.

86. Kumar M, Lu WP, Ragsdale SW. Binding of carbon disufide to the site of acetyl-CoA synthesis by the nickel-iron-sulfur protein, carbon monoxide dehydrogenase, from Clostridium thermoaceticum. Biochemistry. 1994 Aug;33(32):9769-77.

87. Lindahl PA, Munck E, Ragsdale SW. CO dehydrogenase from Clostridium thermoaceticum - EPR and electrochemical studies in CO2 and argon atmosphere. Journal of Biological Chemistry. 1990 Mar;265(7):3873-9.

88. Ragsdale SW, Pierce E. Acetogenesis and the Wood-Ljungdahl pathway of CO(2) fixation. Biochimica et biophysica acta. 2008;1784(12):1873-98.

89. Nevin KP, Hensley SA, Franks AE, Summers ZM, Ou JH, Woodard TL, et al. Electrosynthesis of organic compounds from carbon dioxide is catalyzed by a diversity of acetogenic microorganisms. Applied and Environmental Microbiology. 2011 May;77(9):2882-6.

90. Nevin KP, Woodard TL, Franks AE, Summers ZM, Lovley DR. Microbial electrosynthesis : Feeding microbes electricity to convert carbon dioxide and water to multicarbon extracellular organic compounds. mBio. 2010;1(2).

91. Biegel E, Schmidt S, Müller V. Genetic, immunological and biochemical evidence for a Rnf complex in the acetogen Acetobacterium woodii. Environmental microbiology. 2009;11(6):1438-43.

92. Li F, Hinderberger J, Seedorf H, Zhang J, Buckel W, Thauer RK. Coupled ferredoxin and crotonyl coenzyme A (CoA) reduction with NADH catalyzed by the butyryl-CoA dehydrogenase/Etf complex from Clostridium kluyveri. Journal of bacteriology. 2008;190(3):843-50.

93. Müller V, Imkamp F, Biegel E, Schmidt S, Dilling S. Discovery of a ferredoxin:NAD+-oxidoreductase (Rnf) in Acetobacterium woodii: a novel potential coupling site in acetogens. Annals of the New York Academy of Sciences. 2008;1125:137-46.

94. Herrmann G, Jayamani E, Mai G, Buckel W. Energy conservation via electron-transferring flavoprotein in anaerobic bacteria. Journal of Bacteriology. 2008 Feb;190(3):784-91.

95. Dürre P. Handbook on Clostridia. Dürre P, editor: CRC Press Taylor & Francis Group; 2005.

96. Wang SN, Huang HY, Moll J, Thauer RK. NADP(+) reduction with reduced ferredoxin and NADP(+) reduction with NADH are coupled via an electron-bifurcating enzyme complex in Clostridium kluyveri. Journal of Bacteriology. 2010 Oct;192(19):5115-23.

97. Roberts DL, Jameshagstrom JE, Garvin DK, Gorst CM, Runquist JA, Baur JR, et al. Cloning and expression of the gene-cluster encoding key proteins involved in acetyl-CoA synthesis in Clostridium thermoaceticum - CO dehydrogenase, the corrinoid Fe-S protein, and methyltransferase. Proceedings of the National Academy of Sciences of the United States of America. 1989 Jan;86(1):32-6..

98. Kaster AK, Goenrich M, Seedorf H, Liesegang H, Wollherr A, Gottschalk G, et al. More than 200 genes required for methane formation from H-2 and CO2 and energy conservation are present in Methanothermobacter marburgensis and Methanothermobacter thermautotrophicus. Archaea-an International Microbiological Journal. 2011.

99. Papoutsakis ET, AL-Hinai MA, Jones SW, Indurthi DC, Mitchell DK, Fast A, inventors. Recombinant Clostridia that fix CO2 and CO and uses thereof. U.S.2012.

100. Alsaker KV, Papoutsakis ET. Transcriptional program of early sporulation and stationary-phase events in Clostridium acetobutylicum. Journal of bacteriology. 2005;187(20):7103-18.

101. Köpke M. Genetische Veränderung von Clostridium ljungdahlii zur produktion von 1-butanol aus synthesegas.: Ulm University; 2009.

102. Sakai S, Nakashimada Y, Yoshimoto H, Watanabe S, Okada H, Nishio N. Ethanol production from H-2 and CO2 by a newly isolated thermophilic bacterium, Moorella sp HUC22-1. Biotechnology Letters. 2004 Oct;26(20):1607-12.

103. Sakai S, Nakashimada Y, Inokuma K, Kita M, Okada H, Nishio N. Acetate and ethanol production from H-2 and CO2 by Moorella sp using a repeated batch culture. Journal of Bioscience and Bioengineering. 2005 Mar;99(3):252-8.

104. Köpke M, Mihalcea C, Liew FM, Tizard JH, Ali MS, Conolly JJ, et al. 2,3-butanediol production by acetogenic bacteria, an alternative route to chemical synthesis, using industrial waste gas. Applied and Environmental Microbiology. 2011 Aug;77(15):5467-75. PubMed PMID: WOS:000293224500045.

105. Köpke M, Mihalcea C, Bromley JC, Simpson SD. Fermentative production of ethanol from carbon monoxide. Current Opinion in Biotechnology. 2011 Jun;22(3):320-5..

106. Kundiyana DK, Wilkins MR, Maddipati P, Huhnke RL. Effect of temperature, pH and buffer presence on ethanol production from synthesis gas by "Clostridium ragsdalei". Bioresource technology. 2011;102(10):5794-9.

107. Saxena J, Tanner RS. Effect of trace metals on ethanol production from synthesis gas by the ethanologenic acetogen, Clostridium ragsdalei. Journal of Industrial Microbiology & Biotechnology. 2011 Apr;38(4):513-21.

108. Heiskanen H, Virkajärvi I, Viikari L. The effect of syngas composition on the growth and product formation of Butyribacterium methylotrophicum. Enzyme and Microbial Technology. 2007;41(3):362-7.

109. Noelling J. Genome sequence and comparative analysis of the solvent-producing bacterium Clostridium acetobutylicum. Journal of. 2001;183(16):4823-38.

110. Huang H, Liu H, Gan Y-R. Genetic modification of critical enzymes and involved genes in butanol biosynthesis from biomass. Biotechnology advances. 2010;28(5):651-7.

111. Ukpong MN, Atiyeh HK, De Lorme MJM, Liu K, Zhu X, Tanner RS, et al. Physiological response of Clostridium carboxidivorans during conversion of synthesis gas to solvents in a gas-fed bioreactor. Biotechnology and bioengineering. 2012;Accepted article.

112. Kundiyana DK, Huhnke RL, Wilkins MR. Syngas fermentation in a 100-L pilot scale fermentor: Design and process considerations. Journal of Bioscience and Bioengineering. 2010 May;109(5):492-8.

113. Phillips JR, Klasson KT, Claussen EC, Gaddy JL. Biological production of ethanol from coal synthesis gas. Applied Biochemistry And Biotechnology. 1993;39(1):0-7.

114. Lee SY, Park JH, Jang SH, Nielsen LK, Kim J, Jung KS. Fermentative butanol production by clostridia. Biotechnology and Bioengineering. 2008 Oct;101(2):209-28.

115. Sauer U, Durre P. Differential induction of genes related to solvent formation during the shift from acidogenesis to solventogenesis in continuous culture of Clostridium acetobutylicum. Fems Microbiology Letters. 1995 Jan;125(1):115-20.

116. Shi Z, Blaschek HP. Transcriptional analysis of Clostridium beijerinckii NCIMB 8052 and the hyper-butanol-producing mutant BA101 during the shift from acidogenesis to solventogenesis. Applied and Environmental Microbiology. 2008 Dec;74(24):7709-14.

117. Köpke M, Dürre P. Biochemical production of biobutanol. In: Luque R, Campelo J, Clark JH, editors. Camebridge, UK: Woodhead Publishing Ltd; 2011. p. 221-57.

118. Gheshlaghi R, Scharer JM, Moo-young M, Chou CP. Metabolic pathways of clostridia for producing butanol. Biotechnology advances. 2009;27(6):764-81.

119. Seedorf H, Fricke WF, Veith B, Bruggemann H, Liesegang H, Strittimatter A, et al. The genome of Clostridium kluyveri, a strict anaerobe with unique metabolic features. Proceedings of the National Academy of Sciences of the United States of America. 2008 Feb;105(6):2128-33.

120. Inui M, Suda M, Kimura S, Yasuda K, Suzuki H, Toda H, et al. Expression of Clostridium acetobutylicum butanol synthetic genes in Escherichia coli. Applied microbiology and biotechnology. 2008;77(6):1305-16.

121. Tomas CA, Beamish J, Papoutsakis ET. Transcriptional analysis of butanol stress and tolerance in Clostridium acetobutylicum. Journal of Bacteriology. 2004 Apr;186(7):2006-18.

122. Jones DT, Woods DR. Acetone-butanol fermentation revisited. Microbiological reviews. 1986;50(4):484-524.

123. Paul D, Austin FW, Arick T, Bridges SM, Burgess SC, Dandass YS, et al. Genome sequence of the solvent-producing bacterium Clostridium carboxidivorans strain P7T. Journal of bacteriology. 2010;192(20):5554-5.

124. Ismaiel aa, Zhu CX, Colby GD, Chen JS. Purification and characterization of a primary-secondary alcohol dehydrogenase from two strains of Clostridium beijerinckii. Journal of bacteriology. 1993;175(16):5097-105.

125. Ramachandriya KD, Wilkins MR, Delorme MJM, Zhu X, Kundiyana DK, Atiyeh HK, et al. Reduction of acetone to isopropanol using producer gas fermenting microbes. Biotechnology and bioengineering. 2011;108(10):2330-8.

126. Toth J, Ismaiel aa, Chen JS. The ald gene, encoding a coenzyme A-acylating aldehyde dehydrogenase, distinguishes Clostridium beijerinckii and two other solvent-producing clostridia from Clostridium acetobutylicum. Applied and environmental microbiology. 1999;65(11):4973-80.

127. Bertram J, Kuhn A, Durre P. Tn916-induced mutants of Clostridium acetobutylicum defective in regulation of solvent formation. Archives of Microbiology. 1990 Mar;153(4):373-7.

128. Durre P, Fischer RJ, Kuhn A, Lorenz K, Schreiber W, Sturzenhofecker B, et al. Solventogenic enzymes of Clostridium acetobutylicum - catalytic properties, genetic organization, and transcriptional regulation. Fems Microbiology Reviews. 1995 Oct;17(3):251-62.

129. Thauer RK, Jungerma.Ka, Kirchnia.Fh. Properties and function of pyruvate-formate-lyase reaction in Clostridiae. European Journal of Biochemistry. 1972;27(2):282-90.

130. Hemme CL, Mouttaki H, Lee Y-J, Zhang G, Goodwin L, Lucas S, et al. Genome Announcement - Sequencing of multiple Clostridia genomes related to biomass conversion and biofuels production. Journal of bacteriology. 2010;192(24):6494-6.

131. Wang Y, Li X, Mao Y, Blaschek HP. Genome-wide dynamic transcriptional profiling in Clostridium beijerinckii NCIMB 8052 using single-nucleotide resolution RNA-Seq. BMC genomics. 2012;13(1):102-.

132. Janssen H, Döring C, Ehrenreich A, Voigt B, Hecker M, Bahl H, et al. A proteomic and transcriptional view of acidogenic and solventogenic steady-state cells of Clostridium acetobutylicum in a chemostat culture. Applied microbiology and biotechnology. 2010;87(6):2209-26.

133. Haus S, Jabbari S, Millat T, Janssen H, Fischer R-J, Bahl H, et al. A systems biology approach to investigate the effect of pH-induced gene regulation on solvent production by Clostridium acetobutylicum in continuous culture. BMC systems biology. 2011;5(1):10-.

134. Alsaker KV, Paredes C, Papoutsakis ET. Metabolite stress and tolerance in the production of biofuels and chemicals: gene-expression-based systems analysis of butanol, butyrate, and acetate stresses in the anaerobe Clostridium acetobutylicum. Biotechnology and bioengineering. 2010;105(6):1131-47.

135. Allcock ER, Reid SJ, Jones DT, Woods DR. Autolytic activity of an autolysin-deficient mutant of Clostridium acetobutylicum. Appllied Environmental Microbiology. 1981;42:929-35.

136. Hermann M, Fayolle F, Marchal R, Podvin L, Sebald M, Vandecasteele JP. Isolation and characterization of butanol-resistant mutants of Clostridium acetobutylicum. Applied Environmental Microbiology. 1985;50:1238-43.

137. Vanderwesthuizen A, Jones DT, Woods DR. Autolytic activity and butanol tolerance of Clostridium acetobutylicum. Applied and Environmental Microbiology. 1982;44(6):1277-81..

138. Baer SH, Blaschek HP, Smith TL. Effect of butanol challenge and temperature on lipid composition and membrane fluidity of butanol-tolerant Clostridium acetobutylicum. Applied and environmental microbiology. 1987;53(12):2854-61.

139. Soucaille P, Joliff G, Izard A, Goma G. Butanol tolerance and autobacteriocin production by Clostridium acetobutylicum. Current Microbiology. 1987;14(5):295-9.

140. Mermelstein LD, Welker NE, Bennett GN, Papoutsakis ET. Expression of cloned homologous fermentative genes in Clostridium acetobutylicum ATCC 824. Bio-Technology. 1992 Feb;10(2):190-5.

141. Mermelstein LD, Papoutsakis ET, Petersen DJ, Bennett GN. Metabolic engineering of Clostridium acetobutylicum ATCC 824 for increased solvent production by enhancement of acetone formation enzyme activities using a synthetic acetone operon. Biotechnology and Bioengineering. 1993 Nov;42(9):1053-60.

142. Walter KA, Mermelstein LD, Papoutsakis ET. Studies of recombinant Clostridium acetobutylicum with increased dosages of butyrate formation genes. Recombinant DNA Technology Ii. 1994;721:69-72.

143. Desai RP, Papoutsakis ET. Antisense RNA strategies for metabolic engineering of Clostridium acetobutylicum. Applied and Environmental Microbiology. 1999 Mar;65(3):936-45.

144. Sillers R, A-Hinai MA, Papoutsakis ET. Aldehyde-alcohol dehydrogenase and/or thiolase overexpression coupled with CoA transferase downregulation lead to higher alcohol titers and selectivity in Clostridium acetobutylicum fermentations. Biotechnology and Bioengineering. 2009 Jan 1;102(1):38-49.

145. Tummala SB, Junne SG, Papoutsakis ET. Antisense RNA downregulation of coenzyme A transferase combined with alcohol-aldehyde dehydrogenase overexpression leads to predominantly alcohologenic Clostridium acetobutylicum fermentations. Journal of Bacteriology. 2003 Jun;185(12):3644-53.

146. Green EM, Boynton ZL, Harris LM, Rudolph FB, Papoutsakis ET, Bennett GN. Genetic manipulation of acid formation pathways by gene inactivation in Clostridium acetobutylicum ATCC 824. Microbiology-Uk. 1996 Aug;142:2079-86.

147. Green EM, Bennett GN. Inactivation of an aldehyde/alcohol dehydrogenase gene from Clostridium acetobutylicum ATCC 824. Applied biochemistry and biotechnology. 1996;57-58:213-21.

148. Wilkinson SR, Young M. Targeted integration of genes into the Clostridium acetobutylicum genome. Microbiology-Uk. 1994 Jan;140:89-95.

149. Shimizu T, Bathein W, Tamaki M, Hayashi H. The virr gene, a member of a class of 2-component response regulators, regulates the production of perfringolysin-o, collagenase, and hemagglutinin in Clostridium perfringens. Journal of Bacteriology. 1994 Mar;176(6):1616-23.

150. Liyanage H, Kashket S, Young M, Kashket ER. Clostridium beijerinckii and Clostridium difficile detoxify methylglyoxal by a novel mechanism involving glycerol dehydrogenase. Applied and Environmental Microbiology. 2001 May;67(5):2004-10.

151. O'Connor JR, Lyras D, Farrow KA, Adams V, Powell DR, Hinds J, et al. Construction and analysis of chromosomal Clostridium difficile mutants. Molecular Microbiology. 2006 Sep;61(5):1335-51.

152. Wilkinson SR, Young DI, Morris JG, Young M. Molecular genetics and the initiation of solventogenesis in Clostridium beijerinckii (formerly Clostridium acetobutylicum) NCIMB-8052. Fems Microbiology Reviews. 1995 Oct;17(3):275-85.

153. Harris LM, Welker NE, Papoutsakis ET. Northern, morphological, and fermentation analysis of spo0A inactivation and overexpression in Clostridium acetobutylicum ATCC 824. Journal of Bacteriology. 2002 Jul;184(13):3586-97.

154. Sarker MR, Carman RJ, McClane BA. Inactivation of the gene (cpe) encoding Clostridium perfringens enterotoxin eliminates the ability of two cpe-positive C-perfringens type A human gastrointestinal disease isolates to affect rabbit ileal loops. Molecular Microbiology. 1999 Sep;33(5):946-58.

155. Soucaille P, Rainer F, Christian C, inventors. Process for chromosomal integration and DNA sequence replacement in clostridia. EU. 2008.

156. Argyros DA, Tripathi SA, Barrett TF, Rogers SR, Feinberg LF, Olson DG, et al. High ethanol titers from cellulose by using metabolically engineered thermophilic, anaerobic microbes. Applied and Environmental Microbiology. 2011 Dec;77(23):8288-94.

157. Karberg M, Guo HT, Zhong J, Coon R, Perutka J, Lambowitz AM. Group II introns as controllable gene targeting vectors for genetic manipulation of bacteria. Nature Biotechnology. 2001 Dec;19(12):1162-7.

158. Heap JT, Pennington OJ, Cartman ST, Carter GP, Minton NP. The ClosTron: A universal gene knock-out system for the genus Clostridium. Journal of Microbiological Methods. 2007 Sep;70(3):452-64.

159. Heap JT, Kuehne Sa, Ehsaan M, Cartman ST, Cooksley CM, Scott JC, et al. The ClosTron: Mutagenesis in Clostridium refined and streamlined. Journal of microbiological methods. 2010;80(1):49-55.

160. Camiade E, Peltier J, Bourgeois I, Couture-Tosi E, Courtin P, Antunes A, et al. Characterization of Acp, a Peptidoglycan hydrolase of Clostridium perfringens with N-acetylglucosaminidase activity That is implicated in cell separation and stress-induced autolysis. Journal of Bacteriology. 2010 May;192(9):2373-84.

161. Heap JT, Ehsaan M, Cooksley CM, Ng YK, Cartman ST, Winzer K, et al. Integration of DNA into bacterial chromosomes from plasmids without a counter-selection marker. Nucleic Acids Res. 2012 Apr 1;40(8):e59.

162. Keis S, Shaheen R, Jones DT. Emended descriptions of Clostridium acetobutylicum and Clostridium beijerinckii, and descriptions of Clostridium saccharoperbutylacetonicum sp nov and Clostridium saccharobutylicum sp nov. International Journal of Systematic and Evolutionary Microbiology. 2001 Nov;51:2095-103.

163. Bertram J, Durre P. Conjugal transfer and expression of Streptococcal transposons in Clostridium acetobutylicum. Archives of Microbiology. 1989 May;151(6):551-7. PubMed PMID: WOS:A1989AA14000015.

164. Woolley RC, Pennock A, Ashton RJ, Davies A, Young M. Transfer of Tn1545 and Tn916 to Clostridium acetobutylicum. Plasmid. 1989 Sep;22(2):169-74.

165. Blouzard JC, Valette O, Tardif C, de Philip P. Random mutagenesis of Clostridium cellulolyticum by using a Tn1545 derivative. Applied and Environmental Microbiology. 2010 Jul;76(13):4546-9.

166. Cartman ST, Minton NP. A mariner-based transposon system for in vivo random mutagenesis of Clostridium difficile. Appl Environ Microbiol. 2010 Feb;76(4):1103-9. PubMed PMID: 20023081. Pubmed Central PMCID: 2820977. Epub 2009/12/22. eng.

167. Koepke M, Liew FM, inventors. Recombinant microorganism and methods of production thereof. U.S.2011.

168. Simpson SD, Koepke M, Liew FM, inventors. Recombinant microorganisms and methods of use thereof. U.S.2011.

169. Lederle S. Heterofermentative acetonproduktion: University of Ulm, Germany; 2011.

170. Becker U, Grund G, Orschel M, Doderer K, Löhden G, Brand G, et al., inventors. Zellen und Verfahren zur Herstellung von Aceton. Germany. 2009.

171. Harris LM, Desai RP, Welker NE, Papoutsakis ET. Characterization of recombinant strains of the Clostridium acetobutylicum butyrate kinase inactivation mutant: need for new phenomenological models for solventogenesis and butanol inhibition? Biotechnology and bioengineering. 2000;67(1):1-11.

172. Lehmann D, Lutke-Eversloh T. Switching Clostridium acetobutylicum to an ethanol producer by disruption of the butyrate/butanol fermentative pathway. Metabolic Engineering. 2011 Sep;13(5):464-73.

173. Kuit W, Minton NP, Lopez-Contreras AM, Eggink G. Disruption of the acetate kinase (ack) gene of Clostridium acetobutylicum results in delayed acetate production. Applied Microbiology and Biotechnology. 2012 May;94(3):729-41.

174. Yu MR, Zhang YL, Tang IC, Yang ST. Metabolic engineering of Clostridium tyrobutyricum for n-butanol production. Metabolic Engineering. 2011 Jul;13(4):373-82.

175. Li YC, Tschaplinski TJ, Engle NL, Hamilton CY, Rodriguez M, Liao JC, et al. Combined inactivation of the Clostridium cellulolyticum lactate and malate dehydrogenase genes substantially increases ethanol yield from cellulose and switchgrass fermentations. Biotechnology for Biofuels. 2012 Jan;5.

176. Lee J, Jang YS, Choi SJ, Im JA, Song H, Cho JH, et al. Metabolic engineering of Clostridium acetobutylicum ATCC 824 for isopropanol-butanol-ethanol fermentation. Applied and Environmental Microbiology. 2012 Mar;78(5):1416-23.

177. Higashide W, Li YC, Yang YF, Liao JC. Metabolic Engineering of Clostridium cellulolyticum for Production of isobutanol from cellulose. Applied and Environmental Microbiology. 2011 Apr;77(8):2727-33.

178. Cai XP, Bennett GN. Improving the Clostridium acetobutylicum butanol fermentation by engineering the strain for co-production of riboflavin. Journal of Industrial Microbiology & Biotechnology. 2011 Aug;38(8):1013-25.

179. Zhu LJ, Dong HJ, Zhang YP, Li Y. Engineering the robustness of Clostridium acetobutylicum by introducing glutathione biosynthetic capability. Metabolic Engineering. 2011 Jul;13(4):426-34.

180. Tomas CA, Welker NE, Papoutsakis ET. Overexpression of groESL in Clostridium acetobutylicum results in increased solvent production and tolerance, prolonged metabolism, and changes in the cell's transcriptional program. Applied and Environmental Microbiology. 2003 Aug;69(8):4951-65.

181. Nakayama S-i, Kosaka T, Hirakawa H, Matsuura K, Yoshino S, Furukawa K. Metabolic engineering for solvent productivity by downregulation of the hydrogenase gene cluster hupCBA in Clostridium saccharoperbutylacetonicum strain N1-4. Applied Microbiology and Biotechnology. 2008 Mar;78(3):483-93.

182. Atsumi S, Hanai T, Liao J. Non-fermentative pathways for synthesis of branched-chain higher alcohols as biofuels. Nature. 2008;451(7174):86-9.

183. Clomburg JM, Gonzalez R. Biofuel production in Escherichia coli: the role of metabolic engineering and synthetic biology. Applied Microbiology and Biotechnology. 2010 Mar;86(2):419-34.

184. Peralta-Yahya PP, Keasling JD. Advanced biofuel production in microbes. Biotechnology Journal. 2010 Feb;5(2):147-62.

185. Kapic A, Jones ST, Heindel TJ. Carbon monoxide mass transfer in a syngas mixture. Industrial & Engineering Chemistry Research. 2006 Dec;45(26):9150-5.

186. Mohammadi M, Younesi H, Najafpour G, Mohamed AR. Sustainable ethanol fermentation from synthesis gas by Clostridium ljungdahlii in a continuous stirred tank bioreactor. Journal of Chemical Technology and Biotechnology. 2012 Jun;87(6):837-43.

187. Bredwell MD, Srivastava P, Worden RM. Reactor design issues for synthesis-gas fermentations. Biotechnology Progress. 1999 Sep-Oct;15(5):834-44. PubMed PMID: WOS:000082936200007.

188. Ahmed A, Cateni BG, Huhnke RL, Lewis RS. Effects of biomass-generated producer gas constituents on cell growth, product distribution and hydrogenase activity of Clostridium carboxidivorans P7T. Biomass and Bioenergy. 2006;30(7):665-72.

189. Xu D, Lewis RS. Syngas fermentation to biofuels: Effects of ammonia impurity in raw syngas on hydrogenase activity. Biomass and Bioenergy. 2012.

190. Ahmed A, Lewis RS. Fermentation of biomass-generated synthesis gas: Effects of nitric oxide. Biotechnology and Bioengineering. 2007 Aug;97(5):1080-6.

191. Brogren C, Karlsson HT, Bjerle I. Absorption of NO in an alkaline solution of KMnO4. Chemical Engineering & Technology. 1997 Aug;20(6):396-402.

192. Chu H, Chien TW, Li SY. Simultaneous absorption of SO2 and NO from flue gas with KMnO4/NaOH solutions. Science of the Total Environment. 2001 Jul;275(1-3):127-35.

193. Klasson KT, Ackerson MD, Clausen EC, Gaddy JL. Biological conversion of coal and coal-derived synthesis gas. Fuel. 1993 Dec;72(12):1673-8.

194. Vega JL, Holmberg VL, Clausen EC, Gaddy JL. Fermentation parameters of Peptostreptococcus productus on gaseous substrates (CO, H2/CO2). Archives of Microbiology. 1989;151(1):65-70.

195. Hurst KM, Lewis RS. Carbon monoxide partial pressure effects on the metabolic process of syngas fermentation. Biochemical Engineering Journal. 2010;48(2):159-65.

196. Najafpour G, Younesi H. Ethanol and acetate synthesis from waste gas using batch culture of Clostridium ljungdahlii. Enzyme and Microbial Technology. 2006 Jan;38(1-2):223-8.

197. Gaddy JL, Chen GJ, inventors. Bioconversion of waste biomass to useful products.1998.

198. Genthner BRS, Bryant MP. Growth of Eubacterium limosum with carbon monoxide as the energy source. Applied and Environmental Microbiology. 1982;43(1):70-4.

199. Demler M, Weuster-Botz D. Reaction Engineering analysis of hydrogenotrophic production of ccetic ccid by Acetobacterium woodii. Biotechnology and Bioengineering. 2011 Feb;108(2):470-4.

200. Cotter JL, Chinn MS, Grunden AM. Ethanol and acetate production by Clostridium ljungdahlii and Clostridium autoethanogenum using resting cells. Bioprocess Biosyst Eng. 2009;32(3):369-80.

201. Kanchanatawee S, Maddox IS. Effect of phosphate and ammonium ion concentrations on solvent production in defined medium by Clostridium acetobutylicum. Journal of Industrial Microbiology. 1990 Jul;5(5):277-82.

202. Guo Y, Xu JL, Zhang Y, Xu HJ, Yuan ZH, Li D. Medium optimization for ethanol production with Clostridium autoethanogenum with carbon monoxide as sole carbon source. Bioresource Technology. 2010 Nov;101(22):8784-9.

203. Kundiyana DK, Huhnke RL, Maddipati P, Atiyeh HK, Wilkins MR. Feasibility of incorporating cotton seed extract in Clostridium strain P11 fermentation medium during synthesis gas fermentation. Bioresource Technology. 2010 Dec;101(24):9673-80.

204. Sim JH, Kamaruddin AH. Optimization of acetic acid production from synthesis gas by chemolithotrophic bacterium - Clostridium aceticum using statistical approach. Bioresource Technology. 2008 May;99(8):2724-35..

205. Vega JL, Prieto S, Elmore BB, Clausen EC, Gaddy JL. The biological production of ethanol from synthesis gas. Applied Biochemistry and Biotechnology. 1989 Jan-Aug;20-1:781-97.

206. Hu P, Jacobsen LT, Horton JG, Lewis RS. Sulfide assessment in bioreactors with gas replacement. Biochemical Engineering Journal. 2010 May;49(3):429-34.

207. Do YS, Smeenk J, Broer KM, Kisting CJ, Brown R, Heindel TJ, et al. Growth of Rhodospirillum rubrum on synthesis gas: Conversion of CO to H-2 and poly-beta-hydroxyalkanoate. Biotechnology and Bioengineering. 2007 Jun;97(2):279-86.

208. Devi MP, Mohan SV, Mohanakrishna G, Sarma PN. Regulatory influence of CO2 supplementation on fermentative hydrogen production process. International Journal of Hydrogen Energy. 2010 Oct;35(19):10701-9.

209. Grethlein AJ, Worden RM, Jain MK, Datta R. Continuous production of mixed alcohol and acids from carbon monoxide. Applied Biochemistry and Biotechnology. 1990 Spr-Sum;24-5:875-84.

210. Girbal L, Croux C, Vasconcelos I, Soucaille P. Regulation of metabolic shifts in Clostridium acetobutylicum ATCC 824. Fems Microbiology Reviews. 1995 Oct;17(3):287-97.

211. Kundiyana DK, Wilkins MR, Maddipati P, Huhnke RL. Effect of temperature, pH and buffer presence on ethanol production from synthesis gas by "Clostridium ragsdalei". Bioresource Technology. 2011 May;102(10):5794-9.

212. Gaddy JL, Clausen EC, inventors. Clostridium ljungdahlii, an anaerobic ethanol and acetate producing microorganism. 1992.

213. Cotter JL, Chinn MS, Grunden AM. Influence of process parameters on growth of Clostridium ljungdahlii and Clostridium autoethanogenum on synthesis gas. Enzyme and Microbial Technology. 2009 May;44(5):281-8.

214. Frohlich H, Villian L, Melzner D, Strube J. Membrane technology in bioprocess science. Chemie Ingenieur Technik. 2012 Jun;84(6):905-17.

215. Rushton A, Ward AS, Holdich RG. Solid-liquid filtration and separation technology. Germany: VCH Publishers, Inc.; 1996.

216. Kargupta K, Datta S, Sanyal SK. Analysis of the performance of a continuous membrane bioreactor with cell recycling during ethanol fermentation. Biochemical Engineering Journal. 1998 Jan;1(1):31-7.

217. Brown SD, Guss AM, Karpinets TV, Parks JM, Smolin N, Yang S, et al. Mutant alcohol dehydrogenase leads to improved ethanol tolerance in Clostridium thermocellum. Proceedings of the National Academy of Sciences of the United States of America. 2011;108(33):13752-7.

218. Shao XJ, Raman B, Zhu MJ, Mielenz JR, Brown SD, Guss AM, et al. Mutant selection and phenotypic and genetic characterization of ethanol-tolerant strains of Clostridium thermocellum. Applied Microbiology and Biotechnology. 2011 Nov;92(3):641-52.

219. Ezeji T, Milne C, Price ND, Blaschek HP. Achievements and perspectives to overcome the poor solvent resistance in acetone and butanol-producing microorganisms. Applied Microbiology and Biotechnology. 2010 Feb;85(6):1697-712.

220. Borden JR, Papoutsakis ET. Dynamics of genomic-library enrichment and identification of solvent tolerance genes for Clostridium acetobutylicum. Applied and environmental microbiology. 2007;73(9):3061-8.

221. Ezeji TC, Li Y. Advanced product recovery technologies. Hoboken: Wiley; 2009.

222. Ezeji TC, Qureshi N, Blaschek HP. Bioproduction of butanol from biomass: from genes to bioreactors. Current opinion in biotechnology. 2007;18(3):220-7.

223. Jauregui-Haza UJ, Pardillo-Fontdevila EJ, Wilhelm AM, Delmas H. Solubility of hydrogen and carbon monoxide in water and some organic solvents. Latin American Applied Research. 2004 Apr;34(2):71-4.

224. Karcher P, Ezeji TC, Qureshi N, Blaschek HP. Microbial production of butanol: product recovery by extraction. New Delhi: IK International Publishing House Pvt Ltd; 2005.

225. Ezeji TC, Li Y. Advanced product recovery technologies. Hoboken: Wiley; 2009.

226. Ezeji TC, Qureshi N, Blaschek HP. Production of acetone, butanol and ethanol by Clostridium beijerinckii BA101 and in situ recovery by gas stripping. World Journal of Microbiology & Biotechnology. 2003 Aug;19(6):595-603.

227. Wei L, Pordesimo LO, Igathinathane C, Batchelor WD. Process engineering evaluation of ethanol production from wood through bioprocessing and chemical catalysis. Biomass & Bioenergy. 2009 Feb;33(2):255-66.

228. Piccolo C, Bezzo F. A techno-economic comparison between two technologies for bioethanol production from lignocellulose. Biomass & Bioenergy. 2009 Mar;33(3):478-91.

229. Yoneda N, Kusano S, Yasui M, Pujado P, Wilcher S. Recent advances in processes and catalysts for the production of acetic acid. Applied Catalysis a-General. 2001 Nov;221(1-2):253-65.

230. Zigova J, Sturdik E, Vandak D, Schlosser S. Butyric acid production by Clostridium butyricum with integrated extraction and pertraction. Process Biochemistry. 1999 Oct;34(8):835-43.

231. Gaddy JL, inventor. Biological production of ethanol from waste gases with Clostridium ljungdahlii. US. 2000.

232. Guzman DD. INEOS Bio looks to start US cellulosic ethanol in 2012. ICIS Chemical Business. 2011:22-3.

233. Bio I. INEOS Bio - Pioneering waste-to-biofuel technology attracts cross-party support in UK 2011. Available from: http://www.ineosbio.com/76-Press_releases-16.htm.

234. Coskata. Coskata, Inc.'s semi-commercial facility demonstrates two years of successful operation 2011. Available from:

235. Coskata. About us 2012. Available from: http://www.coskata.com/company/.

236. Huhnke RL, Lewis RS, Tanner RS, inventors. Isolation and characterization of novel Clostridial species. US. 2010.

237. Kundiyana DK, Huhnke RL, Wilkins MR. Effect of nutrient limitation and two-stage continuous fermentor design on productivities during "Clostridium ragsdalei" syngas fermentation. Bioresource Technology. 2011 May;102(10):6058-64.

238. Zahn JA, Saxena J, inventors. Novel ethanologenic Clostridium species, Clostridium coskatii. US. 2012.

239. Kuo I. Coskata grabs biggest slice of USDA's $405M in biofuels loan guarantees: Reuters; 2011. Available from: http://www.reuters.com/article/2011/01/21/idUS126007139820110121.

240. Lane J. Coskata switches focus from biomass to natural gas; to raise $100M in natgas-oriented private placement: Biofuels Digest; 2012. Available from: http://www.biofuelsdigest.com/bdigest/2012/07/20/coskata-switches-from-biomass-to-natural-gas-to-raise-100m-in-natgas-oriented-private-placement/.

241. Wernau J. Biofuel firm Coskata drops initial public offering plans amid volatile IPO market: Chicago Tribute Business; 2012. Available from: http://www.chicagotribune.com/business/ct-biz-0721-ipos-20120721,0,1092013.story.

242. Lane J. The intrepid investor: How are biofuels IPOs performing and why? : Biofuels Digest; 2012. Available from: http://www.biofuelsdigest.com/bdigest/2012/02/09/the-intrepid-investor-part-i/.

243. Guzman DD, Chang J. Bio-based 2,3 BDO set for 2014 sales. ICIS Chemical Business. 2012:24.

244. LanzaTech. Media Releases. Available from: http://lanzatech.com/media/media-releases.

CHAPTER 9

Assessment of Feedstock Options for Biofuels Production in Ghana

Francis Kemausuor, Joseph Oppong Akowuah, Emmanuel Ofori

Department of Agricultural Engineering, Kwame Nkrumah University of Science and Technology, Kumasi, Ghana

ABSTRACT

In the wake of climate change and increasing fossil fuel prices, biofuels are becoming attractive to agricultural dependent economies in sub-Saharan Africa and other regions of the world. This study evaluates the energy production potential of biomass resources grown on the available arable agricultural land under two principal scenarios: using 2.5% and 5% of the available arable land for energy crop expansion. Using conservative biofuel yields from crops in the sub-region, a 2.5% of uncultivated arable land dedicated to four traditional crops grown in Ghana namely maize, cassava, sweet sorghum and oil palm could potentially replace 9.3% and 7.2% of transportation fuels by 2020 and 2030 respectively. Using 5% of the uncultivated arable land to cultivate the above four crops and jatropha could potentially produce biofuel to replace 17.3% of transport fuels by 2020 and 13.3% by 2030. In order to enrol such a scheme, government is encouraged to put in place appropriate structures to ensure that, the industry meet international sustainability standards.

KEYWORDS

Biofuels; Feedstock; Ghana

1. INTRODUCTION

As energy demand grows, self-sufficiency in energy requirement has become critical to the socio-economic development of most developing countries. The growth in energy demand due to rapid industrialization and socioeconomic growth require huge investments to meet them. For decades now, fossil oil has contributed immensely to meeting these demands. However, increasing concerns about emissions from fossil fuels and the impending peak oil theory [1-4] have stimulated interest for alternative energy sources. An alternative fuel must be technically feasible, economically competitive, environmentally acceptable, and readily available [5]. Biofuel produced from energy crops and other resources has been touted as one of the viable sources of energy for the foreseeable future in dealing with these worldwide socioeconomic and environmental concerns [6,7]. Global production of biofuel has been growing rapidly in recent years, rising from about 18 billion litres in 2000 to 105 billion litres in 2010 [8]. Despite this exponential increase, biofuel still represents a very small share of global energy consumption. Biofuels provided 2.7% of all global fuel for road transportation in 2010—an increase from 2% in 2009. As usual, supply is dominated by bioethanol, which accounted for approximately 82% of total biofuel production in 2010. Ethanol's contribution is largely responsible for the increase in biofuel production. Together, Brazil and the US contributed about 90% of global ethanol production with the European Union contributing about 53% of all biodiesel in 2010. Global demand for biofuel is projected to increase to about 183.8 billion litres by 2015 with ethanol expected to contribute about 80% of the total [9]. Already, several countries have adopted policies to promote liquid biofuel development led by the US, the EU, Brazil, Canada, Australia and Japan [10]. A growing number of developing countries, including a few African countries have also started to introduce similar policies. Notable countries in Africa with biofuel strategies in place are South Africa, Mozambique, Ethiopia, Senegal and Mali.

In Ghana, a draft bioenergy policy of 2010 [11] aims to substitute national petroleum fuels consumption with 10% biofuel by 2020 and 20% by 2030. The policy intends to make use of the country's vast biomass potential and resources in modern applications for the production of transport fuels and electricity. Biomass already dominates the energy consumption pattern in Ghana, accounting for over 63% of total energy consumed [12] but used principally in traditional cookstoves as firewood and charcoal. Liquid biofuel use is not commercialised in the country but there are plans to begin commercial production as soon as the infrastructure is in place. Plans for biodiesel production are at a more advanced stage as opposed to those for ethanol production. The draft bioenergy policy is open to the commercial scale production of biofuel feedstocks to sustain supply but is not targeting any specific energy crop(s) to meet the demand targets. In the past, jatropha was thought to be the best feedstock due to previous impression that it was drought resistant. But current research has shown that jatropha may not be the wonder crop that it was thought to be [13].

With conflicting reports on the prospects or potential of jatropha and recent revelations that it performs poorly in marginal lands [13], there is the need to perform an analysis of possible biofuel feedstock choices in order to recommend the best choices to investors and policy makers. The aim of this study was to perform an analysis of the technical potential of first-generation biofuels from energy crops in Ghana. The study aims to contribute knowledge to the call from Ghana's Energy Commission for research support to select the best biofuel feedstock options for the country. Knowledge about the technical potentials would enable a good estimation of land occupancy to attain biofuel targets proposed by the Energy Commission. This paper is structured into 5 sections. As part of the introduction, a review of the agricultural resources in the country has been presented. The methodology for conducting the analysis follows the introduction after which the results is presented and discussed. Section 4 presents a brief description of the recipes for a successful biofuel industry in Ghana after which Section 5 summarises the conclusions.

1.1 Agricultural Resources in Ghana

Ghana is administratively divided into ten regions with a total land area of approximately 238,537 sq. km. There are 212 Districts/ Municipal/Metropolitan Areas (MMDAs), which increased from 138 in year 2000[1]. A recent population census in 2010 estimates Ghana's population at 24.6 million [14]. A regional population density map is shown in **Figure 1**.

The coastal and southern parts have higher population density than the northern parts of the country. The Greater Accra region which also houses the nation's capital has the highest population density whereas the Northern region has the lowest. The Northern region has an estimated 35 persons per square kilometre with vast uncultivated arable lands. In relation to agriculture, there are six agro-ecological zones in Ghana defined on the basis of climate, reflected by the natural vegetation and influenced by the soil type (**Figure 2**).

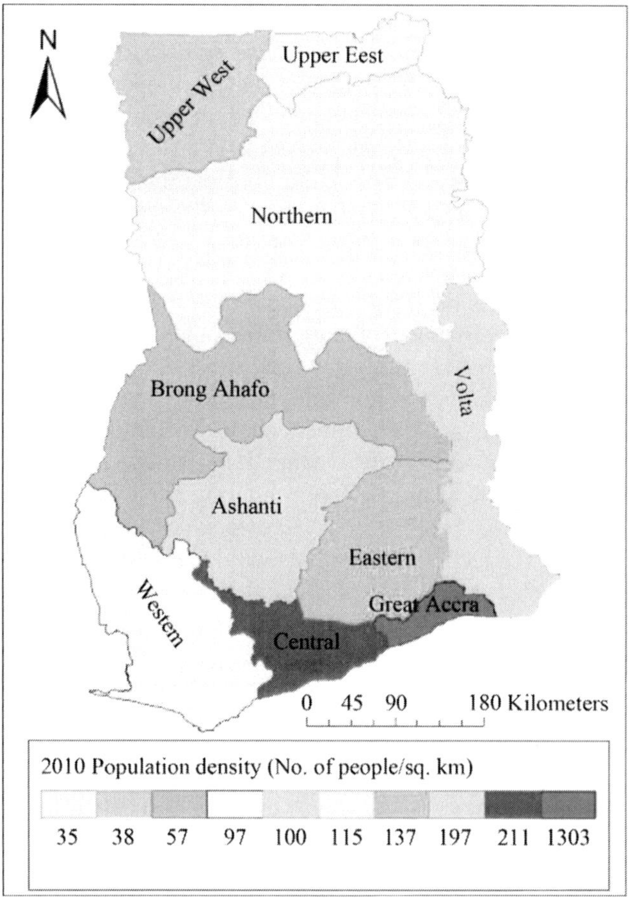

Figure 1. Regional population density map of Ghana for year 2010 [14].

These are Rain Forest, Deciduous Forest, Transitional Zone, Coastal Savannah, Guinea Savannah and Sudan Savannah. The characteristics of the various agro-ecological zones are shown in **Table 1**. Annual rainfall ranges from about 800 mm along the coastal savannah to 2200 mm in the Rain forest. Annual average temperatures range from 26.1°C near the coast to 28.9°C in the northern parts of the country. About 155,000 sq. km., approximately 65% of the total land area, is classified as agricultural land area [15]. The agricultural sector is characterized by a large number of dispersed small-scale producers, employing manual cultivation techniques dependent on rainfall with little or no purchased inputs but providing over 90% of the food needs of the country [16]. Farming systems vary with the six agro-ecological zones (**Table 1**). However, certain general features are discernible throughout the country. According to [17], crop production in Ghana is hampered by land degradation, improper field development, use of low-yielding varieties, lack of organised seed production and distribution systems, and inadequate storage structures. Situated in a tropical climate, Ghana has the advantage of growing the most desirable energy crops for biofuel production. Table 2 shows a summary of key characteristics of some of these crops.

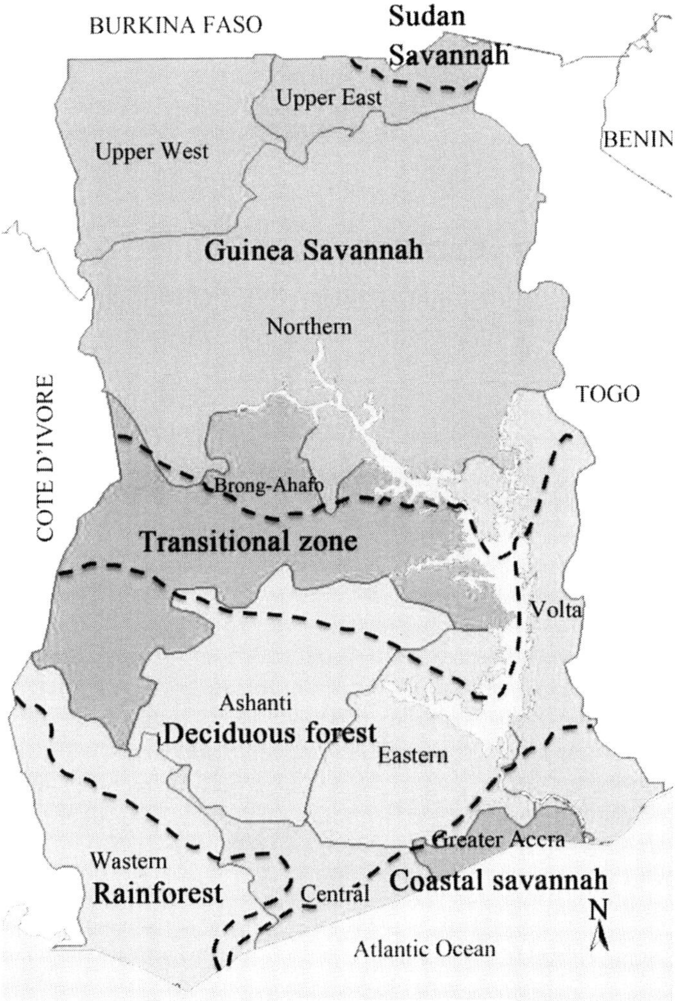

Figure 2. Agro-ecological zone classification in Ghana.

With regards to Food security, 5% of Ghana's population are food insecure. The food security problem exists in all the agro-ecological zones and in both urban and rural areas. There have been many agricultural projects implemented to address the food security problem particularly in relation to stable food crops such as maize, rice, sorghum and soybean. Root and tuber crops such as cassava, yam and cocoyam have also been considered. The government is introducing block farm programmes to make it possible for government sponsored mechanisation and extension services to cover large number of farmers. The Ministry of Food and Agriculture has also set up the National Food Buffer Stock Company (NAFCO) to ensure food security and to insulate farmers against losses resulting from anticipated increases in production.

2. METHODOLOGY

2.1. Crop Cultivation Pattern in Ghana

In 2010, approximately 7.8 million hectares, or 57.6% of the available arable land for agriculture in Ghana was under agricultural cultivation [16] leaving approximately 5.8 million (or 42.4%) which was classified as uncultivated. **Table 3** shows the cultivated area and production of major agricultural crops in Ghana in the year 2010. Potential crops for bioenergy that dominate the land area are maize, cassava, oil palm and sorghum. Other potential crops such as soybean, sunflower, sugarcane and coconut occupy less land areas. Maize is the most cultivated non-export crop in Ghana followed by cassava. In 2010, approximately 1.9 million tonnes of maize was harvested from an area of 992 thousand ha. Maize is produced in all districts in Ghana, making it the most popular food crop in the country. Over 13.5 million tonnes of cassava was produced from an area of 875 thousand hectares in the same year. About 2 million tonnes of oil palm fruit and 0.32 million tonnes of sorghum were produced in 2010 from an area of 360 thousand hectares and 253 thousand hectares respectively. With regards to the trend in the cultivated area of these four crops, the pattern shows a fairly even trend over the last decade, with marginal increase or decrease in some cases (**Figure 3**). There were notable increases in the cultivated areas of oil palm (96%) and maize (39%) and a fairly even trend in the cultivated area for sorghum (**Figure 3**).

Cassava is the dominant food crop in Ghana in terms of production output and has remained so for the last decade (**Figure 4**). It is cultivated in eight of the ten regions in the country. The large production of cassava is partly the result of a cassava improvement programme— the "Root and Tuber Improvement and Marketing Programme" (RTIMP)[2]—supported by the International Fund for Agricultural Development (IFAD) and the Government of Ghana (GoG) through the Ministry of Food and Agriculture. Generally, oil palm cultivation is carried out in varying scales such as smallholder farms, and medium to large-scale plantations. Of the 360 thousand hectares of oil palm cultivated in 2010, 60 thousand hectares were commercial plantations and the rest (over 80%) are small-scale farms scattered over the southern parts of the country [16].

Based on land area already occupied by potential bioenergy crops, four crops namely, maize, cassava, oil palm, and sorghum have been considered in this paper as possible crops for biofuel production in Ghana. Production trends were similar to the area cultivated for these crops. There are notable increases in the production of maize (99.5%), oil palm (81%) and cassava (51%) which corresponds somewhat to the increases in their cultivated area, indicating that there has been little improvement in the yields of these four crops over the period.

2.2. Scenarios Development

Two principal scenarios were adopted in the assessment of feedstock options for biofuel production in Ghana. A percentage of the uncultivated arable agricultural land available in Ghana is dedicated to growing the selected potential bioenergy crops in the different scenarios to determine the resulting biofuel yields. The biofuel produced in the different scenarios is compared to projected petroleum fuel demand for transportation to determine the percentage of fuel it can replace. Crop data used for the assessment was crop production data from the Statistics, Research and Information Directorate (SRID) of Ministry of Food and Agriculture and also from the Food and Agriculture Organisation. Fuel consumption data and projections were from the National Petroleum Authority and the Energy Commission of Ghana. In the analysis, a portion of uncultivated arable agricultural land in Ghana is dedicated to cultivating the four potential energy crops identified, i.e. maize, cassava, oil palm and sweet sorghum in a proportion equal to the proportion in which they were cultivated in 2010. Besides being the most cultivated bioenergy crops in terms of land area, Ghanaian farmers are familiar with their cultivation and the crops are also well suited to the climatic and soil conditions in Ghana. With regards to sorghum, we expect the "sweet sorghum" variety to be cultivated for energy crop purposes. In the first scenario, 2.5% of unused agricultural land (or 144,541 ha) is used to expand the cultivation of these four crops for biofuel purposes. In 2010, approximately 2480 hectares of land was dedicated to these four crops in Ghana. Of this amount, maize and cassava occupied about 75% as shown in **Table 4**. In scenario one, the assumption was made that, 2.5% of uncultivated agricultural land in 2010, which is 144,541 ha is shared among the four crops in the percentages shown in **Table 4** which is the percentage occupied by each of these crops as a total of the area they occupied in 2010. Cultivation is assumed to start gradually and the 2.5% or 144,541 hectares is expected to be fully cultivated in the year 2020. Beyond 2020, no additional land is expected to be allocated to biofuel crops but rather expect further research conducted to improve upon the conservative yields that were used in the initial analysis up to year 2030. The timelines adopted is in line with the Energy Commission's adoption of these dates to realize biofuel blends in transportation fuels in Ghana.

Table 1. Characteristics of agro-ecological zones in Ghana.

Zone	Area (1000 ha)	Rainfall (mm/yr)	Length of growing season (days)	Dominant land use systems	Main food crops
Rain forest	750	2200	Major season: 150 - 160 Minor season: 100	Forest, plantations	Cassava, plantain, maize, oil palm
Deciduous forest	740	1500	Major season: 150 - 160 Minor season: 90	Forest, plantations	Cocoa, cassava, plantain, maize, oil palm
Transition Zone	6630	1300	180 - 200	Annual food and cash crops	Maize, cassava, yam, taro (cocoyam), plantain, groundnut, cowpea, maize
Guinea savannah	14.790	1100	180 - 200	Annual food and cash crops, livestock	Sorghum, maize, groundnut, millet, yam, cowpea, maize
Sudan Savannah	190	1000	150 - 160	Annual food crops, livestock	Millet, sorghum, cowpea, groundnut, yam, maize
Coastal savannah	580	800	Major season: 100 - 110 Minor season: 50	Annual food crops	Cassava, maize

Source: [18].

Table 2. Overview of principal biofuel feedstocks.

Feedstock	Conditions for best performance	Fuel yield per hectare of crop (l/ha)
Sugarcane	Up to 1600 m above sea level	5800 l/ha in Brazil, 4000 l/ha for Africa
Corn (maize)	Grows everywhere but often needs irrigation	3000 l/ha in advanced agricultural systems, 700 l/ha for Africa due to poor yields
Sweet sorghum	Dryer tropics and temperate regions up to 2500 m altitude	3000 - 6000 l/ha
Cassava	Tropical climates up to 1000 m altitude	5400 l/ha advanced agricultural systems, 1750 l/ha in Africa due to low efficiency in current production methods
Oil Palm	Humid tropic areas up to 700 m altitude	6000 l/ha in Malaysia, Africa plantations average less than half
Jatropha	Tropical & semi-arid regions at altitudes up to 500 m. Rainfall from as low as 300 mm	400 - 2200 l/ha of pure plant oil

Source: Compiled from [19].

Table 3. Cultivated area and production of major crops in Ghana, 2010.

Crop type	Area harvested 2010 (ha)	Production 2010 (t)	Yield (t/ha)
Cocoa beans	1,625,000	632,037	0.39
Maize	991,669	1,871,700	1.89
Cassava	875,013	13,504,100	15.44
Yams	384,942	5,960,490	15.48
Groundnuts	353,376	530,887	1.50
Oil palm fruit	352,800	2,004,300	5.68
Plantains	327,870	3,537,730	10.79
Sorghum	252,555	324,422	1.28
Cocoyam	205,342	1,354,800	6.60
Rice, paddy	181,228	491,603	2.71
Millet	176,600	218,952	1.24
Vegetables	165,010	818,615	4.9610

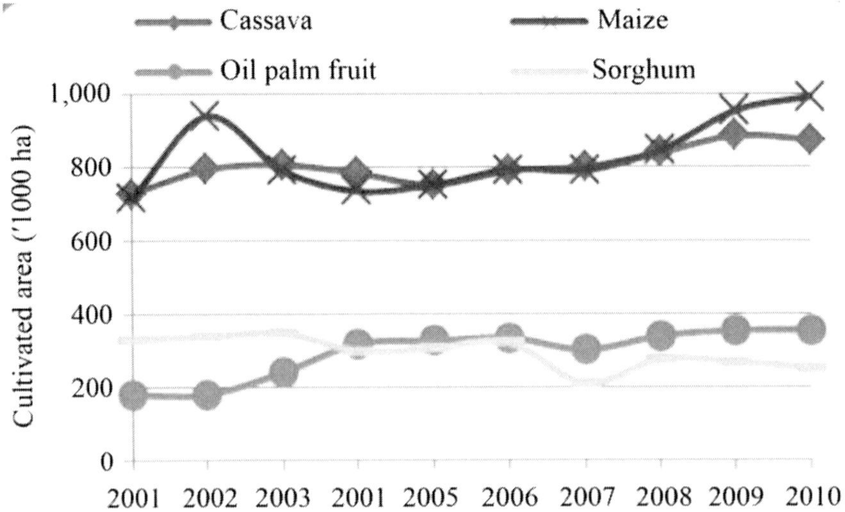

Figure 3. Pattern of land use for selected crops in Ghana. Source: Data from [20].

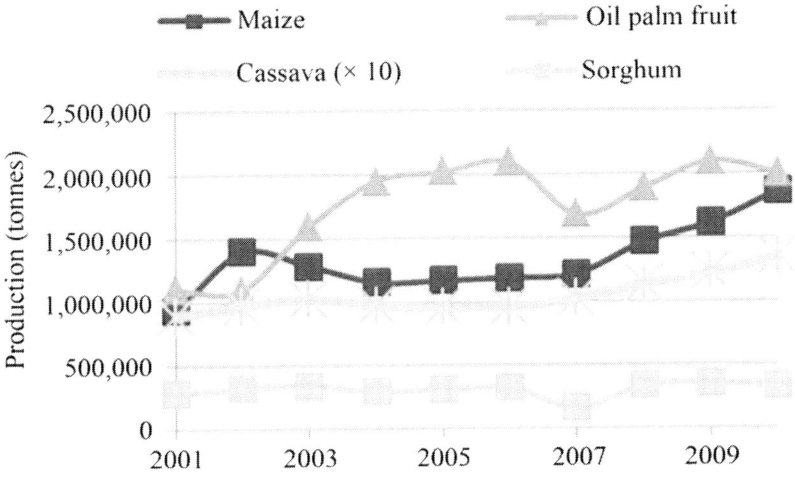

Figure 4. Pattern of production of selected crops in Ghana. Source: Data from [20].

Having allocated land space to each of these crops, biofuel production potential was estimated based on conservative yields achieved in most African countries as reported by [19] and also presented in **Table 2**. These yield figures are based on current agricultural practices in African countries with low irrigation and low fertiliser input. For example, whereas ethanol from maize could yield as much as 3000 litres/hectare in developed agricultural systems, the reported figure for Africa is 700 litres/hectare due to low yields. Also in the case of cassava, about 5400 litres/hectare could be realised in countries such as Thailand but only 1750 litres/hectare is reported for Africa. In order to make a case for research into agronomic improvements, the amount of land space that could be saved is estimated if yields in Ghana were to improve comparable to the current yield in advanced agricultural economies and assuming the same amount of biofuel is to be produced. The resulting biofuel yields are then compared to transportation fuel demand projected for the years 2020 and 2030.

The second scenario considered 5% of the uncultivated agricultural land available in Ghana dedicated to growing the identified potential bioenergy crops for biofuel. Scenario 2 differs from Scenario 1 with the addition of jatropha to the four original crops. In the land allocations, 2.5% of the uncultivated land is still allocated to the four crops in the same proportion as in Scenario 1 and an additional 2.5% dedicated to cultivating only jatropha. The 2.5% allocation to jatropha is admittedly high; but this is based on the fact that cassava and especially oil palm are, to a large extent, not cultivated in the northern parts of the country where there is very low population density with large uncultivated lands. The northern parts of the country occupy close to 40% of Ghana's land size. Annual rainfall amounts of 800 to 1100 mm make it feasible to cultivate jatropha in the northern parts of the country. It was therefore assumed that, the bulk of the land dedicated to jatropha in Scenario 2 would be sourced from the northern parts of the country where jatropha is already cultivated on commercial scale.

Table 4. Agricultural land allocation to crops in Scenario 1.

Crop	Land area in 2010 (ha)	Fraction of total land area occupied by 4 crops	Portion allocated from 2.5% uncultivated land
Maize	992	0.4	57816.3
Cassava	875	0.35	50589.2
Oil palm	360	0.15	21681.1
Sorghum	253	0.1	14454.1
TOTAL	2480	1	144540.7

Again the assumption was made that biofuels produced in both scenarios are to be used locally while the trend in food crop production remains unchanged. It is expected that, improved agronomic practices in the energy crop plantations would trickle down to smallholder food crop programmes which should bring about marginal increases in yields.

3. RESULTS AND DISCUSSION

This section presents the results and discussion of the assessment of feedstock options for biofuel production in Ghana.

3.1. Results from Scenario 1

Table 5 presents the results for biofuel production from Scenario 1. Using conservative yields for biofuels from the various crops under consideration, about 303 million litres can be produced from 2.5% of uncultivated agricultural land in Ghana. This is made up of about 219 million litres bioethanol and 84 million litres biodiesel. About 85% of the total land allocated is dedicated to bioethanol crops and only 15% to biodiesel crops. **Table 6** shows the amount of land area that can be saved if the same amount of biofuel is produced but with high crop yields. In **Table 6**, it is assumed that the country intends to produce the same amount of biofuel (303,194,517 litres) as in **Table 5** but invests in agronomic research that improves yields comparable to the current levels in advanced agricultural economies. It also further takes into consideration, the fact that, high agricultural productivity is achieved with inputs that enhance crop growth and with irrigation schemes where necessary that ensure that crops are not rainfall dependent. Using yield figures from **Table 2**, the targeted biofuel is produced with 57,756 hectares of land and saves over 86,000 hectares of land space in the process compared with conservative yields.

Table 5. Biofuel production under Scenario 1.

Crop	Land allocated (ha)	Biofuel yield (l/ha)	Biofuel production (l)	Fuel type
Maize	57816.3	700	40,471,410	Bioethanol
Cassava	50997.2	3000	152,931,600	Bioethanol
Sweet Sorghum	14745.5	1750	25,804,625	Bioethanol
Oil Palm	20981.7	4000	83,926,800	Biodiesel
Total	**144540.7**		**303,134,435**	

Table 6. Land area saved with high yields from Scenario 1.

Crop	Desired biofuel production (l)	High biofuel yield (l/ha)	New land area to cultivate (ha)	Land area saved
Maize	40,471,396	3000	13,490	44,326
Cassava	152,991,668	6000	25,499	25,499
Sweet Sorghum	25,804,595	5400	4779	9967
Oil Palm	83,926,858	6000	13,988	6994
Total	**303,194,517**		**57,756**	**86,786**

3.2. Results from Scenario 2

Table 7 shows biofuel production in Scenario 2 where an extra 2.5% of the land is used for jatropha cultivation alone. In this scenario, about 563 million litres of biofuel is produced in the country. Bioethanol production under this scenario remains at the same level as in Scenario 1, at 219 million litres. However, biodiesel production increases from the original 84 million litres in scenario 1 to 344 million litres. Again in Scenario 2, if the same amount of biofuel mentioned above is to be produced from high yielding crops, the land savings expected is estimated. In this case achieving high yields from biofuel crops would save the country about 113 thousand hectares of land (**Table 8**) which is more than a third of the land used in Scenario 2 for conservative yields. To achieve such high biofuel yields, cassava, oil palm, sweet sorghum and maize yields would have to increase two-fold, three-fold, four-fold and five-fold respectively compared to existing yields.

Table 7. Biofuels production under Scenario 2.

Crop	Land allocated (ha)	Biofuel yield (l/ha)	Biofuel production (l)	Fuel type
Maize	57.816	700	40.471.396	Bioethanol
Cassava	50.997	3000	152.991.668	Bioethanol
Sweet Sorghum	14.745	1750	25.804.595	Bioethanol
Oil Palm	20.982	4000	83.926.858	Biodiesel
Jatropha	144.541	1800	260.173.260	Biodiesel
Total	289.081		563.367.777	

3.3. Fossil Fuel Replacement

Petrol and diesel consumption for the years 2020 and 2030 was estimated using correlation with population growth to determine portions of road transport fuels that can be substituted with biofuel. Demand for petrol and diesel is expected to reach 1.28 billion litres and 1.97 billion litres respectively by the year 2020. This increases to 1.66 billion litres and 2.56 billion litres for petrol and diesel respectively in 2030. **Table 9** shows the percentage of petroleum fuels that can be substituted with biofuel produced in Scenarios 1 and 2. In 2020, there is a potential of replacing 9.3% of overall diesel and petrol demand with biofuel from Scenario 1. In Scenario

2, there is the potential to substitute 17.3% of petroleum fuels. In Scenario 2, Ghana can substitute 17.1% of petrol with ethanol and 17.5% of fossil diesel with biodiesel. In 2030, Ghana would be able to substitute 7.2% of its petrol and diesel consumption in Scenario 1 and 13.3% in Scenario 2.

Table 8. Land area saved with high yields in Scenario 2.

Crop	Desired biofuel production (l)	High biofuel yield (l/ha)	New land area to cultivate (ha)	Land area saved
Maize	40.471.396	3000	13.490	44.326
Cassava	152.991.668	6000	25.499	25.499
Sweet Sorghum	25.804.595	5400	4779	9967
Oil Palm	83.926.858	6000	13.988	6994
Jatropha	260.173.260	2200	118.261	26.280
Total	563.367.777		**176,016**	**113,066**

3.4. Discussion

Ghana's draft bioenergy policy of 2010 [11] aims for the substitution of national petroleum fuels consumption with biofuel by 10% by 2020 and 20% by 2030. If Ghana were to adopt Scenario 1, the country would be able to replace 9.3% and 7.2% respectively by 2020 and 2030.

In effect, Scenario 1, which uses 2.5% of uncultivated arable land at conservative yield estimates, comes close to meeting the 2020 target but fall short of the 2030 target. A 2.5% of uncultivated land space can only meet EC targets if there is a significant increase in crop yields. For both target years, i.e. 2020 and 2030, Scenario 2 presents relatively high percentage replacements. However, in terms of actual energy content, more bioethanol and biodiesel would be required to meet its equivalent petroleum fuel product. This is because the energy content of bioethanol is about two thirds that of petrol and biodiesel has an energy content of about 12% less than petroleumbased diesel fuel on a mass basis [21]. Scenario 2 serve a better option in both target years for meeting the targets of the draft bioenergy policy.

Table 9. Transport fuel substitution with biofuel from conservative yields.

	Petroleum consumption (l)	Biofuel production (l)	Petroleum replaceable (%)
Scenario 1 (2020)			
Bioethanol/Petrol	1.280.303.721	219.207.635	17.1
Biodiesel/Diesel	1.970.649.110	83.926.800	4.3
Total	**3,250,952,831**	**303,134,435**	**9.3**
Scenario 2 (2020)			
Bioethanol/Petrol	1.280.303.721	219.207.635	17.1
Biodiesel/Diesel	1.970.649.110	344.100.060	17.5
Total	**3,250,952,831**	**563,307,695**	**17.3**
Scenario 1 (2030)			
Bioethanol/Petrol	1.666.282.480	219.207.635	13.2
Biodiesel/Diesel	2.564.749.310	83.926.800	3.3
Total	**4,231,031,790**	**303,134,435**	**7.2**
Scenario 2 (2030)			
Bioethanol/Petrol	1.666.282.480	219.207.635	13.2
Biodiesel/Diesel	2.564.749.310	344.100.060	13.4
Total	**4,231,031,790**	**563,307,695**	**13.3**

The decision to increase cultivation of certain crops would demand better technological input into the cultivation of crops and in the end, enhance food crop production in the country. Already, existing bioenergy companies in the country are investing into heavy agricultural machinery and agronomic improvements in order to boost their yields and improve on the economics of production. It is therefore anticipated that, the spill over effects would enhance crop yields from small scale farmers as they learn to embrace new and improved variety of crops. Increased agricultural activities, even with high mechanization would create more employment for the country and the extra biofuel could be exported to other countries in the early stages until the infrastructure is ready to support biofuel consumption in Ghana. Presently, the infrastructure for retailing biofuel blends are not in place as biofuel consumption is not mandated by government.

In order to achieve targets of land use, energy and land use regulators, i.e. the Energy Commission and Lands Commission, must ensure that regulatory systems are enforced in order to protect the environment and livelihood. A weakened system and/or regulator would expose the system to exploitation and the perceived benefits would not be achieved. Land acquisition guidelines which factors in all stakeholders (and not chiefs alone as the case is presently), should be in place to ensure a wider participation. Representatives of community members as well as district agricultural and planning officers should be actively involved. Once the structures are in place and lands are demarcated, the Lands Commission must ensure that only lands meant for energy crop production are used for the purpose.

The local production and use of biofuel as substitute for fossil-based fuels offers many attractive benefits for Ghana, but they could also impact negatively if not managed properly. Even though a detailed study on the impact of biofuels falls beyond the scope of this paper, there are notable points that must be taken into consideration in any serious biofuel programme of the kind suggested in this paper. The production of biofuel in Ghana presents an opportunity for local and foreign investments in Ghana, as well as offers the opportunity to reduce crude oil import and reduce the country's vulnerability to supply disruptions. Pursuing Ghana's biofuel potentials would boost agricultural and technological development and create jobs that would enhance the quality of life. The cultivation, harvesting and processing of energy crops could create an additional income avenue for farmers especially during the off-season when they are not engaged in any farming activities. Besides the financial savings that biofuel production of this scale may offer the country, there could also be environmental benefits which include the offset of greenhouse gases (GHGs) associated with the burning of fossil fuels. Exactly how much GHG savings would be realised is still debatable as there is no global agreement on the issue as different computing methodologies have yielded different results.

Notwithstanding the benefits, such a large scale production of energy crops for biofuel could bring with it negative socio-economic and environmental implications. First of all, wages and working conditions could be poor [22,23] and jobs may be given to people living outside plantation communities which could create conflicts. There is also the fear of environmental degradation if regulations are not enforced. This calls for effective regulatory and monitoring agencies actively engaged with all stakeholders.

4. RECIPE FOR SUCCESS

In order to realise a successful biofuel industry, there is the need for important policy issues to be put in place and for government to support the industry, especially during its nascent stage. Some of the more critical issues include:

4.1. Enactment of Policies and Regulatory Frameworks

Government support, legislation and regulation are critical to a successful biofuel industry in Ghana. Even though Ghana has targets for overall biofuels consumption in the near future, there are no mandates in place for biofuel blends (e.g. E5, E10, B5, B10, etc.) which could discourage investment plans, especially by the private sector. It could also lead to a situation where companies might be forced to export seeds/oil with a justification that there is no mandate for the use of biofuel. In order to succeed in introducing biofuel consumption in larger quantities in Ghana, the Energy Commission must set blending ratios similar to what countries such as Brazil has done. For a programme of this nature, biofuel agencies or parallel organisations should be established to oversee biofuel development programmes. Such an agency would be able to determine the appropriate support that the industry needs through interactions with various stakeholders and identify the needed infrastructure necessary to attract local and foreign investors into the industry. Clear responsibilities for Environmental and Social Impact Assessments (ESIAs) should be given to the Environmental Protection Agency. ESIAs will be crucial for ensuring socio-economic and environmental sustainability of biofuel projects. Strong government regulatory agencies will be required to undertake ESIAs and also to force their implementation. One critical issue here is the mapping of lands for food and energy crops, as well as for other purposes. Land use maps must be prepared to cover all agro-ecological zones in the country using GIS as appropriate, in order to lessen or eliminate exploitation of food crop lands by biofuel producers. Such maps should inform decision makers on which areas in the country must be reserved only for food crop production and which areas could be used for energy crop cultivation. This would go a long way towards ensuring that poor farmers are not deprived of their farm lands in the biofuel boom.

4.2. Support for R & D and Technology Transfer

The process of transferring biofuel technology and expertise involves the flow of knowledge, experience and equipment among different stakeholders, including governments, the private sector, financial institutions, NGOs, research and educational institutions and labour unions [23]. One of the biggest challenges, for biofuels in Ghana, at the moment has to do with achieving high yields from energy crops and this can be done through partnership with the countries that have achieved higher yields in biofuel feedstocks, such as Brazil for sugarcane, Malaysia and Indonesia for oil palm and Thailand for cassava. These countries have already gone through the experimental stages of agronomical development and there are opportunities for Ghana to partner with them to transfer the knowledge acquired to the country. Apart from improvements in feedstock, processing equipment is another important area that needs development. Biofuel plant providers could build plants on a "build-operatetransfer" basis in partnership with Ghana in the beginning. This would bring about improved local capacity to address the technology issue in future projects.

For biofuel technology transfer to be successful, government must be actively involved. Government support for technology transfer may include direct financial incentives, including grants for Research, Development and Demonstration (R, D & D). Policies supportive of international joint ventures would also help provide domestic companies in Ghana with access to intellectual property owned by international companies. With a natural favourable climate for energy crop production, partners in such joint ventures might contribute host sites for demonstrations and first commercial plants, as well as avenues for entering local biofuel markets. R & D support from the research community is needed in developing high yielding energy crops, standards for blending fuels, the development and improvements in production technology and processing, market development and consumer use. It is important that R & D is linked to technology transfer in such a way that it feeds directly into the scale up of biofuel in the country. R & D efforts need to tackle the practical problems encountered by biofuel technology transfer, not just at the technological level, but at all the critical levels as dictated by the experience from the field. An effective R & D programme should really cover Research, Development, Demonstration and Diffusion (R, D, D & D) and the activities should be closely linked with innovation in the market and education both in its formal and informal/non-formal aspects.

4.3. Ensuring Social Equity and Ecological Sustainability

The production of biofuel in Ghana must be done to ensure social equity. Incentives should be provided for the establishment of small scale biofuel conversion technologies in rural communities so that higher-value products may be produced and sold directly from the rural industries. Thus, Prescribed Obligatory Purchases (POPs) requiring private biofuel companies to purchase stipulated percentages of their feedstock or crude oil requirements from local farmers should be enshrined in the national biofuel policy. Another mechanism that could be used to address multiple objectives for the direct benefit of producers at the lowest and local levels is the Biofuel Export Levy (BEL). The objectives of the BEL could include investment in social services, particularly educational and health facilities, in biofuel producing areas; funding for R & D in improving yields of indigenous feedstocks and purchasing of improved feedstock varieties for small-scale local producers. A BEL could be set at a fixed percentage, up to about 10%, of revenues from biofuel exports and this should be explicitly shared among the different objectives agreed for the BEL.

The production and use of biofuel may have serious environmental impacts such as the use of large amounts of water, destruction of forests, reduction in food production and increase in soil degradation. It is important for Ghana to recognise the ecological challenges associated with biofuel and to address these challenges in order to ensure a sustainable industry. In this regard, it is necessary to put in place sustainability standards to address specific environmental and social issues such as soil erosion; damage to water and soil from the application of pesticides and fertilisers; increased use of freshwater sources; ecological impacts of monoculture crop plantations; the loss of biodiversity and wildlife habitat; potential impacts on agricultural and rural incomes; access to biofuel markets by small landholders and indigenous groups; job availability and quality; potential use of child labour; and access to education and healthcare for workers. The residues generated during the harvesting and processing of the crops could be used to produce biogas which can be used as cooking fuels, thereby reducing deforestation and improving public health.

5. CONCLUSION

Ghana has the potential to cultivate energy crops for biofuel production. In this assessment, two scenarios were used: Scenario 1 had 2.5% of uncultivated arable land dedicated to only four potential energy crops and Scenario 2 builds on Scenario 1 with an additional 2.5% of uncultivated arable land dedicated to only jatropha. The 5% of the uncultivated agricultural land was allocated to the energy crops in the following percentages: 21.2% for maize, 2.5% for sweet sorghum, 18.7% for cassava, 7.6% for oil palm and 50% for jatropha. Using conservative biofuel yields of 700 l/ha for maize, 3000 l/ha for cassava, 1750 l/ha for sweet sorghum, 4000 l/ha for oil palm and 1800 l/ha for jatropha, a 2.5% of uncultivated arable land dedicated to the four crops presented in Scenario 1 would yield 303 million litres of biofuel. Using the same yields in Scenario 1 and adding 2.5% of uncultivated arable land for jatropha cultivation yields 563 million litres of biofuel. The volume of biofuel produced in Scenario 1 could substitute 9.3% of transportation fuels by 2020 and 7.2% by 2030. In the second scenario, Ghana could replace 17.3% of transportation fuels by 2020 and 13.3% by 2030. At such conservative yield, the country would have to dedicate more land to energy crops in order to meet projections for biofuel demand. Other ways of meeting demand are to raise yields of these crops and/or supplement with production from lignocellulosic materials.

REFERENCES

1. G. Maggio and G. G. Cacciola, "When Will Oil, Natural Gas, and Coal Peak?" Fuel, Vol. 98, 2012, pp. 111-123.

2. M. Aftabuzzaman and E. Mazloumi, "Achieving Sustainable Urban Transport Mobility in Post Peak Oil Era," Transport Policy, Vol. 18, No. 5, 2011, pp. 695-702.

3. A. Zecca and L. Chiari, "Fossil-Fuel Constraints on Global Warming," Energy Policy, Vol. 38, No. 1, 2010, pp. 1-3.

4. N. A. Owen, O. R. Inderwildi and D. A. King, "The Status of Conventional World Oil Reserves-Hype or Cause for Concern?" Energy Policy, Vol. 38, No. 8, 2010, pp. 4743-4749.

5. J. Singh and S. Gu, "Biomass Conversion to Energy in India—A Critique," Renewable and Sustainable Energy Reviews, Vol. 14, 2010, pp. 1367-1378.

6. J. J. Cheng and G. R. Timilsina, "Status and Barriers of Advanced Biofuel Technologies: A Review," Renewable Energy, Vol. 36, No. 12, 2011, pp. 3541-3549.

7. A. Demirbas, "Biofuels Securing the Planet's Future Energy Needs," Energy Conversion and Management, Vol. 50, 2009, pp. 2239-2249.

8. Worldwatch Institute, "Biofuels Make a Comeback Despite Tough Economy," Worldwatch Institute, 2011. http://www.worldwatch.org/biofuels-make-comeback-despite-tough-economy

9. HART/GBC, "Global Biofuels Outlook 2009-2015," Hart Energy Consulting/Hart's Global Biofuels Center, Texas, 2009.

10. C. Mandil and A. Shihab-Eldin, "Assessment of Biofuels Potential and Limitations," A Report Commissioned by the International Energy Forum, 2010. http://www.ief.org/_resources/files/ content/news/presentations/ief-report-biofuels-potentials-and-limitations-february-2010.pdf

11. Energy Commission, “Draft Bioenergy Policy of Ghana,” Energy Commission, Accra, 2010.

12. F. Kemausuor, G. Y. Obeng, A. Brew-Hammond and A. Duker, “A Review of Trends, Policies and Plans for Increasing Energy Access in Ghana,” Renewable and Sustainable Energy Reviews, Vol. 15, No. 9, 2011, pp. 5143- 5154.

13. Africa Biofuel Network, “Biofuels—A Failure for Africa,” 2010. http://www.africanbiodiversity.org/sites/default/files/PDFs/Biofuels%20-%20A%20Failure%20for%20Africa%20(ABN,%20Dec%202010).pdf

14. Ghana Statistical Service, “Population and Housing Census Summary of Final Results, 2012,” 2012. http://www.statsghana.gov. gh/docfiles/2010phc/2010_popution_and_housing_census(view_summary_of_final_results).pdf

15. Food and Agricultural Organisation, “Ghana Overview, 2009.” http://faostat3.fao.org/home/index.html#VISUALIZE_BY_AREA

16. Ministry of Food and Agriculture, “Agriculture in Ghana: Facts and Figures 2010,” Ministry of Food and Agriculture, Statistics, Research and Information Directorate (SRID), 2011.

17. M. H. Duku, S. Gu and E. B. Hagan, “A Comprehensive Review of Biomass Resources and Biofuels Potential in Ghana,” Renewable and Sustainable Energy Reviews, Vol. 15, No. 1, 2011, pp. 404-415.

18. Food and Agricultural Organisation, “Ghana Aquastat 2005.” http://www.fao.org/fileadmin/user_upload/aquastat/pdf_files/GHA_tables.pdf

19. S. Sielhorst, J. W. Molenaar and D. Offermans, “Biofuels in Africa: An Assessment of Risks and Benefits for African Wetlands,” Wetlands International, Wageningen, 2008.

20. Food and Agricultural Organisation, “Crop production Ghana 2010.” http://faostat.fao.org/site/567/default.aspx

21. K. M. Rahman, M. Mashud, M. Roknuzzaman and A. Al Galib, “Biodiesel from Jatropha Oil as an Alternative Fuel for Diesel Engine,” International Journal of Mechanical & Mechatronics, Vol. 10, No. 3, 2010, pp. 1-6.

22. A. Milbrandt, "Assessment of Biomass. Resources in Liberia," Prepared for the US Agency for International Development (USAID) under the Liberia Energy Assistance Program (LEAP), Technical Report NREL/TP-6A2- 44808, 2009.

23. Worldwatch Institute, "Biofuels for Transport: Global Potential and Implications for Energy and Agriculture," Earthscan Publishing, 2007.

Index

I

L

M

P

R

S

T